Elements for Physics

To Maria

Albert Tarantola

Elements for Physics

Quantities, Qualities, and Intrinsic Theories

With 44 Figures (10 in colour)

Professor Albert Tarantola
Institut de Physique du Globe de Paris
4, place Jussieu
75252 Paris Cedex 05
France

Library of Congress Control Number: 2005935447

ISBN-10 3-540-25302-5 Springer Berlin Heidelberg New York
ISBN-13 978-3-540-25302-0 Springer Berlin Heidelberg New York

Springer is a part of Springer Science+Business Media.

springeronline.com

Printed in Germany

Typesetting: Data prepared by the Author using a Springer TEX macro package
Cover design: *design & production* GmbH, Heidelberg

Printed on acid-free paper SPIN 11406990 57/3141/SPI 5 4 3 2 1 0

Preface

Physics is very successful in describing the world: its predictions are often impressively accurate. But to achieve this, physics limits terribly its scope. Excluding from its domain of study large parts of biology, psychology, economics or geology, physics has concentrated on quantities, i.e., on notions amenable to accurate measurement.

The meaning of the term physical 'quantity' is generally well understood (everyone understands what it is meant by "the frequency of a periodic phenomenon", or "the resistance of an electric wire"). It is clear that behind a set of *quantities* like temperature – inverse temperature – logarithmic temperature, there is a qualitative notion: the 'cold–hot' *quality*. Over this one-dimensional quality space, we may choose different 'coordinates': the temperature, the inverse temperature, etc. Other quality spaces are multidimensional. For instance, to represent the properties of an ideal elastic medium we need 21 coefficients, that can be the 21 components of the elastic stiffness tensor $c_{ijk\ell}$, or the 21 components of the elastic compliance tensor (inverse of the stiffness tensor), or the proper elements (six eigenvalues and 15 angles) of any of the two tensors, etc. Again, we are selecting coordinates over a 21-dimensional quality space. On this space, each point represents a particular elastic medium.

So far, the consideration is trivial. What is important is that *it is always possible to define the distance between two points of any quality space, and this distance is —inside a given theoretical context— uniquely defined*. For instance, two periodic phenomena can be characterized by their periods, T_1 and T_2, or by their frequencies, ν_1 and ν_2. The only definition of distance that respects some clearly defined invariances is $D = |\log(T_2/T_1)| = |\log(\nu_2/\nu_1)|$.

For many vector and tensor spaces, the distance is that associated with the ordinary norm (of a vector or a tensor), but some important spaces have a more complex structure. For instance, 'positive tensors' (like the electric permittivity or the elastic stiffness) are not, in fact, elements of a linear space, but oriented geodesic segments of a curved space. The notion of *geotensor* ("geodesic tensor") is developed in chapter 1 to handle these objects, that are like tensors but that do not belong to a linear space.

The first implications of these notions are of mathematical nature, and a point of view is proposed for understanding Lie groups as metric manifolds with curvature and torsion. On these manifolds, a sum of geodesic segments can be introduced that has the very properties of the group. For instance, in the manifold representing the group of rotations, a 'rotation vector' is not a vector, but a geodesic segment of the manifold, and the composition of rotations is nothing but the geometric sum of these segments.

More fundamental are the implications in physics. As soon as we accept that behind the usual physical quantities there are quality spaces, that usual quantities are only special 'coordinates' over these quality spaces, and that there is a metric in each space, the following question arises: Can we do physics intrinsically, i.e., can we develop physics using directly the notion of physical quality, and of metric, and without using particular coordinates (i.e., without any particular choice of physical quantities)? For instance, Hooke's law $\sigma_{ij} = c_{ij}{}^{k\ell}\, \varepsilon_{k\ell}$ is written using three quantities, stress, stiffness, and strain. But why not using the exponential of the strain, or the inverse of the stiffness? One of the major theses of this book is that physics can, and must, be developed independently of any particular choice of coordinates over the quality spaces, i.e., independently of any particular choice of physical quantities to represent the measurable physical qualities.

Most current physical theories, can be translated so that they are expressed using an intrinsic language. Other theories (like the theory of linear elasticity, or Fourier's theory of heat conduction) cannot be written intrinsically. I claim that these theories are inconsistent, and I propose their reformulation.

Mathematical physics strongly relies on the notion of derivative (or, more generally, on the notion of tangent linear mapping). When taking into account the geometry of the quality spaces, another notion appears, that of *declinative*. Theories involving nonflat manifolds (like the theories involving Lie group manifolds) are to be expressed in terms of declinatives, not derivatives. This notion is explored in chapter 2.

Chapter 3 is devoted to the analysis of some spaces of physical qualities, and attempts a classification of the more common types of physical quantities used on these spaces. Finally, chapter 4 gives the definition of an intrinsic physical theory and shows, with two examples, how these intrinsic theories are built.

Many of the ideas presented in this book crystallized during discussions with my colleagues and students. My friend Bartolomé Coll deserves special mention. His understanding of mathematical structures is very deep. His logical rigor and his friendship have made our many discussions both a pleasure and a source of inspiration. Some of the terms used in this book have been invented during our discussions over a cup of coffee at Café Beaubourg, in Paris. Special thanks go to my professor Georges Jobert, who introduced me to the field of inverse problems, with dedication and rigor. He has contributed to this text with some intricate demonstrations. Another friend, Klaus Mosegaard, has been of great help, since the time we developed together Monte Carlo methods for the resolution of inverse problems. With probability one, he defeats me in chess playing and mathematical problem solving. Discussions with Peter Basser, João Cardoso, Guillaume Evrard, Jean Garrigues, José-Maria Pozo, John Scales, Loring Tu, Bernard Valette, Peiliang Xu, and Enrique Zamora have helped shape some of the notions presented in this book.

Paris, August 2005 *Albert Tarantola*

[illegible] Monte Carlo methods for the resolution of inverse problems. With probability one, [illegible] mathematical problem [illegible]

[illegible] 2005 [illegible]

Contents

List of Appendices

Overview

One-dimensional Quality Spaces

Consider a one-dimensional space, each point $\mathcal{N}$ of it representing a musical note. This line has to be imagined infinite in its two senses, with the infinitely acute tones at one "end" and the infinitely grave tones at the other "end". Musicians can immediately give the *distance* between two points of the space, i.e., between two notes, using the octave as unit. To express this distance by a formula, we may choose to represent a note by its frequency, ν, or by its period, τ. The distance between two notes $\mathcal{N}_1$ and $\mathcal{N}_2$ is[1]

$$D_{\text{music}}(\mathcal{N}_1, \mathcal{N}_2) = |\log_2 \frac{\nu_2}{\nu_1}| = |\log_2 \frac{\tau_2}{\tau_1}| \quad . \tag{1}$$

This distance is the only one that has the following properties:

- its expression is identical when using the positive quantity $\nu = 1/\tau$ or its inverse, the positive quantity $\tau = 1/\nu$;
- it is additive, i.e., for any set of three ordered points $\{\mathcal{N}_1, \mathcal{N}_2, \mathcal{N}_3\}$, the distance from point $\mathcal{N}_1$ to point $\mathcal{N}_2$, plus the distance from point $\mathcal{N}_2$ to point $\mathcal{N}_3$, equals the distance from point $\mathcal{N}_1$ to point $\mathcal{N}_3$.

This one-dimensional space (or, to be more precise, this one-dimensional *manifold*) is a simple example of a *quality space*. It is a metric manifold (the distance between points is defined). The quantities frequency ν and period τ are two of the *coordinates* that can be used on the quality space of the musical notes to characterize its points. Infinitely many more coordinates are, of course, possible, like the logarithmic frequency $\nu^* = \log(\nu/\nu_0)$, the cube of the frequency, $\eta = \nu^3$, etc. Given the expression for the distance in some coordinate system, it is easy to obtain an expression for it using another coordinate system. For instance, it follows from equation (1) that the distance between two musical notes is, in terms of the logarithmic frequency, $D_{\text{music}}(\mathcal{N}_1, \mathcal{N}_2) = |\nu_2^* - \nu_1^*|$.

There are many quantities in physics that share three properties: (*i*) their range of variation is $(0, \infty)$, (*ii*) they are as commonly used as their in-

[1]To obtain the distance in octaves, one must use base 2 logarithms.

verses, and (*iii*) they display the Benford effect.[2] Examples are the frequency ($\nu = 1/\tau$) and period ($\tau = 1/\nu$) pair, the temperature ($T = 1/\beta$) and thermodynamic parameter ($\beta = 1/T$) pair, or the resistance ($R = 1/C$) and conductance ($C = 1/R$) pair. These quantities typically accept the expression in formula (1) as a natural definition of distance. In this book we say that we have a pair of *Jeffreys quantities*.

For instance, before the notion of *temperature*[3] was introduced, physicists followed Aristotle in introducing the *cold–hot (quality) space*. Even if a particular coordinate over this one-dimensional manifold was not available, physicists could quite precisely identify many of its points: the point $\mathcal{Q}_1$ corresponding to the melting of sulphur, the point $\mathcal{Q}_2$ corresponding to the boiling of water, etc. Among the many coordinates today available in the cold–hot space (like the Celsius or the Fahrenheit temperatures), the pair absolute temperature $T = 1/\beta$ and thermodynamic parameter $\beta = 1/T$ are obviously a Jeffreys pair. In terms of these coordinates, the natural distance between two points of the cold–hot space is (using natural logarithms)

$$D_{\text{cold–hot}}(\mathcal{Q}_1, \mathcal{Q}_2) \;=\; |\, \log \frac{T_2}{T_1} \,| \;=\; |\, \log \frac{\beta_2}{\beta_1} \,| \;=\; |\, T_2^* - T_1^* \,| \;=\; |\, \beta_2^* - \beta_1^* \,| \quad , \qquad (2)$$

where, for more completeness, the logarithmic temperature T^* and the logarithmic thermodynamic parameter β^* have also been introduced. An expression using other coordinates is deduced using any of those equivalent expressions. For instance, using Celsius temperatures, $D_{\text{cold–hot}}(\mathcal{Q}_1, \mathcal{Q}_2) = |\, \log(\, (t_2 + T_0)/(t_1 + T_0)\,)\,|$, where $T_0 = 273.15\,\mathrm{K}$.

At this point, without any further advance in the theory, we could already ask a simple question: if the tone produced by a musical instrument depends on the position of the instrument in the cold–hot space (using ordinary language we would say that the 'frequency' of the note depends on the 'temperature', but we should not try to be specific), what is the simplest dependence that we can imagine? Surely a *linear* dependence. But as both spaces, the space of musical notes and the cold–hot space, are metric, the only *intrinsic* definition of linearity is a proportionality between the *distances* in the two spaces,

$$D_{\text{music}}(\mathcal{N}_1, \mathcal{N}_2) \;=\; \alpha \; D_{\text{cold–hot}}(\mathcal{Q}_1, \mathcal{Q}_2) \quad , \qquad (3)$$

where α is a positive real number. Note that we have just expressed a physical law without being specific about the many possible physical quantities

[2]The Benford effect is an uneven probability distribution for the first digit in the numerical expression of a quantity: when using a base K number system, the probability that the first digit is n is $p_n = \log_K (n+1)/n$. For instance, in the usual base 10 system, about 30% of the time the first digit is one, while for only 5% of the time is the first digit a nine. See details in chapter 3.

[3]Before Galileo, the quantity 'temperature' was not defined. Around 1592, he invented the first thermometer, using air.

that one may use in each of the two quality spaces. Choosing, for instance, temperature T in the cold–hot space, and frequency ν in the space of musical notes, the expression for the linear law (3) is

$$\nu_2 / \nu_1 = (T_2 / T_1)^\alpha \quad . \tag{4}$$

Note that the linear law takes a *formally* linear aspect only if logarithmic frequency (or logarithmic period) and logarithmic temperature (or logarithmic thermodynamic parameter) are used. An expression like $\nu_2 - \nu_1 = \alpha\,(T_2 - T_1)$ although formally linear, is not a linear law (as far as we have agreed on given metrics in our quality spaces).

Multi-dimensional Quality Spaces

Consider a homogeneous piece of a linear elastic material, in its unstressed state. When a (homogeneous) stress $\sigma = \{\sigma^{ij}\}$ is applied, the body experiences a strain $\varepsilon = \{\varepsilon^{ij}\}$ that is related to the stress through any of the two equivalent equations (Hooke's law)

$$\varepsilon^{ij} = d^{ij}{}_{k\ell}\, \sigma^{k\ell} \quad ; \quad \sigma^{ij} = c^{ij}{}_{k\ell}\, \varepsilon^{k\ell} \quad , \tag{5}$$

where $\mathbf{d} = \{d^{ij}{}_{k\ell}\}$ is the *compliance* tensor, and $\mathbf{c} = \{c^{ij}{}_{k\ell}\}$ is the *stiffness* tensor. These two tensors are positive definite and are mutually inverse, $d^{ij}{}_{k\ell}\, c^{k\ell}{}_{mn} = c^{ij}{}_{k\ell}\, d^{k\ell}{}_{mn} = \delta^i_m\, \delta^j_n$, and one can use any of the two to characterize the elastic medium.

In elementary elasticity theory one assumes that the compliance tensor has the symmetries $d_{ijk\ell} = d_{jik\ell} = d_{k\ell ij}$, with an equivalent set of symmetries for the stiffness tensor. An easy computation shows that (in 3D media) one is left with 21 degrees of freedom, i.e., 21 quantities are necessary and sufficient to characterize a linear elastic medium. We can then introduce an abstract 21-dimensional manifold $\mathfrak{E}$, such that each point $\mathcal{E}$ of $\mathfrak{E}$ corresponds to an elastic medium (and vice versa). This is the (quality) space of elastic media.

Which sets of 21 quantities can we choose to represent a linear elastic medium? For instance, we can choose 21 independent components of the compliance tensor $d^{ij}{}_{k\ell}$, or 21 independent components of the stiffness tensor $c^{ij}{}_{k\ell}$, or the six eigenvalues and the 15 proper angles of the one or the other. Each of the possible choices corresponds to choosing a coordinate system over $\mathfrak{E}$.

Is the manifold $\mathfrak{E}$ metric, i.e., is there a natural definition of distance between two of its points? The requirement that the distance must have the same expression in terms of compliance, $\mathbf{d}$, and in terms of stiffness, $\mathbf{c}$, that it must have an invariance of scale (multiplying all the compliances or all the stiffnesses by a given factor should not alter the distance), and that it should depend only on the invariant scalars of the compliance or of the stiffness tensor leads to a unique expression. The distance between the elastic medium $\mathcal{E}_1$, characterized by the compliance tensor $\mathbf{d}_1$ or the stiffness tensor $\mathbf{c}_1$, and the elastic medium $\mathcal{E}_2$ characterized by the compliance tensor $\mathbf{d}_2$ or the stiffness tensor $\mathbf{c}_2$, is

$$D_{\mathfrak{E}}(\mathcal{E}_1,\mathcal{E}_2) = \| \log(\mathbf{d}_2\, \mathbf{d}_1^{-1}) \| = \| \log(\mathbf{c}_2\, \mathbf{c}_1^{-1}) \| \quad . \tag{6}$$

In this equation, the logarithm of an adimensional, positive definite tensor $\mathbf{T} = \{T^{ij}{}_{k\ell}\}$ can be defined through the series[4]

$$\log \mathbf{T} = (\mathbf{T} - \mathbf{I}) - \tfrac{1}{2}(\mathbf{T} - \mathbf{I})^2 + \dots \quad . \tag{7}$$

Alternatively, the logarithm of an adimensional, positive definite tensor can be defined as the tensor having the same proper angles as the original tensor, and whose eigenvalues are the logarithms of the eigenvalues of the original tensor. Also in equation (6), the norm of a tensor $\mathbf{t} = \{t^{ij}{}_{k\ell}\}$ is defined through

$$\| \mathbf{t} \| = \sqrt{t^{ij}{}_{k\ell}\, t^{k\ell}{}_{ij}} \quad . \tag{8}$$

It can be shown (see chapter 1) that the finite distance in equation (6) does derive from a metric, in the sense of the term in differential geometry, i.e., it can be deduced from a quadratic expression defining the distance element ds^2 between two infinitesimally close points.[5] An immediate question arises: is this 21-dimensional manifold flat? To answer this question one must evaluate the Riemann tensor of the manifold, and when this is done, one finds that this tensor is different from zero: *the manifold of elastic media has curvature.*

Is this curvature an artefact, irrelevant to the physics of elastic media, or is this curvature the sign that the quality spaces here introduced have a nontrivial geometry that may allow a geometrical formulation of the equations of physics? This book is here to show that it is the second option that is true. But let us take a simple example: the three-dimensional rotations.

A rotation $\mathcal{R}$ can be represented using an orthogonal matrix $\mathbf{R}$. The *composition* of two rotations is defined as the rotation $\mathcal{R}$ obtained by first applying the rotation $\mathcal{R}_1$, then the rotation $\mathcal{R}_2$, and one may use the notation

$$\mathcal{R} = \mathcal{R}_2 \circ \mathcal{R}_1 \quad . \tag{9}$$

It is well known that when rotations are represented by orthogonal matrices, the composition of two rotations is obtained as a matrix product:

$$\mathbf{R} = \mathbf{R}_2\, \mathbf{R}_1 \quad . \tag{10}$$

But there is a second useful representation of a rotation, in terms of a rotation pseudovector $\boldsymbol{\rho}$, whose axis is the rotation axis and whose norm equals the

[4] I.e., $(\log \mathbf{T})^{ij}{}_{k\ell} = (T^{ij}{}_{k\ell} - \delta^i_k\,\delta^j_\ell) - \frac{1}{2}(T^{ij}{}_{rs} - \delta^i_r\,\delta^j_s)(T^{rs}{}_{k\ell} - \delta^r_k\,\delta^s_\ell) + \dots$.

[5] This distance is closely related to the "Cartan metric" of Lie group manifolds.

rotation angle. As pseudovectors are, in fact, antisymmetric tensors, let us denote by $\mathbf{r}$ the antisymmetric matrix related to the components of the pseudovector ρ through the usual duality,[6] $r_{ij} = \epsilon_{ijk}\,\rho^k$. For instance, in a Euclidean space, using Cartesian coordinates,

$$\mathbf{r} = \begin{pmatrix} r_{xx} & r_{xy} & r_{xz} \\ r_{yx} & r_{yy} & r_{yz} \\ r_{zx} & r_{zy} & r_{zz} \end{pmatrix} = \begin{pmatrix} 0 & \rho^z & -\rho^y \\ -\rho^z & 0 & \rho^x \\ \rho^y & -\rho^x & 0 \end{pmatrix} \quad . \tag{11}$$

We shall sometimes call the antisymmetric matrix $\mathbf{r}$ the rotation "vector".

Given an orthogonal matrix $\mathbf{R}$ how do we obtain the antisymmetric matrix $\mathbf{r}$? It can be seen that the two matrices are related via the log–exp duality:

$$\mathbf{r} = \log \mathbf{R} \quad . \tag{12}$$

This is a very simple way for obtaining the rotation vector $\mathbf{r}$ associated to an orthogonal matrix $\mathbf{R}$. Reciprocally, to obtain the orthogonal matrix $\mathbf{R}$ associated to the rotation vector $\mathbf{r}$, we can use

$$\mathbf{R} = \exp \mathbf{r} \quad . \tag{13}$$

With this in mind, it is easy to write the composition of rotations in terms of the rotation vectors. One obtains

$$\mathbf{r} = \mathbf{r}_2 \oplus \mathbf{r}_1 \quad , \tag{14}$$

where the operation $\oplus$ is defined, for any two tensors $\mathbf{t}_1$ and $\mathbf{t}_2$, as

$$\mathbf{t}_2 \oplus \mathbf{t}_1 \equiv \log(\exp \mathbf{t}_2 \, \exp \mathbf{t}_1) \quad . \tag{15}$$

The two expressions (10) and (14) are two different representations of the abstract notion of composition of rotations (equation 9), respectively in terms of orthogonal matrices and in terms of antisymmetric matrices (rotation vectors). Let us now see how the operation $\oplus$ in equation (14) can be interpreted as a sum, provided that one takes into account the geometric properties of the space of rotations.

It is well known that the rotations form a group, the Lie group SO(3). Lie groups are manifolds, in fact, quite nontrivial manifolds, having curvature and torsion.[7] In the (three-dimensional) Lie group manifold SO(3), the orthogonal matrices $\mathbf{R}$ can be seen as the points of the manifold. When the identity matrix $\mathbf{I}$ is taken as the origin of the manifold, an antisymmetric matrix $\mathbf{r}$ can be interpreted as the oriented geodesic segment going from the origin $\mathbf{I}$ to the point $\mathbf{R} = \exp \mathbf{r}$. Then, let two rotations be represented by the two antisymmetric matrices $\mathbf{r}_2$ and $\mathbf{r}_1$, i.e., by two oriented geodesic

[6]Here, ϵ_{ijk} is the totally antisymmetric symbol.

[7]And such that autoparallel lines and geodesic lines coincide.

segments of the Lie group manifold. It is demonstrated in chapter 1 that the *geometric sum* of the two segments (performed using the curvature and torsion of the manifold) exactly corresponds to the operation $\mathbf{r}_2 \oplus \mathbf{r}_2$ introduced in equations (14) and (15), i.e., *the geometric sum of two oriented geodesic segments of the Lie group manifold* is *the group operation*.

This example shows that the nontrivial geometry we shall discover in our quality spaces is fundamentally related to the basic operations to be performed. One of the major examples of physical theories in this book is, in chapter 4, the theory of ideal elastic media. When acknowledging that the usual 'configuration space' of the body is, in fact, (a submanifold of) the Lie group manifold $GL^+(3)$ (whose 'points' are all the 3×3 real matrices with positive determinant), one realizes that the *strain* is to be defined as a geodesic line joining two configurations: the strain is not an element of a linear space, but a geodesic of a Lie group manifold. This, in particular, implies that the proper definition of strain is *logarithmic*.

This is one of the major lessons to be learned from this book: the tensor equations of properly developed physical theories, usually contain logarithms and exponentials of tensors. The conspicuous absence of logarithms and exponentials in present-day physics texts suggests that there is some basic aspect of mathematical physics that is not well understood. I claim that a fundamental invariance principle should be stated that is not yet recognized.

Invariance Principle

Today, a physical theory is seen as relating different physical quantities. But we have seen that *physical quantities* are nothing but *coordinates* over spaces of *physical qualities*. While present tensor theories assure invariance of the equations with respect to a change of coordinates over the physical space (or the physical space-time, in relativity), we may ask if there is a formulation of the tensor theories that assure invariance with respect to *any* choice of coordinates over *any* space, including the spaces of physical qualities (i.e., invariance with respect to any choice of physical quantities that may represent the physical qualities).

The goal of this book is to demonstrate that the answer to that question is positive.

For instance, when formulating Fourier's law of heat conduction, we have to take care to arrive at an equation that is independent of the fact that, over the cold–hot space, we may wish to use as coordinate the temperature, its inverse, or its cube. When doing so, one arrives at an expression (see equation 4.21) that has no immediate resemblance to the original Fourier's law. This expression does not involve specific quantities; rather, it is valid for any possible choice of them. When being specific and choosing, for instance, the (absolute) temperature T the law becomes

$$\phi_i = -\kappa \frac{1}{T} \frac{\partial T}{\partial x^i} \quad , \tag{16}$$

where $\{x^i\}$ is any coordinate system in the physical space, ϕ_i is the heat flux, and κ is a constant. This is not Fourier's law, as there is an extra factor $1/T$. Should we write the law using, instead of the temperature, the thermodynamic parameter $\beta = 1/T$, we would arrive at

$$\phi_i = +\kappa \frac{1}{\beta} \frac{\partial \beta}{\partial x^i} \quad . \tag{17}$$

It is the symmetry between these two expressions of the law (a symmetry that is not satisfied by the original Fourier's law) that suggests that the equations at which we arrive when using our (strong) invariance principle may be more physically meaningful than ordinary equations. In fact, nothing in the arguments of Fourier's work (1822) would support the original equation, $\phi_i = -\kappa\, \partial T / \partial x^i$, better than our equation (16). In chapter 4, it is suggested that, quantitatively, equations (16) and (17) are at least as good as Fourier's law, and, qualitatively, they are better.

In the case of one-dimensional quality spaces, the necessary invariance of the expressions is achieved by taking seriously the notion of *one-dimensional linear space*. For instance, as the cold–hot quality space is a one-dimensional metric manifold (in the sense already discussed), once an arbitrary origin is chosen, it becomes a linear space. Depending on the particular coordinate chosen over the manifold (temperature, cube of the temperature), the natural basis (a single vector) is different, and vectors on the space have different components. Nothing is new here with respect to the theory of linear spaces, but this is not the way present-day physicists are trained to look at one-dimensional qualities.

In the case of multi-dimensional quality spaces, one easily understands that physical theories do not relate particular quantities but, rather, they relate the geometric properties of the different quality spaces involved. For instance, the law defining an ideal elastic medium can be stated as follows: when a body is subjected to a linear change of stress, its configuration follows a geodesic line in the configuration space.[8]

Mathematics

To put these ideas on a clear basis, we need to develop some new mathematics.

[8]More precisely, as we shall see, an ideal elastic medium is defined by a 'geodesic' mapping' between the (linear) stress space and the submanifold of the Lie group manifold $GL^+(3)$ that is geodesically connected to the origin of the group (this submanifold is the configuration space).

Our quality spaces are manifolds that, in general, have curvature and torsion (like the Lie group manifolds). We shall select an origin on the manifold, and consider the collection of all 'autoparallel' or 'geodesic' segments with that common origin. Such an oriented segment shall be called an *autovector*. The sum of two autovectors is defined using the parallel transport on the manifold. Should the manifold be flat, we would obtain the classic structure of linear space. But what is the structure defined by the 'geometric sum' of the autovectors? When analyzing this, we will discover the notion of *autovector space*, which will be introduced axiomatically. In doing so, we will find, as an intermediary, the *troupe* structure (in short, a group without the associativity property).

With this at hand, we will review the basic geometric properties of Lie group manifolds, with special interest in curvature, torsion and parallel transport. While de-emphasizing the usual notion of Lie algebra, we shall study the interpretation of the group operation in terms of the geometric sum of oriented autoparallel (and geodesic) segments. A special term is used for these oriented autoparallel segments, that of *geotensor* (for "geodesic tensor").

Geotensors play an important role in the theory. For many of the objects called "tensors" in physics are improperly named. For instance, as mentioned above, the strain ε that a deforming body may experience *is* a geodesic of the Lie group manifold $GL^+(3)$. As such, it is not an element of a linear space, but an element of a space that, in general, is not flat. Unfortunately, this seems to be more than a simple misnaming: the conspicuous absence of the logarithm and the exponential functions in tensor theories suggests that the geometric structure actually behind some of the "tensors" in physics is not clearly understood. This is why a special effort is developed in this text to define explicitly the main properties of the log–exp duality for tensors.

There is another important mathematical notion that we need to revisit: that of derivative. There are two implications to this. First, when taking seriously the tensor character of the derivative, one does not define the derivative of one *quantity* with respect to another *quantity*, but the derivative of one *quality* with respect to another *quality*. In fact, we have already seen one example of this: in equations (18) and (19) the same derivative is expressed using different coordinates in the cold–hot space (the temperature T and the inverse temperature β). This is the very reason why the law of heat conduction proposed in this text differs from the original Fourier's law.

A less obvious deviation from the usual notion of derivative is when the *declinative* of a mapping is introduced. The declinative differs from the derivative in that the geometrical objects considered are 'transported to the origin'. Consider, for instance, a solid rotating around a point. When characterizing the 'attitude' of the body at some instant t by the (orthogonal) rotation matrix $\mathbf{R}(t)$, we are, in fact defining a mapping from the time axis into the rotation group $SO(3)$. The declinative of this mapping happens to be[9]

$$\dot{\mathbf{R}}(t)\,\mathbf{R}(t)^{-1} \quad , \tag{20}$$

where $\dot{\mathbf{R}}$ is the derivative. The expression $\dot{\mathbf{R}}(t)\,\mathbf{R}(t)^{-1}$ gives, in fact, the instantaneous *rotation velocity,* $\omega(t)$. While the derivative produces $\dot{\mathbf{R}}(t)$, that has no simple meaning, the declinative directly produces the rotation velocity $\omega(t) = \dot{\mathbf{R}}(t)\,\mathbf{R}(t)^{-1} = \dot{\mathbf{R}}(t)\,\mathbf{R}(t)^{*}$ (because the geometry of the rotation group SO(3) is properly taken into account).

Contents

While the mathematics concerning the autovector spaces are developed in chapter 1, those concerning derivatives and declinatives are developed in chapter 2. Chapter 3 gives some examples of identification of the quality spaces behind some of the common physical quantities, and chapter 4 develops two special examples of intrinsic physical theories, the theory of heat conduction and the theory of ideal elastic media. Both theories are chosen because they quantitatively disagree with the versions found in present-day texts.

[9] $\dot{\mathbf{R}}(t)\,\mathbf{R}(t)^{-1}$ is different from $(\log \mathbf{R})^{\cdot} \equiv d(\log \mathbf{R})/dt$.

1 Geotensors

[...] the displacement associated with a small closed path can be decomposed into a translation and a rotation: the translation reflects the torsion, the rotation reflects the curvature.

Les Variétés à Connexion Affine, Élie Cartan, 1923

Even when the physical space (or space-time) is assumed to be flat, some of the "tensors" appearing in physics are not elements of a linear space, but of a space that may have curvature and torsion. For instance, the ordinary sum of two "rotation vectors", or the ordinary sum of two "strain tensors", has no interesting meaning, while if these objects are considered as oriented geodesic segments of a nonflat space, then, the (generally noncommutative) sum of geodesics exactly corresponds to the 'composition' of rotations or to the 'composition' of deformations. It is only for small rotations or for small deformations that one can use a linear approximation, recovering then the standard structure of a (linear) tensor space. The name 'geotensor' (geodesic tensor) is coined to describe these objects that generalize the common tensors.

To properly introduce the notion of geotensor, the structure of 'autovector space' is defined, which describes the rules followed by the sum and difference of oriented autoparallel segments on a (generally nonflat) manifold. At this abstract level, the notions of torsion (defined as the default of commutativity of the sum operation) and of curvature (defined as the default of associativity of the sum operation) are introduced. These two notions are then shown to correspond to the usual notions of torsion and curvature in Riemannian manifolds.

1.1 Linear Space

1.1.1 Basic Definitions and Properties

Consider a set $\mathbb{S}$ with elements denoted $\mathbf{u}, \mathbf{v}, \mathbf{w} \dots$ over which two operations have been defined. First, an internal operation, called *sum* and denoted $+$, that gives to $\mathbb{S}$ the structure of a 'commutative group', i.e., an operation that is associative and commutative,

$$\mathbf{w} + (\mathbf{v} + \mathbf{u}) \;=\; (\mathbf{w} + \mathbf{v}) + \mathbf{u} \qquad ; \qquad \mathbf{w} + \mathbf{v} \;=\; \mathbf{v} + \mathbf{w} \tag{1.1}$$

(for any elements of $\mathbb{S}$), with respect to which there is a zero element, denoted $\mathbf{0}$, that is neutral for any other element, and where any element $\mathbf{v}$ has an opposite element, denoted $\mathbf{-v}$:

$$\mathbf{v} + \mathbf{0} = \mathbf{0} + \mathbf{v} \quad ; \quad \mathbf{v} + (\text{-}\mathbf{v}) = (\text{-}\mathbf{v}) + \mathbf{v} = \mathbf{0} \ . \tag{1.2}$$

Second, a mapping that to any $\lambda \in \Re$ (the field of real numbers) and to any element $\mathbf{v} \in \mathbb{S}$ associates an element of $\mathbb{S}$ denoted $\lambda\,\mathbf{v}$, with the following generic properties,[1]

$$\begin{aligned} 1\,\mathbf{v} &= \mathbf{v} \quad ; \quad (\lambda\,\mu)\,\mathbf{v} = \lambda\,(\mu\,\mathbf{v}) \\ (\lambda + \mu)\,\mathbf{v} &= \lambda\,\mathbf{v} + \mu\,\mathbf{v} \quad ; \quad \lambda\,(\mathbf{w} + \mathbf{v}) = \lambda\,\mathbf{w} + \lambda\,\mathbf{v} \ . \end{aligned} \tag{1.3}$$

Definition 1.1 Linear space. *When the conditions above are satisfied, we shall say that the set* $\mathbb{S}$ *has been endowed with a structure of* linear space, *or* vector space *(the two terms being synonymous). The elements of* $\mathbb{S}$ *are called* vectors, *and the real numbers are called* scalars.

To the sum operation $+$ for vectors is associated a second internal operation, called *difference* and denoted $-$, that is defined by the condition that for any three elements,

$$\mathbf{w} = \mathbf{v} + \mathbf{u} \quad \Longleftrightarrow \quad \mathbf{v} = \mathbf{w} - \mathbf{u} \ . \tag{1.4}$$

The following property then holds:

$$\mathbf{w} - \mathbf{v} = \mathbf{w} + (\text{-}\mathbf{v}) \ . \tag{1.5}$$

From these axioms follow all the well known properties of linear spaces, for instance, for any vectors $\mathbf{v}$ and $\mathbf{w}$ and any scalars λ and μ,

$$\begin{aligned} \mathbf{0} - \mathbf{v} &= \text{-}\mathbf{v} \quad ; \quad \mathbf{0} - (\text{-}\mathbf{v}) = \mathbf{v} \\ \lambda\,\mathbf{0} &= \mathbf{0} \quad ; \quad 0\,\mathbf{v} = \mathbf{0} \ , \\ \mathbf{w} - \mathbf{v} &= \mathbf{w} + (\text{-}\mathbf{v}) \quad ; \quad \mathbf{w} + \mathbf{v} = \mathbf{w} - (\text{-}\mathbf{v}) \\ \mathbf{w} - \mathbf{v} &= \text{-}(\mathbf{v} - \mathbf{w}) \quad ; \quad \mathbf{w} + \mathbf{v} = \text{-}(\,(\text{-}\mathbf{w}) + (\text{-}\mathbf{v})\,) \\ \lambda\,(\text{-}\mathbf{v}) &= \text{-}(\lambda\,\mathbf{v}) \quad ; \quad (\text{-}\lambda)\,\mathbf{v} = \text{-}(\lambda\,\mathbf{v}) \\ (\lambda - \mu)\,\mathbf{v} &= \lambda\,\mathbf{v} - \mu\,\mathbf{v} \quad ; \quad \lambda\,(\mathbf{w} - \mathbf{v}) = \lambda\,\mathbf{w} - \lambda\,\mathbf{v} \ . \end{aligned} \tag{1.6}$$

Example 1.1 *The set of* $p \times q$ *real matrices with the usual sum of matrices and the usual multiplication of a matrix by a real number forms a linear space.*

Example 1.2 *Using the definitions of exponential and logarithm of a square matrix (section 1.4.2), the two operations*

$$\mathbf{M} \boxplus \mathbf{N} \equiv \exp(\log \mathbf{M} + \log \mathbf{N}) \quad ; \quad \mathbf{M}^{\lambda} \equiv \exp(\lambda\,\log \mathbf{M}) \tag{1.7}$$

---

[1] As usual, the same symbol $+$ is used both for the sum of real numbers and the sum of vectors, as this does not generally cause any confusion.

'almost' endow the space of real $n \times n$ matrices (for which the log is defined) with a structure of linear space: if the considered matrices are 'close enough' to the identity matrix, all the axioms are satisfied. With this (associative and commutative) 'sum' and the matrix power, the space of real $n \times n$ matrices is locally *a linear space. Note that this example forces a shift with respect to the additive terminology used above (one does not multiply λ by the matrix $\mathbf{M}$, but raises the matrix $\mathbf{M}$ to the power λ).*

Here are two of the more basic definitions concerning linear spaces, those of subspace and of basis:

Definition 1.2 Linear subspace. *A subset of elements of a linear space $\mathbb{S}$ is called a* linear subspace *of $\mathbb{S}$ if the zero element belongs to the subset, if the sum of two elements of the subset belong to the subset, and if the product of an element of the subset by a real number belongs to the subset.*

Definition 1.3 Basis. *If there is a set of n linearly independent*[2] *vectors $\{\mathbf{e}_1, \dots, \mathbf{e}_n\}$ such that any vector $\mathbf{v} \in \mathbb{S}$ can be written as*[3]

$$\mathbf{v} = v^n \, \mathbf{e}_n + \cdots + v^2 \, \mathbf{e}_2 + v^1 \, \mathbf{e}_1 \equiv v^i \, \mathbf{e}_i \quad , \tag{1.8}$$

we say that $\{\mathbf{e}_1, \dots, \mathbf{e}_n\}$ is a basis *of $\mathbb{S}$, that the* dimension *of $\mathbb{S}$ is n, and that the $\{v^i\}$ are the* components *of $\mathbf{v}$ in the basis $\{\mathbf{e}_i\}$.*

It is easy to demonstrate that the components of a vector on a given basis are uniquely defined.

Let $\mathbb{S}$ be a finite-dimensional linear space. A *form* over $\mathbb{S}$ is a mapping from $\mathbb{S}$ into $\Re$.

Definition 1.4 Linear form. *One says that a form $\mathbf{f}$ is a* linear form, *and uses the notation $\mathbf{v} \mapsto \langle \, \mathbf{f} \, , \, \mathbf{v} \, \rangle$, if the mapping it defines is linear, i.e., if for any scalar and any vectors*

$$\langle \, \mathbf{f} \, , \, \lambda \mathbf{v} \, \rangle = \lambda \, \langle \, \mathbf{f} \, , \, \mathbf{v} \, \rangle \tag{1.9}$$

and

$$\langle \, \mathbf{f} \, , \, \mathbf{v}_2 + \mathbf{v}_1 \, \rangle = \langle \, \mathbf{f} \, , \, \mathbf{v}_2 \, \rangle + \langle \, \mathbf{f} \, , \, \mathbf{v}_1 \, \rangle \quad . \tag{1.10}$$

Definition 1.5 *The product of a linear form $\mathbf{f}$ by a scalar λ is defined by the condition that for any vector $\mathbf{v}$ of the linear space*

$$\langle \, \lambda \mathbf{f} \, , \, \mathbf{v} \, \rangle = \lambda \, \langle \, \mathbf{f} \, , \, \mathbf{v} \, \rangle \quad . \tag{1.11}$$

---

[2] I.e., the relation $\lambda_1 \, \mathbf{e}_1 + \cdots + \lambda_n \, \mathbf{e}_n = \mathbf{0}$ implies that all the λ are zero.

[3] The reverse notation used here is for homogeneity with similar notations to be found later, where the 'sum' is not necessarily commutative.

Definition 1.6 *The sum of two linear forms, denoted* $\mathbf{f}_2 + \mathbf{f}_2$, *is defined by the condition that for any vector* $\mathbf{v}$ *of the linear space*

$$\langle \mathbf{f}_2 + \mathbf{f}_1 , \mathbf{v} \rangle = \langle \mathbf{f}_2 , \mathbf{v} \rangle + \langle \mathbf{f}_1 , \mathbf{v} \rangle \quad . \tag{1.12}$$

We then have the well known

Property 1.1 *With the two operations* (1.11) *and* (1.12) *defined, the space of all linear forms over* $\mathbb{S}$ *is a linear (vector) space. It is called the* dual *of* $\mathbb{S}$, *and is denoted* $\mathbb{S}^*$.

Definition 1.7 Dual basis. *Let* $\{\mathbf{e}_i\}$ *be a basis of* $\mathbb{S}$, *and* $\{\mathbf{e}^i\}$ *a basis of* $\mathbb{S}^*$. *One says that these are* dual bases *if*

$$\langle \mathbf{e}^i , \mathbf{e}_j \rangle = \delta^i{}_j \quad . \tag{1.13}$$

While the components of a vector $\mathbf{v}$ on a basis $\{\mathbf{e}_i\}$, denoted v^i, are defined through the expression $\mathbf{v} = v^i \, \mathbf{e}_i$, the components of a (linear) form $\mathbf{f}$ on the dual basis $\{\mathbf{e}^i\}$, denoted f_i, are defined through $\mathbf{f} = f_i \, \mathbf{e}^i$. The evaluation of $\langle \mathbf{e}^i , \mathbf{v} \rangle$ and of $\langle \mathbf{f} , \mathbf{e}_i \rangle$, and the use of (1.13) immediately lead to

Property 1.2 *The components of vectors and forms are obtained, via the duality product, as*

$$v^i = \langle \mathbf{e}^i , \mathbf{v} \rangle \quad ; \quad f_i = \langle \mathbf{f} , \mathbf{e}_i \rangle \quad . \tag{1.14}$$

Expressions like these seem obvious thanks to the ingenuity of the index notation, with upper indices for components of vectors —and for the numbering of dual basis elements— and lower indices for components of forms —and for the numbering of primal basis elements.—

1.1.2 Tensor Spaces

Assume given a finite-dimensional linear space $\mathbb{S}$ and its dual $\mathbb{S}^*$. A 'tensor space' denoted

$$\mathbb{T} = \underbrace{\mathbb{S} \otimes \mathbb{S} \otimes \cdots \mathbb{S}}_{p \text{ times}} \otimes \underbrace{\mathbb{S}^* \otimes \mathbb{S}^* \otimes \cdots \mathbb{S}^*}_{q \text{ times}} \tag{1.15}$$

is introduced as the set of p linear forms over $\mathbb{S}^*$ and q linear forms over $\mathbb{S}$. Rather than giving here a formal exposition of the properties of such a space (with the obvious definition of sum of two elements and of product of an element by a real number, it is a linear space), let us just recall that an element $\mathbf{T}$ of $\mathbb{T}$ can be represented by the numbers $T^{i_1 i_2 \dots i_p}{}_{j_1 j_2 \dots j_q}$ such that to a set of p forms $\{\mathbf{f}_1, \mathbf{f}_2, \dots, \mathbf{f}_p\}$ and of q vectors $\{\mathbf{v}_1, \mathbf{v}_2, \dots, \mathbf{v}_q\}$ it associates the real number

$$\lambda = T^{i_1 i_2 \ldots i_p}{}_{j_1 j_2 \ldots j_q} \, (\mathbf{f}_1)_{i_1} \, (\mathbf{f}_2)_{i_2} \ldots (\mathbf{f}_p)_{i_p} \, (\mathbf{v}_1)^{j_1} \, (\mathbf{v}_2)^{j_2} \ldots (\mathbf{v}_q)^{j_q} \quad . \tag{1.16}$$

In fact, $T^{i_1 i_2 \ldots i_p}{}_{j_1 j_2 \ldots j_q}$ are the components of $\mathbf{T}$ on the basis induced over $\mathbb{T}$, by the respective (dual) bases of $\mathbb{S}$ and of $\mathbb{S}^*$, denoted $\mathbf{e}_{i_1} \otimes \mathbf{e}_{i_2} \otimes \cdots \mathbf{e}_{i_p} \otimes \mathbf{e}^{j_1} \otimes \mathbf{e}^{j_2} \otimes \cdots \mathbf{e}^{j_p}$, so one writes

$$\mathbf{T} = T^{i_1 i_2 \ldots i_p}{}_{j_1 j_2 \ldots j_q} \, \mathbf{e}_{i_1} \otimes \mathbf{e}_{i_2} \otimes \cdots \mathbf{e}_{i_p} \otimes \mathbf{e}^{j_1} \otimes \mathbf{e}^{j_2} \otimes \cdots \mathbf{e}^{j_p} \quad . \tag{1.17}$$

One easily gives sense to expressions like $w^i = T^{ij}{}_{k\ell} \, f_j \, v^k \, u^\ell$ or $T^i{}_j = S^{ik}{}_{kj}$.

1.1.3 Scalar Product Linear Space

Let $\mathbb{S}$ be a linear (vector) space, and let $\mathbb{S}^*$ be its dual.

Definition 1.8 Metric. *We shall say that the linear space* $\mathbb{S}$ *has a* metric *if there is a mapping* $\mathbf{G}$ *from* $\mathbb{S}$ *into* $\mathbb{S}^*$, *denoted using any of the two equivalent notations*

$$\mathbf{f} = \mathbf{G}(\mathbf{v}) = \mathbf{G}\,\mathbf{v} \quad , \tag{1.18}$$

that is (i) invertible; *(ii)* linear, *i.e., for any real* λ *and any vectors* $\mathbf{v}$ *and* $\mathbf{w}$, $\mathbf{G}(\lambda\,\mathbf{v}) = \lambda\,\mathbf{G}(\mathbf{v})$ *and* $\mathbf{G}(\mathbf{w}+\mathbf{v}) = \mathbf{G}\,\mathbf{w} + \mathbf{G}\,\mathbf{v}$; *(iii)* symmetric, *i.e., for any vectors* $\mathbf{v}$ *and* $\mathbf{w}$, $\langle\, \mathbf{G}\,\mathbf{w} \,,\, \mathbf{v} \,\rangle = \langle\, \mathbf{G}\,\mathbf{v} \,,\, \mathbf{w} \,\rangle$.

Definition 1.9 Scalar product. *Let* $\mathbf{G}$ *be a metric on* $\mathbb{S}$, *the* scalar product *of two vectors* $\mathbf{v}$ *and* $\mathbf{w}$ *of* $\mathbb{S}$, *denoted* $(\,\mathbf{v} \,,\, \mathbf{w}\,)$, *is the real number*[4]

$$(\,\mathbf{v} \,,\, \mathbf{w}\,) = \langle\, \mathbf{G}\,\mathbf{v} \,,\, \mathbf{w} \,\rangle \quad . \tag{1.19}$$

The symmetry of the metric implies the symmetry of the scalar product:

$$(\,\mathbf{v} \,,\, \mathbf{w}\,) = (\,\mathbf{w} \,,\, \mathbf{v}\,) \quad . \tag{1.20}$$

Consider now the scalar $(\,\lambda\,\mathbf{w} \,,\, \mathbf{v}\,)$. We easily construct the chain of equalities $(\,\lambda\,\mathbf{w} \,,\, \mathbf{v}\,) = \langle\, \mathbf{G}(\lambda\,\mathbf{w}) \,,\, \mathbf{v} \,\rangle = \langle\, \lambda\,\mathbf{G}\,\mathbf{w} \,,\, \mathbf{v} \,\rangle = \lambda\,\langle\, \mathbf{G}\,\mathbf{w} \,,\, \mathbf{v} \,\rangle = \lambda\,(\,\mathbf{w} \,,\, \mathbf{v}\,)$. From this and the symmetry property (1.20), it follows that for any vectors $\mathbf{v}$ and $\mathbf{w}$, and any real λ,

$$(\,\lambda\,\mathbf{v} \,,\, \mathbf{w}\,) = (\,\mathbf{v} \,,\, \lambda\,\mathbf{w}\,) = \lambda\,(\,\mathbf{v} \,,\, \mathbf{w}\,) \quad . \tag{1.21}$$

Finally,

$$(\,\mathbf{w} + \mathbf{v} \,,\, \mathbf{u}\,) = (\,\mathbf{w} \,,\, \mathbf{u}\,) + (\,\mathbf{v} \,,\, \mathbf{u}\,) \tag{1.22}$$

and

$$(\,\mathbf{w} \,,\, \mathbf{v} + \mathbf{u}\,) = (\,\mathbf{w} \,,\, \mathbf{v}\,) + (\,\mathbf{w} \,,\, \mathbf{u}\,) \quad . \tag{1.23}$$

[4]As we don't require definite positiveness, this is a 'pseudo' scalar product.

Definition 1.10 Norm. *In a scalar product vector space, the* squared pseudonorm *(or, for short, 'squared norm') of a vector* $\mathbf{v}$ *is defined as* $\|\mathbf{v}\|^2 = (\mathbf{v}, \mathbf{v})$*, and the* pseudonorm *(or, for short, 'norm') as*

$$\|\mathbf{v}\| = \sqrt{(\mathbf{v}, \mathbf{v})} \quad . \tag{1.24}$$

By definition of the square root of a real number, the pseudonorm of a vector may be zero, or positive real or positive imaginary. There may be 'light-like' vectors $\mathbf{v} \neq \mathbf{0}$ such that $\|\mathbf{v}\| = \mathbf{0}$. One has $\|\lambda\mathbf{v}\| = \sqrt{(\lambda\mathbf{v}, \lambda\mathbf{v})} = \sqrt{\lambda^2 (\mathbf{v}, \mathbf{v})} = \sqrt{\lambda^2}\sqrt{(\mathbf{v}, \mathbf{v})}$, i.e., $\|\lambda\mathbf{v}\| = |\lambda|\,\|\mathbf{v}\|$. Taking $\lambda = 0$ in this equation shows that the zero vector has necessarily zero norm, $\|\mathbf{0}\| = 0$ while taking $\lambda = -1$ gives $\|\mathbf{-v}\| = \|\mathbf{v}\|$.

Defining

$$g_{ij} = (\mathbf{e}_i, \mathbf{e}_j) \quad , \tag{1.25}$$

we easily arrive at the relation linking a vector $\mathbf{v}$ and the form $\mathbf{G}\,\mathbf{v}$ associated to it,

$$v_i = g_{ij}\, v^j \quad , \tag{1.26}$$

where, as usual, the same symbol is used to denote the components $\{v^i\}$ of a vector $\mathbf{v}$ and the components $\{v_i\}$ of the associated form. The g_{ij} are easily shown to be the components of the metric $\mathbf{G}$ on the basis $\{\mathbf{e}^i \otimes \mathbf{e}^j\}$. Writing g^{ij} the components of $\mathbf{G}^{-1}$ on the basis $\{\mathbf{e}_i \otimes \mathbf{e}_j\}$, one obtains $g^{ij}\, g_{jk} = \delta^i{}_k$, and the reciprocal of equation (1.26) is then $v^i = g^{ij}\, v_j$. It is easy to see that the duality product of a form $\mathbf{f} = f_i\, \mathbf{e}^i$ by a vector $\mathbf{v} = v^i\, \mathbf{e}_i$ is

$$\langle \mathbf{f}, \mathbf{v} \rangle = f_i\, v^i \quad , \tag{1.27}$$

the scalar product of a vector $\mathbf{v} = v^i\, \mathbf{e}_i$ by a vector $\mathbf{w} = w^i\, \mathbf{e}_i$ is

$$(\mathbf{v}, \mathbf{w}) = g_{ij}\, v^i\, w^j \quad , \tag{1.28}$$

and the (pseudo) norm of a vector $\mathbf{v} = v^i\, \mathbf{e}_i$ is

$$\|\mathbf{v}\| = \sqrt{g_{ij}\, v^i\, v^j} \quad . \tag{1.29}$$

1.1.4 Universal Metric for Bivariant Tensors

Consider an n-dimensional linear space $\mathbb{S}$. If there is a metric g_{ij} defined over $\mathbb{S}$ one may easily define the norm of a vector, and of a form, and, therefore, the norm of a second-order tensor:

Definition 1.11 *The 'Frobenius norm' of a tensor* $\mathbf{t} = \{t^{ij}\}$ *is defined as*

$$\|\mathbf{t}\|_F = \sqrt{g_{ij}\, g_{k\ell}\, t^{ik}\, t^{j\ell}} \quad . \tag{1.30}$$

If no metric is defined over $\mathbb{S}$, the norm of a vector v^i is not defined. But there is a 'universal' way of defining the norm of a 'bivariant' tensor[5] $t^i{}_j$. To see this let us introduce the following

Definition 1.12 Universal metric. *For any two nonvanishing real numbers χ and ψ, the operator with components*

$$g_i{}^j{}_k{}^\ell = \chi\, \delta_i^\ell\, \delta_k^j + \frac{\psi - \chi}{n}\, \delta_i^j\, \delta_k^\ell \tag{1.31}$$

maps the space of bivariant ('contravariant–covariant') tensors into its dual,[6] is symmetric, and invertible. Therefore, it defines a metric over $\mathbb{S} \otimes \mathbb{S}^*$, that we shall call the universal metric.

One may then easily demonstrate the

Property 1.3 *With the universal metric* (1.31), *the (pseudo) norm of a bivariant tensor,* $\| \mathbf{t} \| = \sqrt{g_i{}^j{}_k{}^\ell\, t^i{}_j\, t^k{}_\ell}$ *verifies*

$$\| \mathbf{t} \|^2 = \chi \operatorname{tr} \mathbf{t}^2 + \frac{\psi - \chi}{n} (\operatorname{tr} \mathbf{t})^2 \quad . \tag{1.32}$$

Equivalently,

$$\| \mathbf{t} \|^2 = \chi \operatorname{tr} \tilde{\mathbf{t}}^2 + \psi \operatorname{tr} \bar{\mathbf{t}}^2 \quad . \tag{1.33}$$

where $\bar{\mathbf{t}}$ and $\tilde{\mathbf{t}}$ are respectively the isotropic and the deviatoric parts of $\mathbf{t}$:

$$\bar{t}^i{}_j = \frac{1}{n}\, t^k{}_k\, \delta^i{}_j \quad ; \quad \tilde{t}^i{}_j = t^i{}_j - \frac{1}{n}\, t^k{}_k\, \delta^i{}_j \quad ; \quad t^i{}_j = \bar{t}^i{}_j + \tilde{t}^i{}_j \ . \tag{1.34}$$

Expression (1.33) gives the interpretation of the two free parameters χ and ψ as defining the relative 'weights' with which the isotropic part and the deviatoric part of the tensor enter in its norm.

Defining the inverse (i.e., contravariant) metric by the condition $g_i{}^j{}_p{}^q\, g^p{}_q{}^k{}_\ell = \delta_i^k\, \delta_\ell^j$ gives

$$g^i{}_j{}^k{}_\ell = \alpha\, \delta_\ell^i\, \delta_j^k + \frac{\beta - \alpha}{n}\, \delta_j^i\, \delta_\ell^k \quad ; \quad \alpha = \frac{1}{\chi} \quad ; \quad \beta = \frac{1}{\psi} \ . \tag{1.35}$$

It follows from expression (1.33) that the universal metric introduced above is the more general expression for an *isotropic* metric, i.e., a metric that respects the decomposition of a tensor into its isotropic and deviatoric parts.

We shall later see how this universal metric relates to the Killing-Cartan definition of metric in the 'algebras' of Lie groups.

---

[5] Here, by *bivariant tensor* we understand a tensor with indices $t^i{}_j$. Similar developments could be made for tensors with indices $t_i{}^j$.

[6] I.e., the space $\mathbb{S}^* \otimes \mathbb{S}$ of 'covariant–contravariant' tensors, via $t_i{}^j \equiv g_i{}^j{}_k{}^\ell\, t^k{}_\ell$.

1.2 Autovector Space

1.2.1 Troupe

A *troupe*, essentially, will be defined as a "group without the associative property". In that respect, the troupe structure is similar, but not identical, to the *loop* structure in the literature, and the differences are fundamental for our goal (to generalize the notion of vector space into that of autovector space). This goal explains the systematic use of the additive notation —rather than the usual multiplicative notation— even when the structure is associative, i.e., when it is a group: in this manner, Lie groups will later be interpreted as local groups of additive geodesics.

As usual, a *binary operation* over a set $\mathbb{S}$ is a mapping that maps every ordered pair of elements of $\mathbb{S}$ into a (unique) element of $\mathbb{S}$.

Definition 1.13 Troupe. *A* troupe *is a set* $\mathbb{S}$ *of elements* $\mathbf{u}, \mathbf{v}, \mathbf{w}, \dots$ *with two internal binary operations, denoted* $\oplus$ *and* $\ominus$*, related through the equivalence*

$$\boxed{\mathbf{w} = \mathbf{v} \oplus \mathbf{u} \quad \Longleftrightarrow \quad \mathbf{v} = \mathbf{w} \ominus \mathbf{u} \quad ,} \tag{1.36}$$

with an element $\mathbf{0}$ *that is neutral for the* $\oplus$ *operation, i.e., such that for any* $\mathbf{v}$ *of* $\mathbb{S}$*,*

$$\boxed{\mathbf{0} \oplus \mathbf{v} = \mathbf{v} \oplus \mathbf{0} = \mathbf{v} \quad ,} \tag{1.37}$$

and such that to any element $\mathbf{v}$ *of* $\mathbb{S}$*, is associated another element, denoted* $\mathbf{-v}$*, and called its* opposite*, satisfying*

$$\boxed{(\mathbf{-v}) \oplus \mathbf{v} = \mathbf{v} \oplus (\mathbf{-v}) = \mathbf{0} \quad .} \tag{1.38}$$

The postulate in equation 1.36 implies that in the relation $\mathbf{w} = \mathbf{v} \oplus \mathbf{u}$, the pair of elements $\mathbf{w}$ and $\mathbf{u}$ determines a unique $\mathbf{v}$ (as $\ominus$ is assumed to be an operation, so that the expression $\mathbf{v} = \mathbf{w} \ominus \mathbf{u}$ determines $\mathbf{v}$ uniquely). It is not assumed that in the relation $\mathbf{w} = \mathbf{v} \oplus \mathbf{u}$ the pair of elements $\mathbf{w}$ and $\mathbf{v}$ determines a unique $\mathbf{u}$ and there are troupes where such a $\mathbf{u}$ is not unique (see example 1.3). It is postulated that there is at least one neutral element satisfying equation (1.37); its uniqueness follows immediately from $\mathbf{v} = \mathbf{0} \oplus \mathbf{v}$, using the first postulate. Also, the uniqueness of the opposite follows immediately from $\mathbf{0} = (\mathbf{-v}) \oplus \mathbf{v}$, while from $\mathbf{0} = \mathbf{v} \oplus (\mathbf{-v})$ follows that the opposite of $\mathbf{-v}$ is $\mathbf{v}$ itself:

$$-(\mathbf{-v}) = \mathbf{v} \quad . \tag{1.39}$$

The expression $\mathbf{w} = \mathbf{v} \oplus \mathbf{u}$ is to be read " $\mathbf{w}$ *is obtained by adding* $\mathbf{v}$ *to* $\mathbf{u}$ ". As this is, in general, a noncommutative sum, the order of the terms matters. Note that interpreting $\mathbf{w} = \mathbf{v} \oplus \mathbf{u}$ as the result of adding $\mathbf{v}$ to a given $\mathbf{u}$ is consistent with the usual multiplicative notation for operators, where

$\mathbf{C} = \mathbf{B}\,\mathbf{A}$ means applying $\mathbf{A}$ first, then $\mathbf{B}$. If there is no risk of confusion, the sentence "$\mathbf{w}$ is obtained by adding $\mathbf{v}$ to $\mathbf{u}$" can be simplified to $\mathbf{w}$ *equals* $\mathbf{v}$ *plus* $\mathbf{u}$ (or, if there is any risk of confusion with a commutative sum, we can say $\mathbf{w}$ *equals* $\mathbf{v}$ *o-plus* $\mathbf{u}$). The expression $\mathbf{v} = \mathbf{w} \ominus \mathbf{u}$ is to be read "$\mathbf{v}$ *is obtained by subtracting* $\mathbf{u}$ *from* $\mathbf{w}$". More simply, we can say $\mathbf{v}$ *equals* $\mathbf{w}$ *minus* $\mathbf{u}$ (or, if there is any risk of confusion, $\mathbf{v}$ *equals* $\mathbf{w}$ *o-minus* $\mathbf{u}$).

Setting $\mathbf{v} = \mathbf{0}$ in equations (1.37), using the equivalence (1.36), and considering that the opposite is unique, we obtain

$$\mathbf{0} \oplus \mathbf{0} = \mathbf{0} \quad ; \quad \mathbf{0} \ominus \mathbf{0} = \mathbf{0} \quad ; \quad -\mathbf{0} = \mathbf{0} \ . \tag{1.40}$$

The most basic properties of the operation $\ominus$ are easily obtained using the equivalence (1.36) to rewrite equations (1.37)–(1.38), this showing that, for any element $\mathbf{v}$ of the troupe,

$$\begin{aligned} \mathbf{v} \ominus \mathbf{v} &= \mathbf{0} \quad ; \quad \mathbf{v} \ominus \mathbf{0} = \mathbf{v} \\ \mathbf{0} \ominus \mathbf{v} &= -\mathbf{v} \quad ; \quad \mathbf{0} \ominus (-\mathbf{v}) = \mathbf{v} \ , \end{aligned} \tag{1.41}$$

all these properties being intuitively expected from a minus operation. Inserting each of the two expressions (1.36) in the other one shows that, for any $\mathbf{v}$ and $\mathbf{w}$ of a troupe,

$$\boxed{(\mathbf{w} \oplus \mathbf{v}) \ominus \mathbf{v} = \mathbf{w} \quad ; \quad (\mathbf{w} \ominus \mathbf{v}) \oplus \mathbf{v} = \mathbf{w} \ ,} \tag{1.42}$$

i.e., one has a *right-simplification property*. While it is clear (using the first of equations (1.41)) that if $\mathbf{w} = \mathbf{v}$, then, $\mathbf{w} \ominus \mathbf{v} = \mathbf{0}$, the reciprocal can also be demonstrated,[7] so that we have the equivalence

$$\mathbf{w} \ominus \mathbf{v} = \mathbf{0} \quad \Longleftrightarrow \quad \mathbf{w} = \mathbf{v} \ . \tag{1.43}$$

Similarly, while it is clear (using the second of equations (1.38)) that if $\mathbf{w} = -\mathbf{v}$, then, $\mathbf{w} \oplus \mathbf{v} = \mathbf{0}$, the reciprocal can also be demonstrated,[8] so that we also have the equivalence

$$\mathbf{w} \oplus \mathbf{v} = \mathbf{0} \quad \Longleftrightarrow \quad \mathbf{w} = -\mathbf{v} \ . \tag{1.44}$$

Another property[9] of the troupe structure may be expressed by the equivalences

$$\mathbf{v} \oplus \mathbf{0} = \mathbf{0} \quad \Longleftrightarrow \quad \mathbf{0} \oplus \mathbf{v} = \mathbf{0} \quad \Longleftrightarrow \quad \mathbf{v} = \mathbf{0} \ , \tag{1.45}$$

[7]From relation (1.36), $\mathbf{w} \ominus \mathbf{v} = \mathbf{0} \Rightarrow \mathbf{w} = \mathbf{0} \oplus \mathbf{v}$, then, using the first of equations (1.37), $\mathbf{w} \ominus \mathbf{v} = \mathbf{0} \Rightarrow \mathbf{w} = \mathbf{v}$.

[8]From relation (1.36), $\mathbf{w} \oplus \mathbf{v} = \mathbf{0} \Rightarrow \mathbf{w} = \mathbf{0} \ominus \mathbf{v}$, then, using the second of equations (1.41), $\mathbf{w} \oplus \mathbf{v} = \mathbf{0} \Rightarrow \mathbf{w} = -\mathbf{v}$.

[9]From relation (1.36) follows that, for any element $\mathbf{v}$, $\mathbf{v} \oplus \mathbf{0} = \mathbf{0} \Leftrightarrow \mathbf{v} = \mathbf{0} \ominus \mathbf{0}$, i.e., using property (1.40), $\mathbf{v} \oplus \mathbf{0} = \mathbf{0} \Leftrightarrow \mathbf{v} = \mathbf{0}$. Also from relation (1.36) follows that, for any element $\mathbf{v}$, $\mathbf{0} \oplus \mathbf{v} = \mathbf{0} \Leftrightarrow \mathbf{0} = \mathbf{0} \ominus \mathbf{v}$, i.e., using property (1.43), $\mathbf{0} \oplus \mathbf{v} = \mathbf{0} \Leftrightarrow \mathbf{v} = \mathbf{0}$.

and there is also a similar series of equivalences for the $\ominus$ operation[10]

$$\mathbf{v}\ominus\mathbf{0} = \mathbf{0} \quad \Longleftrightarrow \quad \mathbf{0}\ominus\mathbf{v} = \mathbf{0} \quad \Longleftrightarrow \quad \mathbf{v} = \mathbf{0} \ . \tag{1.46}$$

To define a particular operation $\mathbf{w}\,\triangle\,\mathbf{v}$ it is sometimes useful to present the results of the operation in a *Cayley table*:

$\triangle$	$\cdots$	$\mathbf{v}$	$\cdots$
$\cdots$	$\cdots$	$\cdots$	$\cdots$
$\mathbf{w}$	$\cdots$	$\mathbf{w}\,\triangle\,\mathbf{v}$	$\cdots$
$\cdots$	$\cdots$	$\cdots$	$\cdots$

,

where we shall use the convention that the element $\mathbf{w}\,\triangle\,\mathbf{v}$ is in the *column* defined by $\mathbf{v}$ and the *row* defined by $\mathbf{w}$. The axiom in equation (1.36) can be translated, in terms of the Cayley tables of the operations $\oplus$ and $\ominus$ of a troupe, by the condition that the elements in every column of the table must all be different.[11]

Example 1.3 *The neutral element* $\mathbf{0}$*, two elements* $\mathbf{v}$ *and* $\mathbf{w}$*, and two elements* $\mathbf{\text{-}v}$ *and* $\mathbf{\text{-}w}$ *(the opposites to* $\mathbf{v}$ *and* $\mathbf{w}$*), submitted to the operations* $\oplus$ *and* $\ominus$ *defined by any of the two equivalent tables*

$\oplus$	-w	-v	0	v	w
w	0	v	w	-w	v
v	w	0	v	w	-v
0	-w	-v	0	v	w
-v	v	-w	-v	0	-w
-w	-v	w	-w	-v	0

$\ominus$	-w	-v	0	v	w
-w	0	-w	-w	w	-v
-v	-w	0	-v	-w	v
0	w	v	0	-v	-w
v	-v	w	v	0	w
w	v	-v	w	v	0

form a troupe.[12] *The operation* $\oplus$ *is not associative,*[13] *as* $\mathbf{v}\oplus(\mathbf{v}\oplus\mathbf{v}) = \mathbf{v}\oplus\mathbf{w} = \mathbf{\text{-}v}$*, while* $(\mathbf{v}\oplus\mathbf{v})\oplus\mathbf{v} = \mathbf{w}\oplus\mathbf{v} = \mathbf{\text{-}w}$*.*

The fact that the set of all oriented geodesic segments (having common origin) on a manifold will be shown to be (locally) a troupe, is what justifies the introduction of this kind of structure. It is easy to see that the sum of oriented geodesic segments does not have the associative property (even locally), so it cannot fit into the more common group structure. Note that in a troupe, in general,

$$\mathbf{w}\ominus\mathbf{v} \neq \mathbf{w}\oplus(\mathbf{\text{-}v}) \tag{1.47}$$

[10] From relation (1.36) it follows that, for any element $\mathbf{v}$, $\mathbf{v}\ominus\mathbf{0} = \mathbf{0} \Leftrightarrow \mathbf{v} = \mathbf{0}\ominus\mathbf{0}$, i.e., using property (1.40), $\mathbf{v}\ominus\mathbf{0} = \mathbf{0} \Leftrightarrow \mathbf{v} = \mathbf{0}$. Also from relation (1.36) it follows that, for any element $\mathbf{v}$, $\mathbf{0}\ominus\mathbf{v} = \mathbf{0} \Leftrightarrow \mathbf{0} = \mathbf{0}\oplus\mathbf{v}$, i.e., using property (1.45), $\mathbf{0}\ominus\mathbf{v} = \mathbf{0} \Leftrightarrow \mathbf{v} = \mathbf{0}$.

[11] In a loop, all the elements in every column and every row must be different (Pflugfelder, 1990).

[12] This troupe is not a loop, as the elements of each row are not all different.

[13] Which means, as we shall see later, that the troupe is not a group.

and, also, in general,

$$\mathbf{w} = \mathbf{v} \oplus \mathbf{u} \quad \not\Rightarrow \quad \mathbf{u} = (\text{-}\mathbf{v}) \oplus \mathbf{w} \quad . \tag{1.48}$$

Although mathematical rigor would impose reserving the term 'troupe' for the pair[14] $(\mathbb{S}, \oplus)$, rather than for the set $\mathbb{S}$ alone (as more than one troupe operation can be defined over a given set), we shall simply say, when there is no ambiguity, " the troupe $\mathbb{S}$ ".

1.2.2 Group

Definition 1.14 First definition of group. *A* group *is a troupe satisfying, for any* $\mathbf{u}$, $\mathbf{v}$ *and* $\mathbf{w}$, *the* homogeneity property

$$\boxed{(\mathbf{v} \ominus \mathbf{w}) \ominus (\mathbf{u} \ominus \mathbf{w}) = \mathbf{v} \ominus \mathbf{u} \quad .} \tag{1.49}$$

From this homogeneity property, it is easy to deduce the extra properties valid in groups (see demonstrations in appendix A.2). First, one sees that for any $\mathbf{u}$ and $\mathbf{v}$ in a group, the *oppositivity property*

$$\mathbf{v} \ominus \mathbf{u} = \text{-}(\mathbf{u} \ominus \mathbf{v}) \tag{1.50}$$

holds. Also, for any $\mathbf{u}$, $\mathbf{v}$ and $\mathbf{w}$ of a group,

$$\mathbf{v} \oplus \mathbf{u} = \mathbf{v} \ominus (\text{-}\mathbf{u}) = \text{-}((\text{-}\mathbf{u}) \oplus (\text{-}\mathbf{v})) \tag{1.51}$$

and

$$\mathbf{v} \ominus \mathbf{u} = \mathbf{v} \oplus (\text{-}\mathbf{u}) = (\mathbf{v} \ominus \mathbf{w}) \oplus (\mathbf{w} \ominus \mathbf{u}) = (\mathbf{v} \oplus \mathbf{w}) \ominus (\mathbf{u} \oplus \mathbf{w}) \quad . \tag{1.52}$$

In a group, also, one has the equivalence

$$\mathbf{w} = \mathbf{v} \oplus \mathbf{u} \quad \Longleftrightarrow \quad \mathbf{v} = \mathbf{w} \oplus (\text{-}\mathbf{u}) \quad \Longleftrightarrow \quad \mathbf{u} = (\text{-}\mathbf{v}) \oplus \mathbf{w} \quad . \tag{1.53}$$

Finally, in a group, one has (see appendix A.2) the following

Property 1.4 *In a group (i.e., in a troupe satisfying the relation (1.49)) the* associativity *property holds, i.e., for any three elements* $\mathbf{u}$, $\mathbf{v}$ *and* $\mathbf{w}$,

$$\boxed{\mathbf{w} \oplus (\mathbf{v} \oplus \mathbf{u}) = (\mathbf{w} \oplus \mathbf{v}) \oplus \mathbf{u} \quad .} \tag{1.54}$$

Better known than this theorem is its reciprocal (the associativity property (1.54) implies the oppositivity property (1.50) and the homogeneity property (1.49)), so we have the equivalent definition:

[14]Or to the pair $(\mathbb{S}, \ominus)$, as one operation determines the other.

Definition 1.15 Second definition of group. *A* group *is an* associative troupe, *i.e., a troupe where, for any three elements* $\mathbf{u}$, $\mathbf{v}$ *and* $\mathbf{w}$, *the property* (1.54) *holds.*

The derivation of the associativity property (1.54) from the homogeneity property (1.49) suggests that there is not much room for algebraic structures that would be intermediate between a troupe and a group.

Definition 1.16 Subgroup. *A subset of elements of a group is called a* subgroup *if it is itself a group.*

Definition 1.17 Commutative group. *A* commutative group *is a group where the operation* $\oplus$ *is commutative, i.e., where for any* $\mathbf{v}$ *and* $\mathbf{w}$, $\mathbf{w}\oplus\mathbf{v} = \mathbf{v}\oplus\mathbf{w}$.

As a group is an associative troupe, we can also define a commutative group as an associative and commutative troupe.

For details on the theory of groups, the reader may consult one of the many good books on the subject, for instance, Hall (1976).

A commutative and associative o-sum $\oplus$ is often an 'ordinary sum', so one can use the symbol $+$ to represent it (but remember example 1.2, where a commutative and associative 'sum' is considered that is not the ordinary sum). The commutativity property then becomes $\mathbf{w}+\mathbf{v} = \mathbf{v}+\mathbf{w}$. Similarly, using the symbol '$-$' for the difference, one has, for instance, $\mathbf{w}-\mathbf{v} = \mathbf{w}+(\text{-}\mathbf{v}) = (\text{-}\mathbf{v})+\mathbf{w}$ and $\mathbf{w}-\mathbf{v} = \text{-}(\mathbf{v}-\mathbf{w})$.

Rather than the additive notation used here for a group, a multiplicative notation is more commonly used. When dealing with Lie groups in later sections of this chapter we shall see that this is not only a matter of notation: Lie groups accept two fundamentally different matrix representations, and while in one of the representations the group operation is the product of matrices, in the second representation, the group operation is a 'noncommutative sum'. For easy reference, let us detail here the basic group equations when a multiplicative representation is used.

Let us denote $\mathbf{A}, \mathbf{B}\ldots$ the elements of a group when using a multiplicative representation.

Definition 1.18 Third definition of group. *A* group *is a set of elements* $\mathbf{A}, \mathbf{B}\ldots$ *endowed with an internal operation* $\mathbf{C} = \mathbf{B}\,\mathbf{A}$ *that has the following three properties:*

- *there is a neutral element, denoted* $\mathbf{I}$ *and called the* identity, *such that for any* $\mathbf{A}$,

$$\mathbf{I}\,\mathbf{A} = \mathbf{A}\,\mathbf{I} = \mathbf{A} \quad ; \tag{1.55}$$

- *for every element* $\mathbf{A}$ *there is an* inverse *element, denoted* $\mathbf{A}^{-1}$, *such that*

$$\mathbf{A}^{-1}\,\mathbf{A} = \mathbf{A}\,\mathbf{A}^{-1} = \mathbf{I} \quad ; \tag{1.56}$$

– *for every three elements, the associative property holds:*

$$\mathbf{C}\,(\mathbf{B}\,\mathbf{A}) \;=\; (\mathbf{C}\,\mathbf{B})\,\mathbf{A} \quad . \tag{1.57}$$

These three axioms are, of course, the immediate translation of properties (1.37), (1.38) and (1.54).

The properties of groups are well known (Hall, 1976). In particular, for any elements, one has (equivalent of equation (1.53))

$$\mathbf{C} = \mathbf{B}\cdot\mathbf{A} \quad \Leftrightarrow \quad \mathbf{B} = \mathbf{C}\cdot\mathbf{A}^{-1} \quad \Leftrightarrow \quad \mathbf{A} = \mathbf{B}^{-1}\cdot\mathbf{C} \tag{1.58}$$

and (equations (1.39), (1.51), and (1.52))

$$(\mathbf{A}^{-1})^{-1} = \mathbf{A} \;\; ; \;\; \mathbf{B}\cdot\mathbf{A} = (\,\mathbf{A}^{-1}\cdot\mathbf{B}^{-1}\,)^{-1} \;\; ; \;\; (\mathbf{B}\cdot\mathbf{C})\cdot(\mathbf{A}\cdot\mathbf{C})^{-1} = \mathbf{B}\cdot\mathbf{A}^{-1}\,. \tag{1.59}$$

A group is called *commutative* if for any two elements, $\mathbf{B}\,\mathbf{A} = \mathbf{A}\,\mathbf{B}$ (for commutative groups the multiplicative notation is usually drop).

1.2.3 Autovector Space

The structure about to be introduced, the "space of autoparallel vectors", is the generalization of the usual structure of (linear) vector space to the case where the sum of elements is not necessarily associative and commutative. If a (linear) vector can be seen as an oriented (straight) segment in a flat manifold, an "autoparallel vector", or 'autovector', represents an oriented autoparallel segment in a manifold that may have torsion and curvature.[15]

Definition 1.19 Autovector space. *Let the set* $\mathbb{S}$*, with elements* $\mathbf{u}, \mathbf{v}, \mathbf{w} \dots$ *, be a linear space with the two usual operations represented as* $\mathbf{w}+\mathbf{v}$ *and* $\lambda\,\mathbf{v}$*. We shall say that* $\mathbb{S}$ *is an* autovector space *if there exists a second internal operation* $\oplus$ *defined over* $\mathbb{S}$*, that is a troupe operation (generally, nonassociative and noncommutative), related to the linear space operation* $+$ *as follows:*

- *the neutral element* $\mathbf{0}$ *for the operation* $+$ *is also the neutral element for the* $\oplus$ *operation;*
- *for colinear elements, the operation* $\oplus$ *coincides with the operation* $+$*;*
- *the operation* $\oplus$ *is* analytic *in terms of* $+$ *inside a finite neighborhood of the origin.*[16]

We say that, while $\{\mathbb{S}, \oplus\}$ *is an autovector space,* $\{\mathbb{S}, +\}$ *is its* tangent linear space. *When considered as elements of* $\{\mathbb{S}, \oplus\}$*, the vectors of* $\{\mathbb{S}, +\}$ *are also called* autovectors.

[15]The notion of autovector has some similarities with the notion of gyrovector, introduced by Ungar (2001) to account for the Thomas precession of special relativity.

[16]I.e., there exists a series expansion written in the linear space $\{\mathbb{S}, +\}$ that, for any elements $\mathbf{v}$ and $\mathbf{w}$ inside a finite neighborhood of the origin, converges to $\mathbf{w} \oplus \mathbf{v}$.

To develop the theory, let us recall that, because we assume that $\mathbb{S}$ is both an autovector space (with the operation $\oplus$) and a linear space (with the operation $+$), all the axioms of a linear space are satisfied, in particular the two first axioms in equations (1.3). They state that for any element $\mathbf{v}$ and for any scalars λ and μ,

$$\boxed{1\,\mathbf{v} = \mathbf{v} \quad ; \quad (\lambda\,\mu)\,\mathbf{v} = \lambda\,(\mu\,\mathbf{v}) \quad .} \tag{1.60}$$

Now, the first of the conditions above means that for any element $\mathbf{v}$,

$$\mathbf{v}\oplus\mathbf{0} = \mathbf{0}\oplus\mathbf{v} = \mathbf{v}+\mathbf{0} = \mathbf{0}+\mathbf{v} = \mathbf{v} \quad . \tag{1.61}$$

The second condition implies that for any element $\mathbf{v}$ and any scalars λ and μ,

$$\mu\,\mathbf{v}\oplus\lambda\,\mathbf{v} = \mu\,\mathbf{v}+\lambda\,\mathbf{v} \quad ; \tag{1.62}$$

i.e., because of the property $\mu\,\mathbf{v}+\lambda\,\mathbf{v} = (\mu+\lambda)\,\mathbf{v}$,

$$\boxed{\mu\,\mathbf{v}\oplus\lambda\,\mathbf{v} = (\mu+\lambda)\,\mathbf{v} \quad .} \tag{1.63}$$

From this, it easily follows[17] that for any vector $\mathbf{v}$ and any real numbers λ and μ,

$$\mu\,\mathbf{v}\ominus\lambda\,\mathbf{v} = (\mu-\lambda)\,\mathbf{v} \quad . \tag{1.64}$$

The analyticity condition imposes that for any two elements $\mathbf{v}$ and $\mathbf{w}$ and for any λ (inside the interval where the operation makes sense), the following series expansion is convergent:

$$\lambda\,\mathbf{w}\oplus\lambda\,\mathbf{v} = \mathbf{c}_0 + \lambda\,\mathbf{c}_1(\mathbf{w},\mathbf{v}) + \lambda^2\,\mathbf{c}_2(\mathbf{w},\mathbf{v}) + \lambda^3\,\mathbf{c}_3(\mathbf{w},\mathbf{v}) + \dots \quad , \tag{1.65}$$

where the $\mathbf{c}_i$ are vector functions of $\mathbf{v}$ and $\mathbf{w}$. As explained in section 1.2.4, the axioms defining an autovector space impose the conditions $\mathbf{c}_0 = \mathbf{0}$ and $\mathbf{c}_1(\mathbf{w},\mathbf{v}) = \mathbf{w}+\mathbf{v}$. Therefore, this series, in fact, starts as $\lambda\,\mathbf{w}\oplus\lambda\,\mathbf{v} = \lambda\,(\mathbf{w}+\mathbf{v}) + \dots$, so we have the property

$$\boxed{\lim_{\lambda\to 0}\frac{1}{\lambda}\,(\lambda\,\mathbf{w}\oplus\lambda\,\mathbf{v}) = \mathbf{w}+\mathbf{v} \quad .} \tag{1.66}$$

This expression shows in which sense the operation $+$ is *tangent* to the operation $\oplus$.

The reader will immediately recognize that the four relations (1.60), (1.63) and (1.66) are those defining a linear space, except that instead of a condition like $\lambda\,\mathbf{w}\oplus\lambda\,\mathbf{v} = \lambda\,(\mathbf{w}\oplus\mathbf{v})$ (not true in an autovector space), we have the relation (1.66). This suggests an alternative definition of an autovector space, less rigorous but much simpler, as follows:

[17]Equation (1.63) can be rewritten $\mu\,\mathbf{v} = (\mu+\lambda)\,\mathbf{v}\ominus\lambda\,\mathbf{v}$, i.e., introducing $\nu = \mu+\lambda$, $(\nu-\lambda)\,\mathbf{v} = \nu\,\mathbf{v}\ominus\lambda\,\mathbf{v}$.

Definition 1.20 Autovector space (alternative definition). *Let* $\mathbb{S}$ *be a set of elements* $\mathbf{u}, \mathbf{v}, \mathbf{w} \ldots$ *with an internal operation* $\oplus$ *that is a troupe operation. We say that the troupe* $\mathbb{S}$ *is an* autovector space *if there also is a mapping that to any* $\lambda \in \Re$ *(the field of real numbers) and to any element* $\mathbf{v} \in \mathbb{S}$ *associates an element of* $\mathbb{S}$ *denoted* $\lambda\,\mathbf{v}$*, satisfying the two conditions* (1.60)*, the condition* (1.63)*, and the condition that the limit on the left in equation* (1.66) *makes sense, this defining a new troupe operation* $+$ *that is both commutative and associative (called the* tangent sum*).*

In the applications considered below, the above definition of autovector space is too demanding, and must be relaxed, as the structure of autovector space is valid only inside some finite region around the origin: when considering large enough autovectors, the o-sum $\mathbf{w} \oplus \mathbf{v}$ may not be defined, or may give an element that is outside the local structure (see example 1.4 below). One must, therefore, accept that the autovector space structures to be examined may have only a local character.

Definition 1.21 Local autovector space. *In the context of definition 1.19 (global autovector space), we say that the defined structure is a* local autovector space *if it is defined only for a certain subset* $\mathbb{S}_0$ *of* $\mathbb{S}$*:*

- *for any element* $\mathbf{v}$ *of* $\mathbb{S}_0$*, there is a finite interval of the real line around the origin such that for any* λ *in the interval, the element* $\lambda\,\mathbf{v}$ *also belongs to* $\mathbb{S}_0$*;*
- *for any two elements* $\mathbf{v}$ *and* $\mathbf{w}$ *of* $\mathbb{S}_0$*, there is a finite interval of the real line around the origin such that for any* λ *and* μ *in the interval, the element* $\mu\,\mathbf{w} \oplus \lambda\,\mathbf{v}$ *also belongs to* $\mathbb{S}_0$*.*

Example 1.4 *When considering a smooth metric manifold with a given origin* $\mathcal{O}$*, the set of oriented geodesic segments having* $\mathcal{O}$ *as origin is locally an autovector space, the sum of two oriented geodesic segments defined using the standard parallel transport (see section 1.3). But the geodesics leaving any point* $\mathcal{O}$ *of an arbitrary manifold shall, at some finite distance from* $\mathcal{O}$*, form caustics (where geodesics cross), whence the locality restriction. The linear tangent space to the local autovector space is the usual linear tangent space at a point of a manifold.*

Example 1.5 *Over the set of all complex squared matrices* $\mathbf{a}$*,* $\mathbf{b} \ldots$*, associate, to a matrix* $\mathbf{a}$ *and a real number* λ*, the matrix* $\lambda\mathbf{a}$*, and consider the operation*[18] $\mathbf{b} \oplus \mathbf{a} = \log(\exp \mathbf{b}\ \exp \mathbf{a})$*. As explained in section 1.4.1.3, they form a local (associative) autovector space, with tangent operation the ordinary sum of matrices* $\mathbf{b} + \mathbf{a}$*.*

To conclude the definition of an autovector space, consider the possibility of defining an 'autobasis'. In equation (1.8) the standard decomposition of a vector on a basis has been considered. With the o-sum operation, a

[18]The exponential and the logarithm of a matrix are defined in section 1.4.2.

different decomposition can be defined, where, given a set of n (auto) vectors $\{\mathbf{e}_1, \mathbf{e}_2, \dots, \mathbf{e}_n\}$, one writes

$$\mathbf{v} = v^n \, \mathbf{e}_n \oplus (\dots \oplus (v^2 \, \mathbf{e}_2 \oplus v^1 \, \mathbf{e}_1) \dots) \quad . \tag{1.67}$$

If any autovector can be written this way, we say that $\{\mathbf{e}_1, \mathbf{e}_2, \dots, \mathbf{e}_n\}$, is an *autobasis*, and that $\{v^1, \dots, v^n\}$, are the *autocomponents* of $\mathbf{v}$ on the autobasis $\{\mathbf{e}_i\}$. The (auto)vectors of an autobasis don't need to be linearly independent.[19]

1.2.4 Series Representations

The analyticity property of the $\oplus$ operation, postulated in the definition of an autovector space, means that inside some finite neighborhood of the origin, the following series expansion makes sense:

$$\begin{aligned}(\mathbf{w} \oplus \mathbf{v})^i &= a^i + b^i{}_j \, w^j + c^i_j \, v^j + d^i{}_{jk} \, w^j \, w^k + e^i{}_{jk} \, w^j \, v^k + f^i_{jk} \, v^j \, v^k \\ &\quad + p^i{}_{jk\ell} \, w^j \, w^k \, w^\ell + q^i{}_{jk\ell} \, w^j \, w^k \, v^\ell + r^i{}_{jk\ell} \, w^j \, v^k \, v^\ell \\ &\quad + s^i{}_{jk\ell} \, v^j \, v^k \, v^\ell + \dots \ ,\end{aligned} \tag{1.68}$$

where only the terms up to order three have been written. Here, $\mathbf{a}$ is some fixed vector and $\mathbf{b}, \mathbf{c}, \dots$ are fixed tensors (i.e., elements of the tensor space introduced in section 1.1.2). We shall see later how this series relates to a well known series arising in the study of Lie groups, called the BCH series. Remember that the operation $\oplus$ is *not* assumed to be associative.

Without loss of generality, the tensors $\mathbf{a}, \mathbf{b}, \mathbf{c} \dots$ appearing in the series (1.68) can be assumed to have the symmetries of the term in which they appear.[20] Introducing into the series the two conditions $\mathbf{w} \oplus \mathbf{0} = \mathbf{w}$ and

---

[19] As explained in appendix A.14, a rotation can be represented by a 'vector' $\mathbf{r}$ whose axis is the rotation axis and whose norm is the rotation angle. While the (linear) sum $\mathbf{r}_2 + \mathbf{r}_1$ of two rotation vectors has no special meaning, if the 'vectors' are considered to be geodesics in a space of constant curvature and constant torsion (see appendix A.14 for details), then, the 'geodesic sum' $\mathbf{r}_2 \oplus \mathbf{r}_1 \equiv \log(\exp \mathbf{r}_2 \, \exp \mathbf{r}_1)$ is identical to the composition of rotations (i.e., to the successive application of rotations). When choosing as a basis for the rotations the vectors $\{\mathbf{c}_1, \mathbf{c}_2, \mathbf{c}_3\} = \{\mathbf{e}_x, \mathbf{e}_y, \mathbf{e}_z\}$, the autocomponents (or, in this context, the 'geocomponents') $\{w^i\}$ defined through $\mathbf{r} = w^3 \, \mathbf{c}_3 \oplus w^2 \, \mathbf{c}_2 \oplus w^1 \, \mathbf{c}_1$ (the rotations form a group, so the parentheses can be dropped) corresponds exactly to the *Cardan angles* in the engineering literature or to the *Brauer angles* in the mathematical literature (Srinivasa Rao, 1988). When choosing as a basis for the rotations the vectors $\{\mathbf{c}_1, \mathbf{c}_2, \mathbf{c}_3\} = \{\mathbf{e}_z, \mathbf{e}_x, \mathbf{e}_z\}$ (note that $\mathbf{e}_z$ is used twice), the geocomponents $\{\varphi, \theta, \psi\}$ defined through $\mathbf{r} = \psi \, \mathbf{c}_3 \oplus \theta \, \mathbf{c}_2 \oplus \varphi \, \mathbf{c}_1 = \psi \, \mathbf{e}_z \oplus \theta \, \mathbf{e}_x \oplus \varphi \, \mathbf{e}_z$ are the standard Euler angles.

[20] I.e., $d^i{}_{jk} = d^i{}_{kj}$, $f^i{}_{jk} = f^i{}_{kj}$, $q^i{}_{jk\ell} = q^i{}_{kj\ell}$, $r^i{}_{jk\ell} = r^i{}_{j\ell k}$, $p^i{}_{jk\ell} = p^i{}_{kj\ell} = p^i{}_{j\ell k}$ and $s^i{}_{jk\ell} = s^i{}_{kj\ell} = s^i{}_{j\ell k}$.

$\mathbf{0} \oplus \mathbf{v} = \mathbf{v}$ (equations 1.37) and using the symmetries just assumed, one immediately obtains $a^i = 0$, $b^i{}_j = c^i{}_j = \delta^i_j$, $d^i{}_{jk} = f^i{}_{jk} = 0$, $p^i{}_{jk\ell} = s^i{}_{jk\ell} = 0$, etc., so the series (1.68) simplifies to

$$(\mathbf{w} \oplus \mathbf{v})^i = w^i + v^i + e^i{}_{jk}\, w^j\, v^k + q^i{}_{jk\ell}\, w^j\, w^k\, v^\ell + r^i{}_{jk\ell}\, w^j\, v^k\, v^\ell + \dots \quad , \qquad (1.69)$$

where $q^i{}_{jk\ell}$ and $r^i{}_{jk\ell}$ have the symmetries

$$q^i{}_{jk\ell} = q^i{}_{kj\ell} \quad ; \quad r^i{}_{jk\ell} = r^i{}_{j\ell k} \quad . \qquad (1.70)$$

Finally, the condition $(\lambda + \mu)\, \mathbf{v} = \lambda\, \mathbf{v} \oplus \mu\, \mathbf{v}$ (equation 1.62) imposes that the circular sums of the coefficients must vanish,[21]

$$\textstyle\sum_{(jk)} e^i{}_{jk} = 0 \quad ; \quad \sum_{(jk\ell)} q^i{}_{jk\ell} = \sum_{(jk\ell)} r^i{}_{jk\ell} = 0 \quad . \qquad (1.71)$$

We see, in particular, that $e^k{}_{ij}$ is necessarily antisymmetric:

$$e^k{}_{ij} = -e^k{}_{ji} \quad . \qquad (1.72)$$

We can now search for the series expressing the $\ominus$ operation. Starting from the property $(\mathbf{w} \ominus \mathbf{v}) \oplus \mathbf{v} = \mathbf{w}$ (second equation of (1.42)), developing the o-sum through the series (1.69), writing a generic series for the $\ominus$ operation, and using the property $\mathbf{w} \ominus \mathbf{w} = \mathbf{0}$, one arrives at a series whose terms up to third order are (see appendix A.3)

$$(\mathbf{w} \ominus \mathbf{v})^i = w^i - v^i - e^i{}_{jk}\, w^j\, v^k - q^i{}_{jk\ell}\, w^j\, w^k\, v^\ell - u^i{}_{jk\ell}\, w^j\, v^k\, v^\ell + \dots \quad , \qquad (1.73)$$

where the coefficients $u^i{}_{jk\ell}$ are given by

$$u^i{}_{jk\ell} = r^i{}_{jk\ell} - (q^i{}_{jk\ell} + q^i{}_{j\ell k}) - \tfrac{1}{2}(e^i{}_{sk}\, e^s{}_{j\ell} + e^i{}_{s\ell}\, e^s{}_{jk}) \quad , \qquad (1.74)$$

and, as easily verified, satisfy $\sum_{(jk\ell)} u^i{}_{jk\ell} = 0$.

1.2.5 Commutator and Associator

In the theory of Lie algebras, the 'commutator' plays a central role. Here, it is introduced using the o-sum and the o-difference, and, in addition to the commutator we need to introduce the 'associator'. Let us see how this can be done.

Definition 1.22 *The* finite commutation *of two autovectors* $\mathbf{v}$ *and* $\mathbf{w}$*, denoted* $\{\mathbf{w}, \mathbf{v}\}$ *is defined as*

$$\boxed{\{\mathbf{w}, \mathbf{v}\} \equiv (\mathbf{w} \oplus \mathbf{v}) \ominus (\mathbf{v} \oplus \mathbf{w}) \quad .} \qquad (1.75)$$

[21] Explicitly, $e^i{}_{jk} + e^i{}_{kj} = 0$, and $q^i{}_{jk\ell} + q^i{}_{k\ell j} + q^i{}_{\ell jk} = r^i{}_{jk\ell} + r^i{}_{k\ell j} + r^i{}_{\ell jk} = 0$.

Definition 1.23 *The* finite association, *denoted* $\{\mathbf{w}, \mathbf{v}, \mathbf{u}\}$ *is defined as*

$$\boxed{\{\mathbf{w}, \mathbf{v}, \mathbf{u}\} \equiv (\mathbf{w} \oplus (\mathbf{v} \oplus \mathbf{u})) \ominus ((\mathbf{w} \oplus \mathbf{v}) \oplus \mathbf{u}) \quad .} \tag{1.76}$$

Clearly, the finite association vanishes if the autovector space is associative. The finite commutation vanishes if the autovector space is commutative.

It is easy to see that when writing the series expansion of the finite commutation of two elements, the first term is a second-order term. Similarly, when writing the series expansion of the finite association of three elements, the first term is a third-order term. This justifies the following two definitions.

Definition 1.24 *The* commutator, *denoted* $[\mathbf{w}, \mathbf{v}]$, *is the lowest-order term in the series expansion of the finite commutation* $\{\mathbf{w}, \mathbf{v}\}$ *defined in equation* (1.75):

$$\boxed{\{\mathbf{w}, \mathbf{v}\} \equiv [\mathbf{w}, \mathbf{v}] + O(3) \quad .} \tag{1.77}$$

Definition 1.25 *The* associator, *denoted* $[\mathbf{w}, \mathbf{v}, \mathbf{u}]$, *is the lowest-order term in the series expansion of the finite association* $\{\mathbf{w}, \mathbf{v}, \mathbf{u}\}$ *defined in equation* (1.76):

$$\boxed{\{\mathbf{w}, \mathbf{v}, \mathbf{u}\} \equiv [\mathbf{w}, \mathbf{v}, \mathbf{u}] + O(4) \quad .} \tag{1.78}$$

Therefore, one has the series expansions

$$\begin{aligned} (\mathbf{w} \oplus \mathbf{v}) \ominus (\mathbf{v} \oplus \mathbf{w}) &= [\mathbf{w}, \mathbf{v}] + \dots \\ (\mathbf{w} \oplus (\mathbf{v} \oplus \mathbf{u})) \ominus ((\mathbf{w} \oplus \mathbf{v}) \oplus \mathbf{u}) &= [\mathbf{w}, \mathbf{v}, \mathbf{u}] + \dots \quad . \end{aligned} \tag{1.79}$$

As explained below, when an autovector space is associative, it is a local Lie group. Then, obviously, the associator $[\mathbf{w}, \mathbf{v}, \mathbf{u}]$ vanishes. As we shall see, the commutator $[\mathbf{w}, \mathbf{u}]$ is then identical to that usually introduced in Lie group theory.

A first property is that the commutator is antisymmetric, i.e., for any autovectors $\mathbf{v}$ and $\mathbf{w}$,

$$[\mathbf{w}, \mathbf{v}] = -[\mathbf{v}, \mathbf{w}] \tag{1.80}$$

(see appendix A.3). A second property is that the commutator and associator are not independent. To prepare the theorem 1.6 below, let us introduce the following

Definition 1.26 *The* Jacobi tensor,[22] *denoted* $\mathbf{J}$, *is defined by its action on any three autovectors* $\mathbf{u}$, $\mathbf{v}$ *and* $\mathbf{w}$:

$$\boxed{\mathbf{J}(\mathbf{u}, \mathbf{v}, \mathbf{w}) \equiv [\mathbf{u}, [\mathbf{v}, \mathbf{w}]] + [\mathbf{v}, [\mathbf{w}, \mathbf{u}]] + [\mathbf{w}, [\mathbf{u}, \mathbf{v}]] \quad .} \tag{1.81}$$

[22]The term 'tensor' means here "element of the tensor space $\mathbb{T}$ introduced in section 1.1.2".

From the antisymmetry property (1.80) follows

Property 1.5 *The Jacobi tensor is totally antisymmetric, i.e., for any three autovectors,*

$$\mathbf{J}(\mathbf{u},\mathbf{v},\mathbf{w}) = -\mathbf{J}(\mathbf{u},\mathbf{w},\mathbf{v}) = -\mathbf{J}(\mathbf{v},\mathbf{u},\mathbf{w}) \quad . \tag{1.82}$$

We can now state

Property 1.6 *As demonstrated in appendix A.3, for any three autovectors,*

$$\boxed{\mathbf{J}(\mathbf{u},\mathbf{v},\mathbf{w}) = 2\,(\,[\,\mathbf{u},\mathbf{v},\mathbf{w}\,] + [\,\mathbf{v},\mathbf{w},\mathbf{u}\,] + [\,\mathbf{w},\mathbf{u},\mathbf{v}\,]\,) \quad .} \tag{1.83}$$

This is a property valid in any autovector space. We shall see later the implication of this property for Lie groups.

Let us come back to the problem of obtaining a series expansion for the o-sum operation $\oplus$. Using the definitions and notations introduced in section 1.2.5, we obtain, up to third order (see the demonstration in appendix A.3),

$$\boxed{\begin{aligned}\mathbf{w}\oplus\mathbf{v} &= (\mathbf{w}+\mathbf{v}) + \tfrac{1}{2}\,[\mathbf{w},\mathbf{v}] + \tfrac{1}{12}\Big(\,[\,\mathbf{v},[\mathbf{v},\mathbf{w}]\,] + [\,\mathbf{w},[\mathbf{w},\mathbf{v}]\,]\Big) + \\ &\quad + \tfrac{1}{3}\Big(\,[\mathbf{w},\mathbf{v},\mathbf{v}] + [\mathbf{w},\mathbf{v},\mathbf{w}] - [\mathbf{w},\mathbf{w},\mathbf{v}] - [\mathbf{v},\mathbf{w},\mathbf{v}]\Big) + \cdots \ .\end{aligned}} \tag{1.84}$$

We shall see below (section 1.4.1.1) that when the autovector space is associative, it is a local Lie group. Then, this series collapses into the well known BCH series of Lie group theory. Here, we have extra terms containing the associator (that vanishes in a group).

For the series expressing the o-difference —that is not related in an obvious way to the series for the o-sum,— one obtains (see appendix A.3)

$$\begin{aligned}\mathbf{w}\ominus\mathbf{v} &= (\mathbf{w}-\mathbf{v}) - \tfrac{1}{2}\,[\mathbf{w},\mathbf{v}] + \tfrac{1}{12}\Big(\,[\,\mathbf{v},[\mathbf{v},\mathbf{w}]\,] - [\,\mathbf{w},[\mathbf{w},\mathbf{v}]\,]\Big) + \\ &\quad - \tfrac{1}{3}\Big(\,[\mathbf{w},\mathbf{v},\mathbf{v}] + [\mathbf{w},\mathbf{v},\mathbf{w}] - [\mathbf{w},\mathbf{w},\mathbf{v}] - [\mathbf{v},\mathbf{v},\mathbf{w}]\Big) + \cdots \quad .\end{aligned} \tag{1.85}$$

1.2.6 Torsion and Anassociativity

Definition 1.27 *The* torsion *tensor* $\mathbf{T}$, *with components* $T^i{}_{jk}$, *is defined through* $[\mathbf{w},\mathbf{v}] = \mathbf{T}(\mathbf{w},\mathbf{v})$, *or, more explicitly,*

$$\boxed{[\mathbf{w},\mathbf{v}]^i = T^i{}_{jk}\, w^j\, v^k \quad .} \tag{1.86}$$

Definition 1.28 *The* anassociativity *tensor* $\mathbf{A}$ *, with components* $A^i{}_{jk\ell}$ *, is defined through* $[\mathbf{w},\mathbf{v},\mathbf{u}] = \frac{1}{2}\,\mathbf{A}(\mathbf{w},\mathbf{v},\mathbf{u})$ *, or, more explicitly,*

$$[\mathbf{w},\mathbf{v},\mathbf{u}]^i = \tfrac{1}{2}\,A^i{}_{jk\ell}\,w^j\,v^k\,u^\ell \quad . \tag{1.87}$$

Property 1.7 *Therefore, using equations* (1.77), (1.78), (1.75), *and* (1.76),

$$\begin{aligned} \left[(\mathbf{w}\oplus\mathbf{v}) \ominus (\mathbf{v}\oplus\mathbf{w})\right]^i &= T^i{}_{jk}\,w^j\,v^k + \dots \\ \left[(\mathbf{w}\oplus(\mathbf{v}\oplus\mathbf{u})) \ominus ((\mathbf{w}\oplus\mathbf{v})\oplus\mathbf{u})\right]^i &= \tfrac{1}{2}\,A^i{}_{jk\ell}\,w^j\,v^k\,u^\ell + \dots \quad . \end{aligned} \tag{1.88}$$

Loosely speaking, the tensors $\mathbf{T}$ and $\mathbf{A}$ give respectively a measure of the default of commutativity and of the default of associativity of the autovector operation $\oplus$. The tensor $\mathbf{T}$ is called the 'torsion tensor' because, as shown below, the autovector space formed by the oriented autoparallel segments on a manifold, corresponds exactly to what is usually called torsion (see section 1.3). We shall also see in section 1.3 that on a manifold with constant torsion, the anassociativity tensor is identical to the Riemann tensor of the manifold (this correspondence explaining the factor 1/2 in the definition of $\mathbf{A}$).

The Jacobi tensor was defined in equation (1.81). Defining its components as

$$\mathbf{J}(\mathbf{w},\mathbf{v},\mathbf{u})^i = J^i{}_{jk\ell}\,w^j\,v^k\,u^\ell \tag{1.89}$$

allows to write its definition in terms of the torsion or of the anassociativity as

$$J^i{}_{jk\ell} = \textstyle\sum_{(jk\ell)} T^i{}_{js}\,T^s{}_{k\ell} = \sum_{(jk\ell)} A^i{}_{jk\ell} \quad . \tag{1.90}$$

From equation (1.80) it follows that the torsion is antisymmetric in its two lower indices:

$$T^i{}_{jk} = -T^i{}_{kj} \quad , \tag{1.91}$$

while equation (1.82) stating the total antisymmetry of the Jacobi tensor now becomes

$$J^i{}_{jk\ell} = -J^i{}_{j\ell k} = -J^i{}_{kj\ell} \quad . \tag{1.92}$$

We can now come back to the two developments (equations (1.69) and (1.73))

$$\begin{aligned} (\mathbf{w}\oplus\mathbf{v})^i &= w^i + v^i + e^i{}_{jk}\,w^j\,v^k + q^i{}_{jk\ell}\,w^j\,w^k\,v^\ell + r^i{}_{jk\ell}\,w^j\,v^k\,v^\ell + \dots \\ (\mathbf{w}\ominus\mathbf{v})^i &= w^i - v^i - e^i{}_{jk}\,w^j\,v^k - q^i{}_{jk\ell}\,w^j\,w^k\,v^\ell - u^i{}_{jk\ell}\,w^j\,v^k\,v^\ell + \dots \quad , \end{aligned} \tag{1.93}$$

with the $u^i{}_{jk\ell}$ given by expression (1.74). Using the definition of torsion and of anassociativity (1.86) and (1.87), we can now express the coefficients of these two series as[23]

$$
\begin{aligned}
e^i{}_{jk} &= \tfrac{1}{2}\, T^i{}_{jk} \\
q^i{}_{jk\ell} &= -\tfrac{1}{12} \textstyle\sum_{(jk)} \big(A^i{}_{jk\ell} - A^i{}_{k\ell j} - \tfrac{1}{2}\, T^i{}_{js}\, T^s{}_{k\ell} \big) \\
r^i{}_{jk\ell} &= \tfrac{1}{12} \textstyle\sum_{(k\ell)} \big(A^i{}_{jk\ell} - A^i{}_{\ell jk} + \tfrac{1}{2}\, T^i{}_{ks}\, T^s{}_{\ell j} \big) \\
u^i{}_{jk\ell} &= \tfrac{1}{12} \textstyle\sum_{(k\ell)} \big(A^i{}_{jk\ell} - A^i{}_{k\ell j} - \tfrac{1}{2}\, T^i{}_{ks}\, T^s{}_{\ell j} \big) \quad ,
\end{aligned}
\tag{1.94}
$$

this expressing terms up to order three of the o-sum and o-difference in terms of the torsion and the anassociativity. A direct check shows that these expressions satisfy the necessary symmetry conditions $\sum_{(jk\ell)} q^i{}_{jk\ell} = \sum_{(jk\ell)} r^i{}_{jk\ell} = \sum_{(jk\ell)} u^i{}_{jk\ell} = 0$.

Reciprocally, we can write[24]

$$
\begin{aligned}
T^i{}_{jk} &= 2\, e^i{}_{jk} \\
\tfrac{1}{2}\, A^i{}_{jk\ell} &= e^i{}_{js}\, e^s{}_{k\ell} + e^i{}_{\ell s}\, e^s{}_{jk} - 2\, q^i{}_{jk\ell} + 2\, r^i{}_{jk\ell} \quad .
\end{aligned}
\tag{1.95}
$$

1.3 Oriented Autoparallel Segments on a Manifold

The major concrete example of an autovector space is of geometric nature. It corresponds to (a subset of) the set of the oriented autoparallel segments of a manifold that have common origin, with the sum of oriented autoparallel segments defined through 'parallel transport'. This example will now be developed.

An n-dimensional *manifold* is a space of elements, called 'points', that accepts in a finite neighborhood of each of its points an n-dimensional system of continuous coordinates. Grossly speaking, an n-dimensional manifold is a space that, locally, "looks like" $\Re^n$. Here, we are interested in the class of smooth manifolds that may or may not be metric, but that have a prescription for the parallel transport of vectors: given a vector at a point (a vector belonging to the linear space tangent to the manifold at the given point), and given a line on the manifold, it is assumed that one is able to transport the vector along the line "keeping the vector always parallel to itself". Intuitively speaking this corresponds to the assumption that there is an "inertial navigation system" on the manifold, analogous to that used in airplanes to keep fixed directions while navigating. The prescription for this parallel transport is not necessarily the one that could be defined using a possible metric (and

[23] The explicit computation is made in appendix A.3.

[24] Expressions (A.44) and (A.45) from the appendix.

'geodesic' techniques), as the considered manifolds may have 'torsion'. In such a manifold, there is a family of privileged lines, the 'autoparallels', that are obtained when constantly following a direction defined by the "inertial navigation system".

If the manifold is, in addition, a metric manifold, then there is a second family of privileged lines, the 'geodesics', that correspond to the minimum length path between any two of its points. It is well known[25] that the two types of lines coincide (the geodesics are autoparallels and vice versa) when the torsion is totally antisymmetric $T_{ijk} = -T_{jik} = -T_{ikj}$.

1.3.1 Connection

We follow here the traditional approach of describing the parallel transport of vectors on a manifold through the introduction of a 'connection'.

Consider the simple situation where some (arbitrary) coordinates $\mathbf{x} \equiv \{x^i\}$ have been defined over the manifold. At a given point $\mathbf{x}_0$ consider the coordinate lines passing through $\mathbf{x}_0$. If $\mathbf{x}$ is a point on any of the coordinate lines, let us denote as $\gamma(\mathbf{x})$ the coordinate line segment going from $\mathbf{x}_0$ to $\mathbf{x}$. The *natural basis* (of the local tangent space) associated to the given coordinates consists of the n vectors $\{\mathbf{e}_1(\mathbf{x}_0), \dots, \mathbf{e}_n(\mathbf{x}_0)\}$ that can formally be denoted as $\mathbf{e}_i(\mathbf{x}_0) = \frac{\partial \gamma}{\partial x^i}(\mathbf{x}_0)$, or, dropping the index 0 ,

$$\mathbf{e}_i(\mathbf{x}) = \frac{\partial \gamma}{\partial x^i}(\mathbf{x}) \quad . \tag{1.96}$$

So, there is a natural basis at every point of the manifold. As it is assumed that a parallel transport exists on the manifold, the basis $\{\mathbf{e}_i(\mathbf{x})\}$ can be transported from a point x^i to a point $x^i + \delta x^i$ to give a new basis that we can denote $\{\mathbf{e}_i(\mathbf{x} + \delta\mathbf{x} \,\|\, \mathbf{x})\}$ (and that, in general, is different from the local basis $\{\mathbf{e}_i(\mathbf{x} + \delta\mathbf{x})\}$ at point $\mathbf{x} + \delta\mathbf{x}$). The *connection* is defined as the set of coefficients $\Gamma^k{}_{ij}$ (that are not, in general, the components of a tensor) appearing in the development

$$\mathbf{e}_j(\mathbf{x} + \delta\mathbf{x} \,\|\, \mathbf{x}) = \mathbf{e}_j(\mathbf{x}) + \Gamma^k{}_{ij}(\mathbf{x})\, \mathbf{e}_k(\mathbf{x})\, \delta x^i + \dots \quad . \tag{1.97}$$

For this first-order expression, we don't need to be specific about the path followed for the parallel transport. For higher-order expressions, the path followed matters (see for instance equation (A.119), corresponding to transport along an autoparallel line).

In the rest of this book, a manifold where a connection is defined is named a *connection manifold*.

[25]See a demonstration in appendix A.11.3.

1.3.2 Oriented Autoparallel Segments

The notion of autoparallel curve is mathematically introduced in appendix A.9.2. It is enough for our present needs to know the main result demonstrated there:

Property 1.8 *A line $x^i = x^i(\lambda)$ is autoparallel if at every point along the line,*

$$\frac{d^2x^i}{d\lambda^2} + \gamma^i{}_{jk} \frac{dx^j}{d\lambda} \frac{dx^k}{d\lambda} = 0 \quad , \tag{1.98}$$

where $\gamma^i{}_{jk}$ is the symmetric part of the connection,

$$\gamma^i{}_{jk} = \tfrac{1}{2} (\Gamma^i{}_{jk} + \Gamma^i{}_{kj}) \quad . \tag{1.99}$$

If there exists a parameter λ with respect to which a curve is autoparallel, then any other parameter $\mu = \alpha \lambda + \beta$ (where α and β are two constants) satisfies also the condition (1.98). Any such parameter associated to an autoparallel curve is called an *affine parameter*.

1.3.3 Vector Tangent to an Autoparallel Line

Let be $x^i = x^i(\lambda)$ the equation of an autoparallel line with affine parameter λ. The *affine tangent vector* $\mathbf{v}$ (associated to the autoparallel line and to the affine parameter λ) is defined, at any point along the line, by

$$v^i(\lambda) = \frac{dx^i}{d\lambda}(\lambda) \quad . \tag{1.100}$$

It is an element of the linear space tangent to the manifold at the considered point. This tangent vector depends on the particular affine parameter being used: when changing from the affine parameter λ to another affine parameter $\mu = \alpha \lambda + \beta$, and defining $\tilde{v}^i = dx^i/d\mu$, one easily arrives at the relation $v^i = \alpha\, \tilde{v}^i$.

1.3.4 Parallel Transport of a Vector

Let us suppose that a vector $\mathbf{w}$ is transported, parallel to itself, along this autoparallel line, and denote $w^i(\lambda)$ the components of the vector in the local natural basis at point λ. As demonstrated in appendix A.9.3, one has

Property 1.9 *The equation defining the parallel transport of a vector $\mathbf{w}$ along the autoparallel line of affine tangent vector $\mathbf{v}$ is*

$$\frac{dw^i}{d\lambda} + \Gamma^i{}_{jk}\, v^j\, w^k = 0 \quad . \tag{1.101}$$

Given an autoparallel line and a vector at any of its points, this equation can be used to obtain the transported vector at any other point along the autoparallel line.

1.3.5 Association Between Tangent Vectors and Oriented Segments

Consider again an autoparallel line $x^i = x^i(\lambda)$ defined in terms of an affine parameter λ. At some point of parameter λ_0 along the curve, we can introduce the affine tangent vector defined in equation (1.100), $v^i(\lambda_0) = \frac{dx^i}{d\lambda}(\lambda_0)$, that belongs to the linear space tangent to the manifold at point λ_0. As already mentioned, changing the affine parameter changes the affine tangent vector.

We could define an association between arbitrary tangent vectors and autoparallel segments characterized using an arbitrary affine parameter,[26] but it is much simpler to proceed through the introduction of a 'canonical' affine parameter. Given an arbitrary vector $\mathbf{V}$ at a point of a manifold, and the autoparallel line that is tangent to $\mathbf{V}$ (at the given point), we can select among all the affine parameters that characterize the autoparallel line, one parameter, say λ, giving $V^i = dx^i/d\lambda$ (i.e., such that the affine tangent vector $\mathbf{v}$ with respect to the parameter λ equals the given vector $\mathbf{V}$). Then, by definition, to the vector $\mathbf{V}$ is associated the oriented autoparallel segment that starts at point λ_0 (the tangency point) and ends at point $\lambda_0 + 1$, i.e., the segment whose "affine length" (with respect to the canonical affine parameter λ being used) equals one. This is represented in figure 1.1.

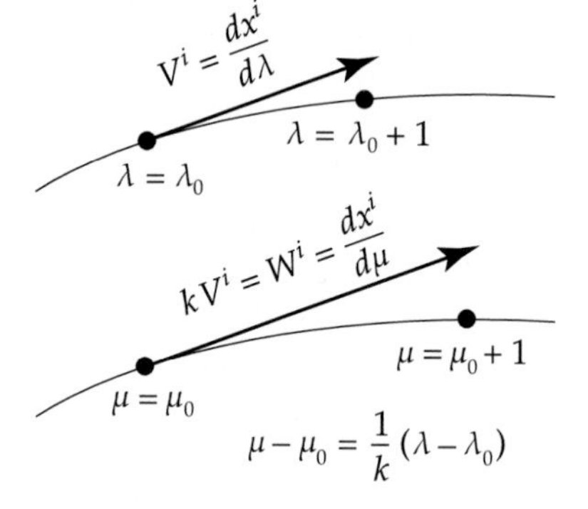

Fig. 1.1. *In a connection manifold (that may or may not be metric), the association between vectors (of the linear tangent space) and oriented autoparallel segments in the manifold is made using a canonical affine parameter.*

Let $\mathcal{O}$ be the point where the vector $\mathbf{V}$ and the autoparallel line are tangent, let $\mathcal{P}$ be the point along the line that the procedure just described associates to the given vector $\mathbf{V}$, and let $\mathcal{Q}$ be the point associated to the vector $\mathbf{W} = k\mathbf{V}$. It is easy to verify (see figure 1.1) that for any affine parameter considered along the line, the increase in the value of the affine parameter

[26]To any point of parameter λ along the autoparallel line we can associate the vector (also belonging to the linear space tangent to the manifold at λ_0) $\mathbf{V}(\lambda;\lambda_0) = ((\lambda - \lambda_0)/(1 - \lambda_0))\,\mathbf{v}(\lambda_0)$. One has $\mathbf{V}(\lambda_0;\lambda_0) = \mathbf{0}$, $\mathbf{V}(1;\lambda_0) = \mathbf{v}(\lambda_0)$, and the more λ is larger than λ_0, the "longer" is $\mathbf{V}(\lambda;\lambda_0)$.

when passing from $\mathcal{O}$ to point $\mathcal{Q}$ is k times the increase when passing from $\mathcal{O}$ to $\mathcal{P}$.

The association so defined between tangent vectors and oriented autoparallel segments is consistent with the standard association between tangent vectors and oriented geodesic segments in metric manifolds without torsion, where the autoparallel lines are the geodesics. The tangent to a geodesic $x^i = x^i(s)$, parameterized by a metric coordinate s, is defined as $v^i = dx^i/ds$, and one has $g_{ij}\, v^i\, v^j = g_{ij}\, (dx^i/ds)\, (dx^j/ds) = ds^2/ds^2 = 1$, this showing that the vector tangent to a geodesic has unit length.

1.3.6 Transport of Oriented Autoparallel Segments

Consider now two oriented autoparallel segments, $\mathbf{u}$ and $\mathbf{v}$ with common origin, as suggested in figure 1.2. To the segment $\mathbf{v}$ we can associate a vector of the tangent space, as we have just seen. This vector can be transported along $\mathbf{u}$ (using equation 1.101) to its tip. The vector obtained there can then be associated to another oriented autoparallel segment, giving the $\mathbf{v}'$ suggested in the figure. So, on a manifold with a parallel transport defined, one can transport not only vectors, but also oriented autoparallel segments.

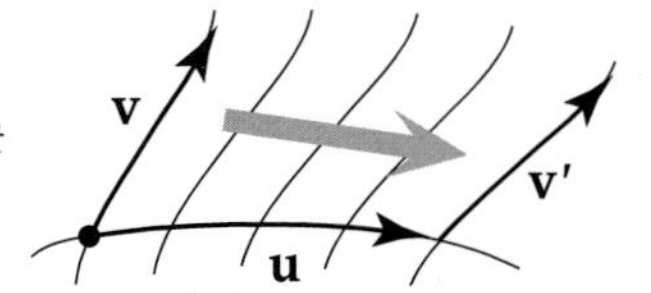

Fig. 1.2. *Transport of an oriented autoparallel segment along another one.*

1.3.7 Oriented Autoparallel Segments as Autovectors

In a sufficiently smooth manifold, take a particular point $\mathcal{O}$ as origin, and consider the set of oriented autoparallel segments, having $\mathcal{O}$ as origin, and belonging to some finite neighborhood of the origin.[27] For the time being let us denote these objects 'autovectors' inside quotes, to be dropped when the demonstration will have been made that they actually form a (local) autovector space. Given two such 'autovectors' $\mathbf{u}$ and $\mathbf{v}$, define the *geometric sum* (or *geosum*) $\mathbf{w} = \mathbf{v} \oplus \mathbf{u}$ by the geometric construction shown in figure 1.3, and given two such 'autovectors' $\mathbf{u}$ and $\mathbf{v}$, define the *geometric difference* (or *geodifference*) $\mathbf{w} = \mathbf{v} \ominus \mathbf{u}$ by the geometric construction shown in figure 1.4.

As the definition of the geodifference $\ominus$ is essentially, the "deconstruction" of the geosum $\oplus$, it is clear that the equation $\mathbf{w} = \mathbf{v} \oplus \mathbf{u}$ can be solved for $\mathbf{v}$:

[27] On an arbitrary manifold, the geodesics leaving a point may form caustics (where the geodesics cross each other). The neighborhood of the origin considered must be small enough to avoid caustics.

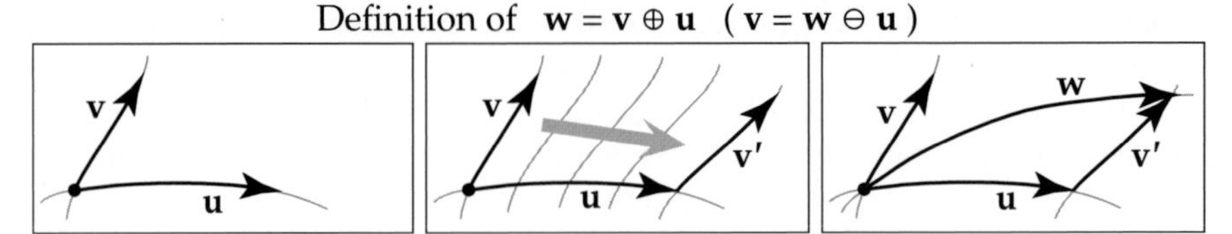

Fig. 1.3. *Definition of the geometric sum of two 'autovectors' at a point $\mathcal{O}$ of a manifold with a parallel transport: the sum* $\mathbf{w} = \mathbf{v} \oplus \mathbf{u}$ *is defined through the parallel transport of* $\mathbf{v}$ *along* $\mathbf{u}$ *. Here,* $\mathbf{v}'$ *denotes the oriented autoparallel segment obtained by the parallel transport of the autoparallel segment defining* $\mathbf{v}$ *along* $\mathbf{u}$ *(as* $\mathbf{v}'$ *does not begin at the origin, it is not an 'autovector'). We may say, using a common terminology that the oriented autoparallel segments* $\mathbf{v}$ *and* $\mathbf{v}'$ *are 'equipollent'. The 'autovector'* $\mathbf{w} = \mathbf{v} \oplus \mathbf{u}$ *is, by definition, the arc of autoparallel (unique in a sufficiently small neighborhood of the origin) connecting the origin* $\mathcal{O}$ *to the tip of* $\mathbf{v}'$ *.*

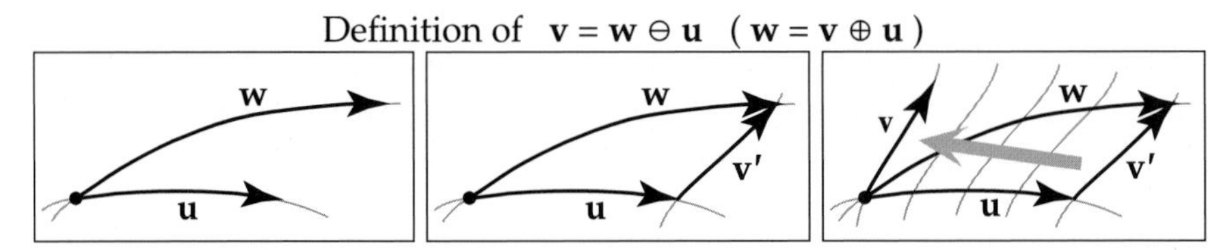

Fig. 1.4. *The geometric difference* $\mathbf{v} = \mathbf{w} \ominus \mathbf{u}$ *of two 'autovectors' is defined by the condition* $\mathbf{v} = \mathbf{w} \ominus \mathbf{u} \;\Leftrightarrow\; \mathbf{w} = \mathbf{v} \oplus \mathbf{u}$ *. This can be obtained through the parallel transport to the origin (along* $\mathbf{u}$ *) of the oriented autoparallel segment* $\mathbf{v}'$ *that " goes from the tip of* $\mathbf{u}$ *to the tip of* $\mathbf{w}$ *". In fact, the transport performed to obtain the difference* $\mathbf{v} = \mathbf{w} \ominus \mathbf{u}$ *is the reverse of the transport performed to obtain the sum* $\mathbf{w} = \mathbf{v} \oplus \mathbf{u}$ *(figure 1.3), and this explains why in the expression* $\mathbf{w} = \mathbf{v} \oplus \mathbf{u}$ *one can always solve for* $\mathbf{v}$ *, to obtain* $\mathbf{v} = \mathbf{w} \ominus \mathbf{u}$ *. This contrasts with the problem of solving* $\mathbf{w} = \mathbf{v} \oplus \mathbf{u}$ *for* $\mathbf{u}$ *, which requires a different geometrical construction, whose result cannot be directly expressed in terms of the two operations* $\oplus$ *and* $\ominus$ *(see the example in figure 1.6).*

Fig. 1.5. *The opposite* $-\mathbf{v}$ *of an 'autovector'* $\mathbf{v}$ *is the 'autovector' opposite to* $\mathbf{v}$ *, and with the same absolute variation of affine parameter as* $\mathbf{v}$ *(or the same length if the manifold is metric).*

$$\mathbf{w} = \mathbf{v} \oplus \mathbf{u} \quad \Longleftrightarrow \quad \mathbf{v} = \mathbf{w} \ominus \mathbf{u} \quad . \tag{1.102}$$

It is obvious that there exists a neutral element $\mathbf{0}$ for the sum of 'autovectors': a segment reduced to a point. For we have, for any 'autovector' $\mathbf{v}$,

$$\mathbf{0} \oplus \mathbf{v} = \mathbf{v} \oplus \mathbf{0} = \mathbf{v} \quad , \tag{1.103}$$

The opposite of an 'autovector' $\mathbf{a}$ is the 'autovector' $-\mathbf{a}$, that is along the same autoparallel line, but pointing towards the opposite direction (see

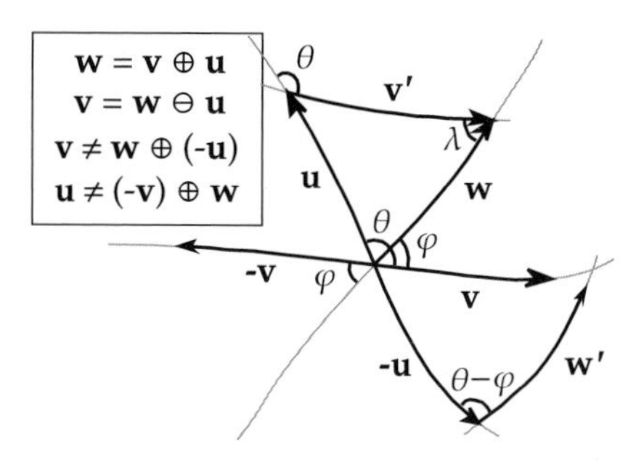

Fig. 1.6. *Over the set of oriented autoparallel segments at a given origin of a manifold we have the equivalence* $\mathbf{w} = \mathbf{v} \oplus \mathbf{u} \Leftrightarrow \mathbf{v} = \mathbf{w} \ominus \mathbf{u}$ *(as the two expressions correspond to the same geometric construction). But, in general,* $\mathbf{v} \neq \mathbf{w} \oplus (\text{-}\mathbf{u})$ *and* $\mathbf{u} \neq (\text{-}\mathbf{v}) \oplus \mathbf{w}$. *For the autovector* $\mathbf{w} \oplus (\text{-}\mathbf{u})$ *is indeed to be obtained by transporting* $\mathbf{w}$ *along* $\text{-}\mathbf{u}$. *There is no reason for the tip of the oriented autoparallel segment* $\mathbf{w}'$ *thus obtained to coincide with the tip of the autovector* $\mathbf{v}$. *Therefore,* $\mathbf{w} = \mathbf{v} \oplus \mathbf{u} \not\Leftrightarrow \mathbf{v} = \mathbf{w} \oplus (\text{-}\mathbf{u})$. *Also, the autovector* $(\text{-}\mathbf{v}) \oplus \mathbf{w}$ *is to be obtained, by definition, by transporting* $\text{-}\mathbf{v}$ *along* $\mathbf{w}$*, and one does not obtain an oriented autoparallel segment that is equal and opposite to* $\mathbf{v}'$ *(as there is no reason for the angles* φ *and* λ *to be identical). Therefore,* $\mathbf{w} = \mathbf{v} \oplus \mathbf{u} \not\Leftrightarrow \mathbf{u} = (\text{-}\mathbf{v}) \oplus \mathbf{w}$. *It is only when the autovector space is associative that all the equivalences hold.*

figure 1.5). The associated tangent vectors are also mutually opposite (in the usual sense). Then, clearly,

$$(\text{-}\mathbf{v}) \oplus \mathbf{v} \;=\; \mathbf{v} \oplus (\text{-}\mathbf{v}) \;=\; \mathbf{0} \quad . \tag{1.104}$$

Equations (1.102)–(1.104) correspond to the three conditions (1.36)–(1.38) defining a troupe. Therefore, with the geometric sum, the considered set of 'autovectors' is a (local) troupe. Let us show that it is an autovector space.

Given an 'autovector' $\mathbf{v}$ and a real number λ, the sense to be given to $\lambda\,\mathbf{v}$ (for any $\lambda \in [\text{-}1, 1]$) is obvious, and requires no special discussion. It is then clear that for any 'autovector' $\mathbf{v}$ and any scalars λ and μ inside some finite interval around zero,

$$(\lambda + \mu)\,\mathbf{v} \;=\; \lambda\,\mathbf{v} \oplus \mu\,\mathbf{v} \quad , \tag{1.105}$$

as this corresponds to translating an autoparallel line along itself.

Whichever method we use to introduce the linear space tangent at the origin $\mathcal{O}$ of the manifold, it is clear that we shall have the property

$$\lim_{\lambda \to 0} \frac{1}{\lambda} (\lambda\,\mathbf{w} \oplus \lambda\,\mathbf{v}) \;=\; \mathbf{w} + \mathbf{v} \quad , \tag{1.106}$$

this linking the geosum to the sum (and difference) in the tangent linear space (through the consideration of the limit of vanishingly small 'autovectors'). Finally, that the operation $\oplus$ is analytical in terms of the operation $+$

in the tangent space can be taken as the very definition of 'smooth' or 'differentiable' manifold. All the conditions necessary for an autovector space are fulfilled (see section 1.2.3), so we have the following

Property 1.10 *On a smooth enough manifold, consider an arbitrary origin* $\mathcal{O}$. *There exists always an open neighborhood of* $\mathcal{O}$ *such that the set of all the oriented autoparallel segments of the neighborhood having* $\mathcal{O}$ *as origin (i.e., the set of 'autovectors'), with the* $\oplus$ *and the* $\ominus$ *operation (defined through the parallel transport of the manifold) forms a* local autovector space. *In a smooth enough (and topologically simple) manifold, the autovector space may be global.*

So we can now drop the quotes and say autovectors, instead of 'autovectors'.

The reader may easily construct the geometric representation of the two properties (1.42), namely, that for any two autovectors, one has $(\mathbf{w} \oplus \mathbf{v}) \ominus \mathbf{v} = \mathbf{w}$ and $(\mathbf{w} \ominus \mathbf{v}) \oplus \mathbf{v} = \mathbf{w}$.

We have seen that the equation $\mathbf{w} = \mathbf{v} \oplus \mathbf{u}$ can be solved for $\mathbf{v}$, to give $\mathbf{v} = \mathbf{w} \ominus \mathbf{u}$. A completely different situation appears when trying to solve $\mathbf{w} = \mathbf{v} \oplus \mathbf{u}$ in terms of $\mathbf{u}$. Finding the $\mathbf{u}$ such that by parallel transport of $\mathbf{v}$ along it one obtains $\mathbf{w}$ correspond to an "inverse problem" that has no explicit geometric solution. It can be solved, for instance using some iterative algorithm, essentially a trial and (correction of) error method.

Note that given $\mathbf{w} = \mathbf{v} \oplus \mathbf{u}$, in general, $\mathbf{u} \neq (-\mathbf{v}) \oplus \mathbf{w}$ (see figure 1.6), the equality holding only in the special situation where the autovector operation is, in fact, a group operation (i.e., it is associative). This is obviously not the case in an arbitrary manifold.

Not only does the associative property not hold on an arbitrary manifold, but even simpler properties are not verified. For instance, let us introduce the following

Definition 1.29 *An autovector space is* oppositive *if for any two autovectors* $\mathbf{u}$ *and* $\mathbf{v}$, *one has* $\mathbf{w} \ominus \mathbf{v} = -(\mathbf{v} \ominus \mathbf{w})$.

Figure 1.7 shows that the surface of the sphere, using the parallel transport defined by the metric, is *not* oppositive.

1.3.8 Torsion and Riemann

From the two operations $\oplus$ and $\ominus$ of an abstract autovector space we have defined the torsion $T^i{}_{jk}$ and the anassociativity $A^i{}_{jk\ell}$. We have seen that the set of oriented autoparallel segments on a manifold forms an autovector space. And we have seen that the geosum and the geodifference on a manifold depend in a fundamental way on the connection $\Gamma^i{}_{jk}$ of the manifold. So we must now calculate expressions for the torsion and the anassociativity, to relate them to the connection. We can anticipate the result: the torsion $T^i{}_{jk}$ (introduced above for abstract autovector spaces) shall match, for the segments on a manifold, the standard notion of torsion (as introduced by Cartan); the anassociativity $A^i{}_{jk\ell}$ shall correspond, for spaces with constant

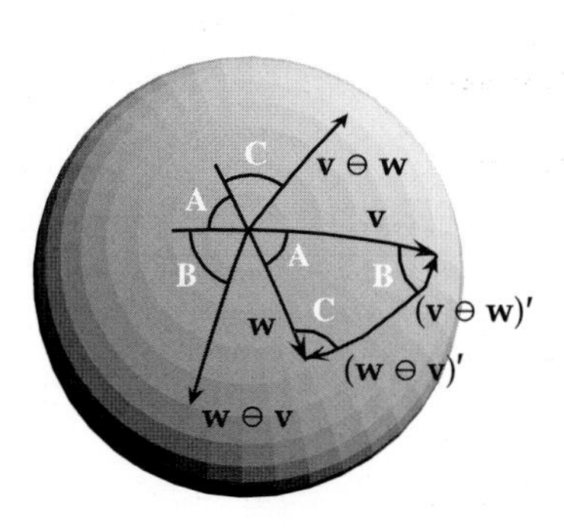

Fig. 1.7. *This figure illustrates the (lack of) oppositivity property for the autovectors on an arbitrary homogeneous manifold (the figure suggests a sphere). The oppositivity property here means that the two following constructions are equivalent. (i) By definition of the operation* $\ominus$*, the oriented geodesic segment* $\mathbf{w} \ominus \mathbf{v}$ *is obtained by considering first the oriented geodesic segment* $(\mathbf{w} \ominus \mathbf{v})'$*, that arrives at the tip of* $\mathbf{w}$ *coming from the tip of* $\mathbf{v}$ *and, then, transporting it to the origin, along* $\mathbf{v}$*, to get* $\mathbf{w} \ominus \mathbf{v}$*. (ii) Similarly, the oriented geodesic segment* $\mathbf{v} \ominus \mathbf{w}$ *is obtained by considering first the oriented geodesic segment* $(\mathbf{v} \ominus \mathbf{w})'$*, that arrives at the tip of* $\mathbf{v}$ *coming from the tip of* $\mathbf{w}$ *and, then, transporting it to the origin, along* $\mathbf{w}$*, to get* $\mathbf{v} \ominus \mathbf{w}$*. We see that, on the surface of the sphere, in general,* $\mathbf{w} \ominus \mathbf{v} \neq -(\mathbf{v} \ominus \mathbf{w})$*.*

torsion, to the Riemann tensor $R^i{}_{jk\ell}$ (and to a sum of the Riemann and the gradient of the torsion for general manifolds).

Remember here the generic expression (1.69) for an o-sum:

$$(\mathbf{w} \oplus \mathbf{v})^i = w^i + v^i + e^i{}_{jk}\, w^j\, v^k + q^i{}_{jk\ell}\, w^j\, w^k\, v^\ell + r^i{}_{jk\ell}\, w^j\, v^k\, v^\ell + \dots \quad . \tag{1.107}$$

With the autoparallel characterized by expression (1.98) and the parallel transport by expression (1.101) it is just a matter of careful series expansion to obtain expressions for $e^i{}_{jk}$, $q^i{}_{jk\ell}$ and $r^i{}_{jk\ell}$ for the geosum defined over the oriented segments of a manifold. The computation is done in appendix A.9.5 and one obtains, in a system of coordinates that is autoparallel at the origin,[28]

$$e^i{}_{jk} = \Gamma^i{}_{jk} \;\; ; \;\; q^i{}_{jk\ell} = -\tfrac{1}{2}\,\partial_\ell \gamma^i{}_{jk} \;\; ; \;\; r^i{}_{jk\ell} = -\tfrac{1}{4}\,\textstyle\sum_{(k\ell)}(\partial_k \Gamma^i{}_{\ell j} - \Gamma^i{}_{ks}\,\Gamma^s{}_{\ell j}) \; . \tag{1.108}$$

The reader may verify (using, in particular, the Bianchi identities mentioned below) that these coefficients $e^i{}_{jk}$, $q^i{}_{jk\ell}$ and $r^i{}_{jk\ell}$, satisfy the symmetries expressed in equation (1.71).

The expressions for the torsion and the anassociativity can then be obtained using equations (1.95). After some easy rearrangements, this gives

$$T^i{}_{jk} = \Gamma^i{}_{jk} - \Gamma^i{}_{kj} \qquad ; \qquad A^i{}_{jk\ell} = R^i{}_{jk\ell} + \nabla_\ell T^i{}_{jk} \quad , \tag{1.109}$$

where

$$R^i{}_{jk\ell} = \partial_\ell \Gamma^i{}_{kj} - \partial_k \Gamma^i{}_{\ell j} + \Gamma^i{}_{\ell s}\,\Gamma^s{}_{kj} - \Gamma^i{}_{ks}\,\Gamma^s{}_{\ell j} \tag{1.110}$$

---

[28]See appendix A.9.4 for details. At the origin of an autoparallel system of coordinates the symmetric part of the connection vanishes (but not its derivatives).

is the Riemann tensor of the manifold,[29] and where $\nabla_\ell T^i{}_{jk}$ is the covariant derivative of the torsion:

$$\nabla_\ell T^i{}_{jk} = \partial_\ell T^i{}_{jk} + \Gamma^i{}_{\ell s} T^s{}_{jk} - \Gamma^s{}_{\ell j} T^i{}_{sk} - \Gamma^s{}_{\ell k} T^i{}_{js} \quad . \tag{1.111}$$

Let us state the two results in equation (1.109) as two explicit theorems.

Property 1.11 *When considering the autovector space formed by the oriented autoparallel segments (of common origin) on a manifold, the torsion is (twice) the antisymmetric part of the connection:*

$$\boxed{T^k{}_{ij} = \Gamma^k{}_{ij} - \Gamma^k{}_{ji} \quad .} \tag{1.112}$$

This result was anticipated when we called the tensor defined in equation (1.86) torsion.

Property 1.12 *When considering the autovector space formed by the oriented autoparallel segments (of common origin) on a manifold, the anassociativity tensor* **A** *is given by*

$$\boxed{A^\ell{}_{ijk} = R^\ell{}_{ijk} + \nabla_k T^\ell{}_{ij} \quad ,} \tag{1.113}$$

where $R^\ell{}_{ijk}$ is the Riemann tensor of the manifold (equation 1.110), and where $\nabla_k T^\ell{}_{ij}$ is the gradient (covariant derivative) of the torsion of the manifold (equation 1.111).

So far, the term 'tensor' has only meant 'element of a tensor space', as introduced in section 1.1.2. In manifolds, one calls tensor an invariantly defined object, i.e., an object that, in a change of coordinates over the manifold (and associated change of natural basis), has its components changed in the standard tensorial way.[30] The connection $\Gamma^i{}_{k\ell}$, for instance, is *not* a tensor. But it is well known that the difference $\Gamma^i{}_{jk} - \Gamma^i{}_{kj}$ is a tensor, and therefore the expression (1.110) defines the components of a tensor:

Property 1.13 *$T^i{}_{jk}$, as expressed in* (1.112)*, are the components of a tensor (the torsion tensor), and $R^i{}_{jk\ell}$, as expressed in* (1.110)*, are the components of a tensor (the Riemann tensor).*

As the covariant derivative of a tensor is a tensor, and the sum of two tensors is a tensor, we have

Property 1.14 *$A^i{}_{jk\ell}$, as expressed in* (1.113)*, are the components of a tensor (the anassociativity tensor).*

---

[29] There are many conventions for the definition of the Riemann tensor in the literature. When the connection is symmetric, this definition corresponds to that of Weinberg (1972).

[30] I.e., $T^{i'j'\dots}{}_{k'\ell'\dots} = \frac{\partial x^{i'}}{\partial x^i} \frac{\partial x^{j'}}{\partial x^j} \cdots \frac{\partial x^k}{\partial x^{k'}} \frac{\partial x^\ell}{\partial x^{\ell'}} \cdots T^{ij\dots}{}_{k\ell\dots}$.

The equations (1.108) are obviously not covariant expressions (they are written at the origin of an autoparallel system of coordinates). But in equations (1.94) we have obtained expressions for $e^i{}_{jk}$, $q^i{}_{jk\ell}$ and $r_{jk\ell}$ in terms of the torsion tensor and the anassociativity tensor. Therefore, equations (1.94) give the covariant expressions of these three tensors.

We can now use here the identity (1.90):

Property 1.15 First Bianchi identity. *At any point*[31] *of a differentiable manifold, the anassociativity and the torsion are linked through*

$$\boxed{\textstyle\sum_{(jk\ell)} A^i{}_{jk\ell} \;=\; \sum_{(jk\ell)} T^i{}_{js}\, T^s{}_{k\ell}} \qquad (1.114)$$

(the common value being the Jacobi tensor $J^i{}_{jk\ell}$ *).*

This is an important identity. When expressing the anassociativity in terms of the Riemann and the torsion (equation 1.113), this is the well known "first Bianchi identity" of a manifold.

The second Bianchi identity is obtained by taking the covariant derivative of the Riemann (as expressed in equation 1.110) and making a circular sum:

Property 1.16 Second Bianchi identity. *At any point of a differentiable manifold, the Riemann and the torsion are linked through*

$$\boxed{\textstyle\sum_{(jk\ell)} \nabla_j R^i{}_{mk\ell} \;=\; \sum_{(jk\ell)} R^i{}_{mjs}\, T^s{}_{k\ell} \quad .} \qquad (1.115)$$

Contrary to what happens with the first identity, no simplification occurs when using the anassociativity instead of the Riemann.

1.4 Lie Group Manifolds

The elements of a Lie group can be interpreted as the points of a manifold. Lie group manifolds have a nontrivial geometry; they are metric spaces with a curvature so strong that whole regions of the manifold may not be joined using geodesic lines. Locally, this curvature is balanced by the existence of a torsion: both curvature and torsion compensate so that there exists an absolute parallelism on the manifold.

Once a point $\mathcal{O}$ of the Lie group manifold has been chosen, one can consider the oriented autoparallel segments having $\mathcal{O}$ as origin. For every parallel transport chosen on the manifold, one can define the geometric sum of two oriented geometric segments, this creating around $\mathcal{O}$ a structure of local autovector space. There is one parallel transport such that the geometric

[31] As any point of a differentiable manifold can be taken as origin of an autovector space.

sum of oriented autoparallel segments happens to be, locally, the group operation.

With the metric over the Lie group manifold properly defined, we shall be able to analyze the relations between curvature and torsion. The definition of metric used here is unconventional: what is called the Killing-Cartan "metric" of a Lie group appears here as the Ricci of the metric.

The 'algebra' of a Lie group plays an important role in conventional expositions of the theory. Its importance is here underplayed, as the emphasis is put on the more general concept of autovector space, and on the notion of additive representation of a Lie group.

Ado's theorem states that any Lie group is, in fact, a subgroup of the 'general linear' group GL(n) (the group of all $n \times n$ real matrices with nonzero determinant), so it is important to understand the geometry of this group. The manifold GL(n) is the disjoint union of two manifolds, representing the matrices having, respectively, a positive and negative determinant. To pass from one submanifold to the other one should pass through a point representing a matrix with zero determinant, but this matrix is not a member of GL(n). Therefore the two submanifolds are *not connected*.

Of these two submanifolds, one is a group, the group $\mathrm{GL}^+(n)$ of all $n \times n$ real matrices with positive determinant (as it contains the identity matrix). As a manifold, it is *connected* (it cannot be divided into two disjoint nonempty open sets whose union is the entire manifold). In fact, it is *simply connected* (it is connected and does not have any "hole"). It is not *compact*.[32]

The autovector structure introduced below will not cover the whole GL(n) manifold but only the part of $\mathrm{GL}^+(n)$ that is connected to the origin through autoparallel paths (that, in fact, are going to also be geodesic paths). For this reason, some of the geometric properties mentioned below are demonstrated only for a finite neighborhood of the origin. But as Lie group manifolds are homogeneous manifolds (any point is identical to any other point), the local properties are valid around any point of the manifold.

Among books studying the geometry of Lie groups, Eisenhart (1961) and Goldberg (1998) are specially recommended. For a more analytical vision, Varadarajan (1984) is clear and complete.

One important topic missing in this text is the study of the set of symmetric, positive definite matrices. It is not a group, as the product of two symmetric matrices is not necessarily symmetric. As this set of matrices is a subset of GL(n) it can also be seen as an $n(n+1)/2$-dimensional submanifold of the Lie group manifold GL(n). These kinds of submanifolds of Lie group manifolds are called *symmetric spaces*.[33] We shall not be much concerned

[32]A manifold is compact if any collection of open sets whose union is the whole space has a finite subcollection whose union is still the whole space. For instance, a submanifold of a Euclidean manifold is compact if it is closed and bounded.

[33]In short, a symmetric space is a Riemannian manifold that has a geodesic-reversing isometry at each of its points.

with symmetric, positive definite matrices in this text, for two reasons. First, when we need to evaluate the distance between two symmetric, positive definite matrices, we can evaluate this distance as if we were working in $GL(n)$ (and we will never need to perform a parallel transport inside the symmetric space). Second, in physics, the symmetry condition always results from a special case being considered (as when the elastic stiffness tensor or the electric permittivity tensors are assumed to be symmetric). In the physical developments in chapter 4, I choose to keep the theory as simple as possible, and I do not impose the symmetry condition. For the reader interested in the theory of symmetric spaces, the highly readable text by Terras (1988) is recommended.

The sections below concern, first, those properties of associative autovector spaces that are easily studied using the abstract definition of autovector space, then the geometric properties of a Lie group manifold. Finally, we will explicitly study the geometry of $GL^+(2)$ (section 1.4.6).

1.4.1 Group and Algebra

1.4.1.1 Local Lie Group

As mentioned at the beginning of section 1.3, an n-dimensional manifold is a space of points, that accepts in a finite neighborhood of each of its points an n-dimensional system of continuous coordinates.

Definition 1.30 *A* Lie group *is a set of elements that (i) is a manifold, and (ii) is a group. The* dimension *of a Lie group is the dimension of its manifold.*

For a more precise definition of a Lie group, see Varadarajan (1984) or Goldberg (1998).

Example 1.6 *By the term 'rotation' let us understand here a geometric construction, independently of any possible algebraic representation. The set of n-dimensional rotations, with the composition of rotations as internal operation, is a Lie group with dimension $n(n-1)/2$. The different possible matrix representations of a rotation define different matrix groups, isomorphic to the group of geometrical rotations.*

Our definition of (local) autovector space has precisely the continuity condition built in (through the existence of the operation that to any element $\mathbf{a}$ and to any real number λ inside some finite interval around zero is associated the element $\lambda\,\mathbf{a}$), and we have seen (section 1.3) that the abstract notion of autovector space precisely matches the geometric properties of manifolds. Therefore, when the troupe operation $\oplus$ is associative (i.e., when it is a group operation), an autovector space is (in the neighborhood of the neutral element) a Lie group:

Property 1.17 *Associative autovector spaces are local Lie groups.*

This is only a local property because the o-sum $\mathbf{b} \oplus \mathbf{a}$ is often defined only for the elements of the autovector space that are "close enough" to the zero element. As we shall see, the difference between an associative autovector space (that is a local structure) and the Lie group (that is a global structure), is that the autovectors of an associative autovector space correspond only to the points of the Lie group that are geodesically connected to the origin.

In any autovector space, the commutator is antisymmetric (see equation 1.80), so the property also holds here: for any two autovectors $\mathbf{v}$ and $\mathbf{w}$ of a Lie group manifold,

$$[\mathbf{w}\,,\,\mathbf{v}] \;=\; -[\mathbf{v}\,,\,\mathbf{w}] \quad . \tag{1.116}$$

The associativity condition precisely corresponds to the condition of vanishing of the finite association $\{\,\mathbf{w}\,,\,\mathbf{v}\,,\,\mathbf{u}\,\}$ introduced in equation (1.76), and it implies, therefore, the vanishing of the associator $[\,\mathbf{w}\,,\,\mathbf{v}\,,\,\mathbf{u}\,]$, as defined in equation (1.78):

Property 1.18 *In associative autovector spaces, the associator always vanishes, i.e., for any three autovectors,*

$$[\,\mathbf{w}\,,\,\mathbf{v}\,,\,\mathbf{u}\,] \;=\; \mathbf{0} \quad . \tag{1.117}$$

Then, using definition (1.81) and theorem (1.83), one obtains

Property 1.19 *In associative autovector spaces, the Jacobi tensor always vanishes,* $\mathbf{J} = \mathbf{0}$ *, i.e., for any three autovectors, one has the* Jacobi property

$$\boxed{[\,\mathbf{w}\,,[\mathbf{v},\mathbf{u}]\,]+[\,\mathbf{u}\,,[\mathbf{w},\mathbf{v}]\,]+[\,\mathbf{v}\,,[\mathbf{u},\mathbf{w}]\,] \;=\; \mathbf{0} \quad .} \tag{1.118}$$

The series (1.84) for the geosum in a general autovector space simplifies here (because of identity (1.117)) to

$$\mathbf{w} \oplus \mathbf{v} \;=\; (\mathbf{w}+\mathbf{v}) + \tfrac{1}{2}\,[\mathbf{w},\mathbf{v}] + \tfrac{1}{12}\Big(\,[\,\mathbf{v}\,,[\mathbf{v},\mathbf{w}]\,]+[\,\mathbf{w}\,,[\mathbf{w},\mathbf{v}]\,]\Big) + \cdots \quad , \tag{1.119}$$

an expression known as the BCH series (Campbell, 1897, 1898; Baker, 1905; Hausdorff, 1906) for a Lie group.[34] As in a group, $\mathbf{w} \ominus \mathbf{v} = \mathbf{w} \oplus (-\mathbf{v})$, the series for the geodifference is easily obtained from the BCH series for the geosum (using the antisymmetry property of the commutator, equation (1.116)).

It is easy to translate the properties represented by equations (1.117), (1.118) and (1.119) using the definitions of torsion and of anassociativity introduced in section 1.2.6: in an associative autovector space (i.e., in a local Lie group) one has

$$A^{\ell}{}_{ijk} \;=\; 0 \qquad ; \qquad T^i{}_{js}\,T^s{}_{k\ell} + T^i{}_{ks}\,T^s{}_{\ell j} + T^i{}_{\ell s}\,T^s{}_{jk} \;=\; 0 \quad , \tag{1.120}$$

and the BCH series becomes

$$(\,\mathbf{w} \oplus \mathbf{v}\,)^i \;=\; (w^i + v^i) \;+\; \tfrac{1}{2}\,T^i{}_{jk}\,w^j\,v^k \;+\; \tfrac{1}{12}\,T^i{}_{js}\,T^s{}_{k\ell}\Big(\,v^j\,v^k\,w^\ell + w^j\,w^k\,v^\ell\,\Big) + \cdots \quad . \tag{1.121}$$

[34] Varadarajan (1984) gives the expression for the general term of the series.

1.4.1.2 Algebra

There are two operations defined on the elements $\mathbf{v}, \mathbf{w}, \ldots$: the geometric sum $\mathbf{w} \oplus \mathbf{v}$ and the tangent operation $\mathbf{w} + \mathbf{v}$. By definition, the commutator $[\,\mathbf{w}\,,\,\mathbf{v}\,]$ gives also an element of the space (definition in equation 1.77). Let us recall here the notion of 'algebra', that plays a central role in the standard presentations of Lie group theory. If one considers the commutator as an operation, then it is antisymmetric (equation 1.116) and satisfies the Jacobi property (equation 1.118). This suggests the following

Definition 1.31 Algebra. *A linear space (where the sum $\mathbf{w} + \mathbf{v}$ and the product of an element by a real number $\lambda\,\mathbf{v}$ are defined, and have the usual properties) is called an* algebra *if a second internal operation $[\,\mathbf{w}\,,\,\mathbf{v}\,]$ is defined that satisfies the two properties* (1.116) *and* (1.118).

Given an associative autovector space (i.e., a local Lie group), with group operation $\mathbf{w} \oplus \mathbf{v}$, the construction of the associated algebra is simple: the linear and the quadratic terms of the BCH series (1.119) respectively define the tangent operation $\mathbf{w} + \mathbf{v}$ and the commutator $[\,\mathbf{w}\,,\,\mathbf{v}\,]$. Although the autovector space (as defined by the operation $\oplus$) may only be local, the linear tangent space is that generated by all the linear combinations $\mu\,\mathbf{w} + \lambda\,\mathbf{v}$ of the elements of the autovector space. The commutator (that is a quadratic operation) can then easily be extrapolated from the elements of the (possibly local) autovector space into all the elements of the linear tangent space.

The reciprocal is also true: given an algebra with a commutator $[\,\mathbf{w}\,,\,\mathbf{v}\,]$ one can build the associative autovector space from which the algebra derives. For the BCH series (1.119) defining the group operation $\oplus$ is written only in terms of the sum and the commutator of the algebra.

Using more geometrical notions (to be developed below), the commutator defines the torsion at the origin of the Lie group manifold. As a group manifold is homogeneous, the torsion is then known everywhere. And a Lie group manifold is perfectly characterized by it torsion.

To check whether a given linear subspace of matrices can be considered as the linear tangent space to a Lie group the condition that the commutator defines an internal operation is the key condition.

Example 1.7 *The set of $n \times n$ real antisymmetric matrices with the operation $[\mathbf{s}, \mathbf{r}] = \mathbf{s}\,\mathbf{r} - \mathbf{r}\,\mathbf{s}$ is an algebra: the commutator $[\mathbf{s}, \mathbf{r}]$ defines an internal operation with the right properties.*[35]

[35] For instance, 3×3 antisymmetric matrices are dual to pseudovectors, $a^i = \frac{1}{2}\,\epsilon^{ijk}\,a_{jk}$. Defining the vector product of two pseudovectors as $(\mathbf{b} \times \mathbf{a})_i = \frac{1}{2}\,\epsilon_{ijk}\,b^j\,a^k$, one can write the commutator of two antisymmetric matrices in terms of the vector product of the associated pseudovectors, $[\mathbf{b}, \mathbf{a}] = -(\mathbf{b} \times \mathbf{a})$. This is obviously an antisymmetric operation. Now, from the formula expressing the double vector product, $\mathbf{c} \times (\mathbf{b} \times \mathbf{a}) = (\mathbf{c} \cdot \mathbf{a})\,\mathbf{b} - (\mathbf{c} \cdot \mathbf{b})\,\mathbf{a}$, it follows that $\mathbf{c} \times (\mathbf{b} \times \mathbf{a}) + \mathbf{a} \times (\mathbf{c} \times \mathbf{b}) + \mathbf{b} \times (\mathbf{a} \times \mathbf{c}) = \mathbf{0}$, that is property (1.118).

Example 1.8 *The set of $n \times n$ real symmetric matrices with the operation $[\mathbf{b}, \mathbf{a}] = \mathbf{b}\,\mathbf{a} - \mathbf{a}\,\mathbf{b}$ is* not *an algebra (the commutator of two symmetric matrices is not a symmetric matrix, so the operation is not internal).*

1.4.1.3 Ado's Theorem

We are about to mention Ados' theorem, stating that Lie groups accept matrix representations. As emphasized in section 1.4.3, in fact, a Lie group accepts two basically different matrix representations. For instance, in example 1.6 we considered the group of 3D (geometrical) rotations, irrespectively of any particular representation. The two basic matrix representations of this group are the following.

Example 1.9 *Let* SO(3) *be the set of all orthogonal 3×3 real matrices with positive (unit) determinant. This is a (multiplicative) group with, as group operation, the matrix product $\mathbf{R}_2\,\mathbf{R}_1$. It is well known that this group is isomorphic to the group of geometrical 3D rotations. A geometrical rotation is then represented by an orthogonal matrix, and the composition of rotations is represented by the product of orthogonal matrices.*

Example 1.10 *Let*[36] $\ell i\,\mathrm{SO}(3)$ *be the set of all 3×3 real antisymmetric matrices $\mathbf{r}$ with $\sqrt{(\mathrm{tr}\,\mathbf{r}^2)/2} < i\pi$, plus certain imaginary matrices (see example 1.15 for details). This is an o-additive group, with group operation $\mathbf{r}_2 \oplus \mathbf{r}_1 = \log(\exp\mathbf{r}_2\ \exp\mathbf{r}_1)$. This group is also isomorphic to the group of geometrical 3D rotations. A rotation is then represented by an antisymmetric matrix $\mathbf{r}$, the dual of which, $\rho_i = \frac{1}{2}\epsilon_{ijk}\,r^{jk}$, is the rotation vector, i.e., the vector whose axis is the rotation axis and whose norm is the rotation angle. The composition of rotations is represented by the o-sum $\mathbf{r}_2 \oplus \mathbf{r}_1$. This operation deserves the name 'sum' (albeit it is a noncommutative one) because for small rotations, $\mathbf{r}_2 \oplus \mathbf{r}_1 \approx \mathbf{r}_2 + \mathbf{r}_1$.*

As we have not yet formally introduced the logarithm and exponential of a matrix, let us postpone explicit consideration of the operation $\oplus$, and let us advance through consideration of the matrix product as group operation.

The sets of invertible matrices are well known examples of Lie groups, (with the matrix product as group operation). It is easy to verify that all axioms are then satisfied.

Lest us make an explicit list of the more common matrix groups.

[36]Given a multiplicative group of matrices $\mathbb{M}$, the notation $\ell i\,\mathbb{M}$, introduced later, stands for 'logarithmic image' of $\mathbb{M}$.

Example 1.11 Usual multiplicative matrix groups.

- *The set of all* $n \times n$ *complex invertible matrices is a* $(2n)^2$*-dimensional multiplicative Lie group, called the* general linear complex *group, and denoted* $\mathrm{GL}(n, \mathbb{C})$.
- *The set of all* $n \times n$ *real invertible matrices is an* n^2*-dimensional multiplicative Lie group, called the* general linear *group, and denoted* $\mathrm{GL}(n)$.
- *The set of all* $n \times n$ *real matrices with positive determinant*[37] *is an* n^2*-dimensional multiplicative Lie group, denoted* $\mathrm{GL}^+(n)$.
- *The set of all* $n \times n$ *real matrices with unit determinant is an* $(n^2 - 1)$*-dimensional multiplicative Lie group, called the* special linear *group, and denoted* $\mathrm{SL}(n)$.
- *The* group of homotheties, $\mathrm{H}^+(n)$, *is the one-dimensional subgroup of* $\mathrm{GL}^+(n)$ *with matrices* $U^\alpha{}_\beta = K\,\delta^\alpha_\beta$ *with* $K > 0$. *One has* $\mathrm{GL}^+(n) = \mathrm{SL}(n) \times \mathrm{H}^+(n)$.
- *The set of all* $n \times n$ *real orthogonal*[38] *matrices with positive determinant (equal to +1) is an* $n(n-1)/2$*-dimensional multiplicative Lie group, called the* special orthogonal *group, and denoted* $\mathrm{SO}(n)$.

In particular, the 1×1 complex "matrices" of $\mathrm{GL}(1, \mathbb{C})$ are just the complex numbers, the zero "matrix" excluded. This two-dimensional (commutative) group $\mathrm{GL}(1, \mathbb{C})$ corresponds to the whole complex plane, excepted the zero (with the product of complex numbers as group operation).

Although the complex matrix groups may seem more general than the real matrix groups, they are not: the group $\mathrm{GL}(n, \mathbb{C})$ can be interpreted as a subgroup of $\mathrm{GL}(2n)$, as the following example shows.

Example 1.12 *When representing complex numbers by* 2×2 *real matrices,*

$$a + i\,b \quad \mapsto \quad \begin{pmatrix} a & b \\ -b & a \end{pmatrix} \quad , \tag{1.122}$$

[37]The *determinant* of a contravariant–covariant tensor $\mathbf{U} = \{U^\alpha{}_\beta\}$ is defined as $\det \mathbf{U} = \frac{1}{n!}\, \underline{\epsilon}_{i_1 i_2 \dots i_n}\, U^{i_1}{}_{j_1}\, U^{i_2}{}_{j_2} \dots U^{i_n}{}_{j_n}\, \overline{\epsilon}^{j_1 j_2 \dots j_n}$, where $\overline{\epsilon}_{i_1 i_2 \dots i_n}$ and and $\underline{\epsilon}_{i_1 i_2 \dots i_n}$ are respectively the *Levi-Civita density* and and the *Levi-Civita capacity* defined as being zero if any index is repeated, equal to one if $\{i_1 i_2 \dots i_n\}$ is an even permutation of $\{1, 2 \dots n\}$ and equal to -1 if the permutation is odd. If the space $\mathbb{E}_n$ has a metric, one can introduce the *Levi-Civita tensor* $\epsilon_{i_1 i_2 \dots i_n}$ related to the Levi-Civita capacity via $\epsilon_{i_1 i_2 \dots i_n} = \sqrt{\det \mathbf{g}}\; \underline{\epsilon}_{i_1 i_2 \dots i_n}$.

[38]Remember here that we are considering mappings $\mathbf{U} = \{U^\alpha{}_\beta\}$ that map $\mathbb{E}_n$ into itself. The transpose of $\mathbf{U}$ is an operator $\mathbf{U}^T$ with components $(\mathbf{U}^T)_\alpha{}^\beta = U^\beta{}_\alpha$. If there is a metric $\mathbf{g} = \{g_{ij}\}$ in $\mathbb{E}_n$, then one can define (see appendix A.1 for details) the adjoint operator $\mathbf{U}^* = \mathbf{g}^{-1}\, \mathbf{U}^T\, \mathbf{g}$. The linear operator $\mathbf{U}$ is called *orthogonal* if its inverse equals its adjoint. Using the equation above this can be written $\mathbf{U}^{-1} = \mathbf{U}^*$, i.e., $\mathbf{g}\, \mathbf{U}^{-1} = \mathbf{U}^T\, \mathbf{g}$. This gives, in terms of components, $g_{ik}\, (\mathbf{U}^{-1})^\mu{}_\beta = U^\mu{}_\alpha\, g_{kj}$, i.e., $(\mathbf{U}^{-1})_{ij} = U_{ji}$. It is important to realize that while the group $\mathrm{GL}(n)$ is defined independently of any possible metric over $\mathbb{E}_n$, the subgroup of orthogonal transformations is defined with respect to a given metric.

it is easy to see that the product of complex numbers is represented by the (ordinary) product of matrices. Therefore, the two-dimensional group GL(1, $\mathbb{C}$) *is isomorphic to the two-dimensional subgroup of* GL(2) *consisting of matrices with the form* (1.122) *(i.e., all real matrices with this form except the zero matrix).*

The (multiplicative) group GL(n) is a good example of a Lie group (we have even seen that GL(n, $\mathbb{C}$) can be considered to be a subgroup of GL($2n$)). In fact it is much more than a simple example, as every Lie group can be considered to be contained by GL(n) :

Property 1.20 (Ado's theorem) *Any Lie group is isomorphic to a real matrix group, i.e., a subgroup of* GL(n) .

Although this is only a free interpretation of the actual theorem,[39] it is more than sufficient for our physical applications. As Iserles et al. (2000) put it, "for practically any concept in general Lie theory there exists a corresponding concept within matrix Lie theory; vice versa, practically any result that holds in the matrix case remains valid within the general Lie theory."

Because of this theorem, we shall now move away from abstract Lie groups (i.e., from abstract autovector spaces), and concentrate on matrix groups. The notations $\mathbf{u}$, $\mathbf{v}$ . . . , that in section 1.4.1.1 represented an abstract element of an autovector space, will, from now on, be replaced by the notation $\mathbf{a}$, $\mathbf{b}$ . . . representing matrices. This allows, for instance, to demonstrate the theorem $[\mathbf{b}, \mathbf{a}] = \mathbf{b}\,\mathbf{a} - \mathbf{a}\,\mathbf{b}$ (see equation 1.148), that makes sense only when the autovectors are represented by matrices. If instead of the additive representation one uses the multiplicative representation, the matrices will be denoted $\mathbf{A}$, $\mathbf{B}$ . . . .

One should remember that the definition of the orthogonal group of matrices depends on the background metric being considered, as the following example highlights.

Example 1.13 *The matrices of* GL(2) *have the form* $\mathbf{U} = \begin{pmatrix} U^1{}_1 & U^1{}_2 \\ U^2{}_1 & U^2{}_2 \end{pmatrix}$ *with real entries such that* $U^1{}_1\, U^2{}_2 - U^1{}_2\, U^2{}_1 \neq 0$. SL(2) *is made by the subgroup with* $U^1{}_1\, U^2{}_2 - U^1{}_2\, U^2{}_1 = 1$. *If the space* $\mathbb{E}_2$ *has a Euclidean metric* $\mathbf{g} = \text{diagonal}(1, 1)$, *the subgroup* SO(2) *of orthogonal matrices corresponds to the matrices* $\begin{pmatrix} U^1{}_1 & U^1{}_2 \\ U^2{}_1 & U^2{}_2 \end{pmatrix} = \begin{pmatrix} \cos\alpha & \sin\alpha \\ -\sin\alpha & \cos\alpha \end{pmatrix}$, *with* $-\pi < \alpha \leq \pi$. *If the space* $\mathbb{E}_2$ *has a Minkowskian metric* $\mathbf{g} = \text{diagonal}(1, -1)$, *the subgroup of orthogonal matrices corresponds to the matrices* $\begin{pmatrix} U^1{}_1 & U^1{}_2 \\ U^2{}_1 & U^2{}_2 \end{pmatrix} = \begin{pmatrix} \cosh\varepsilon & \sinh\varepsilon \\ \sinh\varepsilon & \cosh\varepsilon \end{pmatrix}$, *with* $-\infty < \varepsilon < \infty$.

[39] Any Lie algebra is isomorphic to the Lie algebra of a subgroup of GL(n) . See Varadarajan (1984) for a demonstration.

1.4.2 Logarithm of a Matrix

If a function $z \mapsto f(z)$ is defined for a scalar z, and if $f(\mathbf{M})$ makes sense when z is replaced by a matrix $\mathbf{M}$, then this expression is used to define the function $f(\mathbf{M})$ of the matrix (for a general article about the functions of matrices, see Rinehart, 1955). It is clear, in particular, that we can give sense to any analytical function accepting a series expansion. For instance, the exponential of a square matrix is defined as $\exp \mathbf{M} = \sum_{n=0}^{\infty} \frac{1}{n!} \mathbf{M}^n$. It follows that if one can write a decomposition of a matrix $\mathbf{M}$ as $\mathbf{M} = \mathbf{U}\,\mathbf{J}\,\mathbf{U}^{-1}$, where $\mathbf{J}$ is some simple matrix for which $f(\mathbf{J})$ makes sense, then one defines $f(\mathbf{M}) = \mathbf{U}\, f(\mathbf{J})\, \mathbf{U}^{-1}$. Here below we are mainly concerned with the exponential and the logarithm of a square matrix (in chapter 4 we shall also introduce the square root of a matrix). The exponential and the logarithm of a matrix could have been introduced in section 1.1, where the basic properties of linear spaces were recalled. It seems better to introduce the exponential and the logarithm in this section because, as we are about to see, the natural domain of definition of the logarithm function is a multiplicative group of matrices.

In physical applications, we are not always interested in abstract 'matrices', but, more often, in tensors: the matrices mentioned here usually correspond to components of tensors in a given basis. The reader should note that to give sense to a series containing the components of a tensor, (i) the tensor must be a covariant–contravariant or a contravariant–covariant tensor, and (ii) the tensor must be adimensional (so its successive powers can be added). Here below, the contravariant–covariant notation $M^i{}_j$ is used for the matrices, although the covariant–contravariant notation $M_i{}^j$ could have been used instead. The two types of notation M_{ij} and M^{ij} have no immediate interpretation in terms of components of tensors (for instance, when being used in a series of matrix powers), and are avoided.

1.4.2.1 Analytic Function

For any square complex matrix $\mathbf{M}$ one can write the *Jordan decomposition* $\mathbf{M} = \mathbf{U}\,\mathbf{J}\,\mathbf{U}^{-1}$, where $\mathbf{J}$ is a Jordan matrix.[40] As it is easy to define $f(\mathbf{J})$ for a Jordan matrix, one sets the following

Definition 1.32 Function of a matrix. *Let* $\mathbf{M} = \mathbf{U}\,\mathbf{J}\,\mathbf{U}^{-1}$, *be the Jordan decomposition of the complex matrix* $\mathbf{M}$, *and let* $f(z)$ *be a complex function of the complex variable* z *whose values (and perhaps, the value of some of its derivatives*[41]*) are defined for the eigenvalues of* $\mathbf{M}$. *The matrix* $f(\mathbf{M})$ *is defined as*

[40] A Jordan matrix is a block-diagonal matrix made by Jordan blocks, a Jordan block being a matrix with zeros everywhere, excepted in its diagonal, where there is a constant value λ, and in one of the two adjacent diagonal lines, filled with the number one (see details in appendix A.5).

[41] When the Jordan matrix is not diagonal, the expression $f(\mathbf{J})$ involves derivatives of f. See details in appendix A.5.

$$f(\mathbf{M}) = \mathbf{U} f(\mathbf{J}) \mathbf{U}^{-1} \quad , \tag{1.123}$$

where the function $f(\mathbf{J})$ of a Jordan matrix is defined in appendix A.5. In the particular case where all the eigenvalues of $\mathbf{M}$ are distinct, $\mathbf{J}$ is a diagonal matrix with the eigenvalues $\lambda_1, \lambda_2, \dots$ in its diagonal. Then, $f(\mathbf{J})$ is the diagonal matrix with the values $f(\lambda_1), f(\lambda_2), \dots$ in its diagonal.

1.4.2.2 Exponential

It is easy to see that the above definition of function of a matrix leads, when applied to the exponential function, to the usual exponential series. As this result is general, we can use it as an alternative definition of the exponential function:

Definition 1.33 *For a matrix $\mathbf{m}$, with indices $m^i{}_j$, such that the series $(\exp \mathbf{m})^i{}_j = \delta^i_j + m^i{}_j + \frac{1}{2!} m^i{}_k m^k{}_j + \frac{1}{3!} m^i{}_k m^k{}_\ell m^\ell{}_j + \dots$ makes sense, we shall call the matrix $\mathbf{M} = \exp \mathbf{m}$ the* exponential *of $\mathbf{m}$. The exponential series can be written, more compactly, $\exp \mathbf{m} = \mathbf{I} + \mathbf{m} + \frac{1}{2!} \mathbf{m}^2 + \frac{1}{3!} \mathbf{m}^3 + \dots$, i.e.,*

$$\exp \mathbf{m} = \sum_{n=0}^{\infty} \frac{1}{n!} \mathbf{m}^n \quad . \tag{1.124}$$

Again, for this series to be defined, the matrix (usually representing the components of a tensor) $\mathbf{m}$ has to be adimensional.

As the exponential of a complex number is a periodic function, the matrix exponential is, *a fortiori*, a periodic matrix function. The precise type of periodicity of the matrix exponential will become clear below when analyzing the group SL(2).

Multiplying the series (1.124) by itself n times, one easily verifies the important property

$$(\exp \mathbf{m})^n = \exp(n\,\mathbf{m}) \quad , \tag{1.125}$$

and, in particular, $(\exp \mathbf{m})^{-1} = \exp(-\mathbf{m})$. Another important property of the exponential function is that for any matrix $\mathbf{m}$,

$$\det(\exp \mathbf{m}) = \exp(\operatorname{tr} \mathbf{m}) \quad . \tag{1.126}$$

It may also be mentioned that it follows from the definition of the exponential function that the eigenvalues of $\exp \mathbf{m}$ are the exponential of the eigenvalues of $\mathbf{m}$. Note that, in general,[42]

$$\exp \mathbf{b} \exp \mathbf{a} \neq \exp(\mathbf{b} + \mathbf{a}) \quad . \tag{1.127}$$

The notational abuse

[42]Unless $\exp \mathbf{b} \exp \mathbf{a} = \exp \mathbf{a} \exp \mathbf{b}$.

$$\exp m^i{}_j \equiv (\exp \mathbf{m})^i{}_j \tag{1.128}$$

may be used. It is consistent, for instance, with the common notation $\nabla_i v^j$ for the covariant derivative of a vector that (rigorously) should be written $(\nabla \mathbf{v})_i{}^j$.

1.4.2.3 Logarithm

The logarithm of a matrix (in fact, of a tensor) plays a major role in this book. While in many physical theories involving real scalars, only the logarithm of positive quantities (that is a real quantity) is considered, it appears that most physical theories involving the logarithm of real matrices lead to some special class of complex matrices. Because of this, and because of the periodicity of the exponential function, the definition of the logarithm of a matrix requires some care. It is better to start by recalling the definition of the logarithm of a number, real or complex.

Definition 1.34 *The logarithm of a positive real number* x*, is the (unique) real number, denoted* $y = \log x$*, such that*

$$\exp y = x \quad . \tag{1.129}$$

The log-exp functions define a bijection between the positive part of the real line and the whole real line.

Definition 1.35 *The logarithm of a nonzero complex number* $z = |z|\, e^{i \arg z}$ *is the complex number*

$$\log z = \log |z| + i \arg z \quad . \tag{1.130}$$

As $\log |z|$ is the logarithm of a positive real number, it is uniquely defined. As the argument $\arg z$ of a complex number z is also uniquely defined ($-\pi < \arg z \leq \pi$), it follows that the logarithm of a complex number $z \neq 0$ is uniquely defined. The whole complex plane except the zero, was denoted above $GL(1, \mathbb{C})$, as it is a two-dimensional multiplicative (and commutative) Lie group. It is clear that

$$z \in GL(1, \mathbb{C}) \quad \Rightarrow \quad \exp \log z = z \quad . \tag{1.131}$$

The logarithm function has a discontinuity along the negative part of the imaginary axis, as the logarithm of two points on each immediate side of the imaginary axis differs by 2π. From a geometrical point of view, the logarithm transforms each "radial" line of $GL(1, \mathbb{C})$ into the "horizontal" line of the complex plane whose imaginary coordinate is the angle between the radial line and the real axis. Thus, the logarithmic image of $GL(1, \mathbb{C})$ is a horizontal band of the complex plane, with a width of 2π. Let us denote this band as $\ell i\,\mathrm{GL}(1,\mathbb{C})$ (a notation to be generalized below). It is mapped into $\mathrm{GL}(1,\mathbb{C})$ by the exponential function, so the log-exp functions define a bijection[43] between $\mathrm{GL}(1,\mathbb{C})$ and $\ell i\,\mathrm{GL}(1,\mathbb{C})$. All other similar horizontal bands of the complex plane are mapped by the exponential function into the same $\mathrm{GL}(1,\mathbb{C})$. To the property (1.131) we can therefore add

$$z \in \ell i\,\mathrm{GL}(1,\mathbb{C}) \quad \Rightarrow \quad \log\exp z \;=\; z \quad , \tag{1.132}$$

but one should keep in mind that

$$z \notin \ell i\,\mathrm{GL}(1,\mathbb{C}) \quad \Rightarrow \quad \log\exp z \;\neq\; z \quad . \tag{1.133}$$

To define the logarithm of a matrix, there is no better way than to use the general definition for the function of a matrix (definition 1.32), so let us repeat it here:

Definition 1.36 Logarithm of a matrix. *Let* $\mathbf{M} \;=\; \mathbf{U}\,\mathbf{J}\,\mathbf{U}^{-1}$, *be the Jordan decomposition of an invertible matrix* $\mathbf{M}$. *The matrix* $\log\mathbf{M}$ *is defined as*

$$\log\mathbf{M} \;=\; \mathbf{U}\,(\log\mathbf{J})\,\mathbf{U}^{-1} \quad , \tag{1.134}$$

where the logarithm of a Jordan matrix is defined in appendix A.5. In the particular case where all the eigenvalues of $\mathbf{M}$ *are distinct,* $\mathbf{J}$ *is a diagonal matrix with the eigenvalues* $\lambda_1, \lambda_2, \dots$ *in its diagonal. Then,* $\log\mathbf{J}$ *is the diagonal matrix with the values* $\log\lambda_1, \log\lambda_2, \dots$ *on its diagonal.*[44]

It is well known that, when the series converges, the logarithm of a complex number can be expanded as $\log z = (z-1) - \frac{1}{2}(z-1)^2 + \frac{1}{3}(z-1)^3 + \dots$. It can be shown (e.g., Horn and Johnson, 1999) that this property extends to the matrix logarithm:

Property 1.21 *For a matrix* $\mathbf{M}$ *verifying* $\|\,\mathbf{M} - \mathbf{I}\,\| < 1$, *the following series converges to the logarithm of the matrix:*

$$\log\mathbf{M} \;=\; \sum_{n=1}^{\infty} \frac{(-1)^{n+1}}{n}\,(\,\mathbf{M} - \mathbf{I}\,)^n \quad . \tag{1.135}$$

Explicitly, $\log\mathbf{M} = (\mathbf{M} - \mathbf{I}) - \frac{1}{2}(\mathbf{M} - \mathbf{I})^2 + \frac{1}{3}(\mathbf{M} - \mathbf{I})^3 + \dots$

This is nothing but the extension to matrices of the usual series for the logarithm of a scalar. It cannot be used as a definition of the logarithm of a matrix because it converges only for matrices that are close enough to the

[43] Sometimes the definition of logarithm used here is called the 'principal determination of the logarithm', and any number α such that $e^\alpha = z$ is called 'a logarithm of z' (so all numbers $(\log z) + 2\,n\,i\,\pi$ are 'logarithms' of z). We do not follow this convention here: for any complex number $z \neq 0$, the complex number $\log z$ is uniquely defined.

[44] As the matrix $\mathbf{M}$ is invertible, all the eigenvalues are different from zero.

identity matrix. (Equation (A.72) of the appendix gives another series for the logarithm.)

The uniqueness of the definition of the logarithm of a complex number, leads to the uniqueness of the logarithm of a Jordan matrix, and, from there, the uniqueness of the logarithm of an arbitrary invertible matrix:

Property 1.22 *The logarithm of any matrix of* $\mathrm{GL}(n,\mathbb{C})$ *and, therefore, of any of its subgroups, is defined, and is unique.*

The formulas of this section are not adapted to obtain good analytical expressions of the exponential or the logarithm of a matrix. Because of the Cayley-Hamilton theorem, any matrix function can be reduced to a polynomial of the matrix. Section A.5.5 gives the Sylvester formula, that produces this polynomial.[45]

1.4.2.4 Logarithmic Image

Definition 1.37 Logarithmic image of a multiplicative group of matrices. *Let* $\mathbb{M}$ *be a multiplicative matrix group, i.e., a subgroup of* $\mathrm{GL}(n,\mathbb{C})$. *The image of* $\mathbb{M}$ *through the logarithm function is denoted* $\ell i\,\mathbb{M}$, *and is called the* logarithmic image *of* $\mathbb{M}$. *In particular,*

- $\ell i\,\mathrm{GL}(n,\mathbb{C})$ *is the logarithmic image of* $\mathrm{GL}(n,\mathbb{C})$;
- $\ell i\,\mathrm{GL}(n)$ *is the logarithmic image of* $\mathrm{GL}(n)$;
- $\ell i\,\mathrm{SL}(n)$ *is the logarithmic image of* $\mathrm{SL}(n)$;
- $\ell i\,\mathrm{SO}(n)$ *is the logarithmic image of* $\mathrm{SO}(n)$.

The direct characterization of these different logarithmic images is not obvious, and usually requires some care in the use of the logarithm function. In the two examples below, the (pseudo) norm of a matrix $\mathbf{m}$ is defined as

$$\| \mathbf{m} \| \equiv \sqrt{(\mathrm{tr}\,\mathbf{m}^2)/2} \quad . \tag{1.136}$$

Example 1.14 *While the group* $\mathrm{SL}(2)$ *consists of all* 2×2 *real matrices with unit determinant, its logarithmic image,* $\ell i\,\mathrm{SL}(2)$ *consists (see appendix A.6) of three subsets: (i) the set of all* 2×2 *real traceless matrices* $\mathbf{s}$ *with real norm,* $0 \le \| \mathbf{s} \| < \infty$, *(ii) the set of all* 2×2 *real antisymmetric matrices* $\mathbf{s}$ *with imaginary norm,* $0 < \| \mathbf{s} \| < i\,\pi$, *and (iii) the set of all matrices with form* $\mathbf{t} = \mathbf{s} + i\,\pi\,\mathbf{I}$, *where* $\mathbf{s}$ *is a matrix of the first set.*

---

[45]To obtain rapidly the logarithm $\mathbf{m}$ of a matrix $\mathbf{M}$, one may guess the result, then check, using standard mathematical software, that the condition $\exp \mathbf{m} = \mathbf{M}$ is satisfied. If one is certain of being in the 'principal branch' of the logarithm, the guess is correct.

Example 1.15 *While the group* SO(3) *consists of all* 3×3 *real orthogonal matrices, its logarithmic image,* $\ell i\,\mathrm{SO}(3)$ *consists (see appendix A.7) of two subsets: (i) the set of all* 3 × 3 *real antisymmetric matrices* $\mathbf{r}$ *with*[46] $0 \leq \| \mathbf{r} \| < i\pi$, *and (ii) the set of all imaginary diagonalizable matrices with eigenvalues* $\{0, i\pi, i\pi\}$. *For all the matrices of this set,* $\| \mathbf{r} \| = i\pi$.

Example 1.16 *The set of all complex numbers except the zero was denoted* $\mathrm{GL}(1,\mathbb{C})$ *above. It is a two-dimensional Lie group with respect to the product of complex numbers as group operation. It is, in fact, the group* $\mathrm{GL}(1,\mathbb{C})$. *The set* $\ell i\,\mathrm{GL}(1,\mathbb{C})$ *is (as already mentioned) the band of the complex plane with numbers* $z = a + ib$ *with* a *arbitrary and* $-\pi < b \leq \pi$.

We can now turn to the examination of the precise sense in which the exponential and logarithm functions are mutually inverse. By definition of the logarithmic image of $\mathrm{GL}(n,\mathbb{C})$,

Property 1.23 *For any matrix* $\mathbf{M}$ *of* $\mathrm{GL}(n,\mathbb{C})$,

$$\exp(\log \mathbf{M}) = \mathbf{M} \quad . \tag{1.137}$$

For any matrix $\mathbf{m}$ *of* $\ell i\,\mathrm{GL}(n,\mathbb{C})$,

$$\log(\exp \mathbf{m}) = \mathbf{m} \quad . \tag{1.138}$$

While the condition for the validity of (1.137) only excludes the zero matrix $\mathbf{M} = \mathbf{0}$ (for which the logarithm is not defined), the condition for the validity of (1.138) corresponds to an actual restriction of the domain of matrices $\mathbf{m}$ where this property holds.

Example 1.17 *Let be* $\mathbf{M} = \begin{pmatrix} \cos\alpha & \sin\alpha \\ -\sin\alpha & \cos\alpha \end{pmatrix}$. *One has* $\exp(\log \mathbf{M}) = \mathbf{M}$ *for any value of* α. *Let be* $\mathbf{m} = \begin{pmatrix} 0 & \alpha \\ -\alpha & 0 \end{pmatrix}$. *One has* $\log(\exp \mathbf{m}) = \mathbf{m}$ *only if* $\alpha < \pi$.

Setting $\mathbf{m} = \log \mathbf{M}$ in equation (1.125) gives the expression $[\exp(\log \mathbf{M})]^n = \exp(n \log \mathbf{M})$ that, using equation (1.137), can be written

$$\mathbf{M}^n = \exp(n \log \mathbf{M}) \quad , \tag{1.139}$$

a property valid for any $\mathbf{M}$ in $\mathrm{GL}(n,\mathbb{C})$ and any positive integer n. This can be used to define the real power of a matrix:

Definition 1.38 Matrix power. *For any matrix in* $\mathrm{GL}(n,\mathbb{C})$,

$$\boxed{\mathbf{M}^\lambda \equiv \exp(\lambda \log \mathbf{M}) \quad .} \tag{1.140}$$

---

[46] All these matrices have imaginary norm.

Taking the logarithm of equation (1.140) gives the expression $\log(\mathbf{M}^\lambda) = \log(\exp(\lambda \log \mathbf{M}))$. If $\lambda \log \mathbf{M}$ belongs to $\ell i\,\mathrm{GL}(n,\mathbb{C})$, then, using the property (1.138), this simplifies to $\log(\mathbf{M}^\lambda) = \lambda \log \mathbf{M}$:

$$\lambda \log \mathbf{M} \in \ell i\,\mathrm{GL}(n,\mathbb{C}) \quad \Rightarrow \quad \log(\mathbf{M}^\lambda) = \lambda \log \mathbf{M} \quad . \tag{1.141}$$

In particular, if $-\log \mathbf{M}$ belongs to $\ell i\,\mathrm{GL}(n,\mathbb{C})$, then, $\log(\mathbf{M}^{-1}) = -\log \mathbf{M}$.

Example 1.18 *Let be* $\mathbf{M} = \begin{pmatrix} \cos\alpha & \sin\alpha \\ -\sin\alpha & \cos\alpha \end{pmatrix}$, *with* $\alpha < \pi$. *Then,* $\log \mathbf{M} = \begin{pmatrix} 0 & \alpha \\ -\alpha & 0 \end{pmatrix}$. *As* $\mathbf{M}$ *is a rotation, clearly,* $\mathbf{M}^n = \begin{pmatrix} \cos n\alpha & \sin n\alpha \\ -\sin n\alpha & \cos n\alpha \end{pmatrix}$. *While* $n\alpha < \pi$, *one has* $\log(\mathbf{M}^n) = n \log \mathbf{M}$, *but the property fails if* $n\alpha \geq \pi$.

Setting $\exp \mathbf{m} = \mathbf{M}$ in equation (1.126) shows that for any invertible matrix $\mathbf{M}$, $\det \mathbf{M} = \exp(\mathrm{tr} \log \mathbf{M})$. If $\mathrm{tr} \log \mathbf{M}$ is in the logarithmic image of the complex plane, $\ell i\,\mathrm{GL}(1,\mathbb{C})$, the (scalar) exponential can be inverted:

$$\mathrm{tr}(\log \mathbf{M}) \in \ell i\,\mathrm{GL}(1,\mathbb{C}) \quad \Rightarrow \quad \log(\det \mathbf{M}) = \mathrm{tr}(\log \mathbf{M}) \quad . \tag{1.142}$$

A typical example where the condition $\mathrm{tr}(\log \mathbf{M}) \in \ell i\,\mathrm{GL}(1,\mathbb{C})$ fails, is the 2×2 matrix $\mathbf{M} = -\mathbf{I}$.

One should remember that, in general (unless $\mathbf{B}\,\mathbf{A} = \mathbf{A}\,\mathbf{B}$),

$$\log(\mathbf{B}\,\mathbf{A}) \neq \log \mathbf{B} + \log \mathbf{A} \quad . \tag{1.143}$$

In parallel with the notational abuse (1.128) for the exponential, one may use the notation

$$\log M^i{}_j \equiv (\log \mathbf{M})^i{}_j \quad . \tag{1.144}$$

By no means $\log M^i{}_j$ represents the tensor obtained taking the logarithm of each of the components. Again, this is consistent with the common notational abuse $\nabla_i v^j$ for the covariant derivative of a vector.

1.4.3 Basic Group Isomorphism

By definition of the logarithmic image of a multiplicative group of matrices,

Property 1.24 *The logarithm and exponential functions define a bijection between a set* $\mathbb{M}$ *of matrices that is a Lie group under the matrix product and its image* $\ell i\,\mathbb{M}$ *through the logarithm function.*

Property 1.25 *Let* $\mathbf{A}, \mathbf{B}\ldots$ *be matrices of* $\mathbb{M}$. *Then,* $\mathbf{a} = \log \mathbf{A}$, $\mathbf{b} = \log \mathbf{B}\ldots$ *are matrices of* $\ell i\,\mathbb{M}$. *One has the equivalence*

$$\boxed{\mathbf{C} = \mathbf{B}\,\mathbf{A} \quad \Leftrightarrow \quad \mathbf{c} = \mathbf{b} \oplus \mathbf{a} \quad ,} \tag{1.145}$$

where

$$\boxed{\mathbf{b} \oplus \mathbf{a} \ \equiv \ \log(\exp \mathbf{b} \ \exp \mathbf{a}) \quad .} \tag{1.146}$$

Therefore, $\ell i\,\mathbb{M}$ is also a Lie group, with respect to the operation $\oplus$. The log-exp functions define a group isomorphism between $\mathbb{M}$ and $\ell i\,\mathbb{M}$.

Definition 1.39 *While the group $\mathbb{M}$, with the group operation $\mathbf{C} = \mathbf{B}\,\mathbf{A}$, is called* multiplicative, *the group $\ell i\,\mathbb{M}$, with the (generally noncommutative) group operation $\mathbf{c} = \mathbf{b} \oplus \mathbf{a}$, is called* o-additive.

Using the series for the exponential and for the logarithm, one finds the series expansion

$$\mathbf{b} \oplus \mathbf{a} \ = \ (\mathbf{b} + \mathbf{a}) + \tfrac{1}{2}\,[\mathbf{b}, \mathbf{a}] + \dots \quad , \tag{1.147}$$

where

$$\boxed{[\mathbf{b}, \mathbf{a}] \ = \ \mathbf{b}\,\mathbf{a} - \mathbf{a}\,\mathbf{b} \quad .} \tag{1.148}$$

We thus see that the commutator, as was defined by equation (1.77) for general autovector spaces, contains the usual commutator of Lie group theory.

The symbol $\oplus$ has been introduced in three different contexts. First, in section 1.2 the symbol was introduced as the troupe operation of an abstract autovector space. Second, in section 1.3 the symbol $\oplus$ was introduced for the geometric sum of oriented segments on a manifold. Now in equation (1.146) the symbol $\oplus$ is introduced for an algebraic operation involving the logarithm and the exponential of matrices. These three different introductions are consistent: all correspond to the basic troupe operation in an autovector space (associative or not), and all can be interpreted as an identical sum of oriented segments on a manifold.

1.4.4 Autovector Space of a Group

Given a multiplicative group $\mathbb{M}$ of (square) matrices $\mathbf{A}, \mathbf{B} \dots$, with group operation denoted $\mathbf{B}\,\mathbf{A}$, we can introduce the space $\ell i\,\mathbb{M}$, the logarithmic image of $\mathbb{M}$, with matrices denoted $\mathbf{a}, \mathbf{b} \dots$. It is also a group, with the group operation $\mathbf{b} \oplus \mathbf{a}$ defined by equation (1.146). To have an autovector space we must also define the operation that to any real number and to any element of the given group associates an element of the same group. In the multiplicative representation, this operation is

$$\boxed{\{\lambda, \mathbf{A}\} \quad \mapsto \quad \mathbf{A}^{\lambda}} \tag{1.149}$$

(the matrix exponential having been defined by equation (1.140)), while in the o-additive representation it is

$$\boxed{\{\lambda, \mathbf{a}\} \quad \mapsto \quad \lambda\,\mathbf{a}} \tag{1.150}$$

(the usual multiplication of a matrix by a number).

But for a given multiplicative matrix group $\mathbb{M}$ (resp. a given o-additive matrix group $\ell i \mathbb{M}$) the operation $\mathbf{A}^\lambda$ (resp. the operation $\lambda\, \mathbf{a}$) may not be internal: it may produce a matrix that belongs to a larger group.[47]

This suggests the following definitions.

Definition 1.40 Near-identity subset. *Let* $\mathbb{M}$ *be a multiplicative group of matrices. The subset* $\mathbb{M}_I \subset \mathbb{M}$ *of matrices* $\mathbf{A}$ *such that* $\mathbf{A}^\lambda$ *belongs to* $\mathbb{M}$ *(in fact, to* $\mathbb{M}_I$*) for any* λ *in the interval* $[-1,1]$, *is called the* near-identity subset *of* $\mathbb{M}$.

Definition 1.41 Near-zero subset. *Let* $\mathbb{M}$ *be a multiplicative group of matrices, and* $\ell i \mathbb{M}$ *its logarithmic image. The subset* $\mathfrak{m}_0$ *of matrices of* $\ell i \mathbb{M}$ *such that for any real* $\lambda \in [-1,1]$ *and for any matrix* $\mathbf{a}$ *of the subset,* $\lambda\, \mathbf{a}$ *belongs to* $\ell i \mathbb{M}$ *(in fact, to* $\mathfrak{m}_0$*), is called the* near-zero subset *of* $\ell i \mathbb{M}$.

A schematic illustration of the relations between these subsets, and their basic properties is proposed in figures 1.8 and 1.9. The notation $\mathfrak{m}_0$ for the near-zero subset of $\ell i \mathbb{M}$ is justified because $\mathfrak{m}_0$ is also a subset of the algebra of $\mathbb{M}$ (if a group is denoted $\mathbb{M}$, its algebra is usually denoted $\mathfrak{m}$).

When a matrix $\mathbf{M}$ belongs to $\mathbb{M}_I$, the matrix $\log \mathbf{M}^\lambda$ belongs to $\mathfrak{m}_0$, and

$$\log \mathbf{M}^\lambda = \lambda \log \mathbf{M} \qquad (\mathbf{M} \in \mathbb{M}_I) \quad . \tag{1.151}$$

In particular,

$$\log \mathbf{M}^{-1} = - \log \mathbf{M} \qquad (\mathbf{M} \in \mathbb{M}_I) \quad . \tag{1.152}$$

Example 1.19 *The matrices of the multiplicative group* SL(2) *have two eigenvalues that are both real and positive, or both real and negative, or both complex mutually conjugate. The near-identity subset* $SL(2)_I$ *is made by suppressing from* SL(2) *all the matrices with both eigenvalues real and negative. The matrix algebra* $\mathfrak{sl}(2)$ *consists of* 2×2 *real matrices. The (pseudo)norm* $\| \mathbf{s} \| = \sqrt{(\mathrm{tr}\, \mathbf{s}^2)/2}$ *of these matrices may be positive real or positive imaginary. The near-zero subset* $\mathfrak{sl}(2)_0$ *is made by suppressing from* $\mathfrak{sl}(2)$ *all the matrices with imaginary norm larger than or equal to* $i\pi$. *See section 1.4.6 for the geometrical interpretation of this subset, and appendix A.6 for some analytical details.*

Example 1.20 *The matrices of the multiplicative group* SO(3) *have the three eigenvalues* $\{1, e^{\pm i \alpha}\}$, *where the "rotation angle"* α *is a real number* $0 \le \alpha \le \pi$. *The near-identity subset* $SO(3)_I$ *is made by suppressing the matrices with* $\alpha = \pi$. *The matrix algebra* $\mathfrak{so}(3)$ *consists of* $3{\times}3$ *real antisymmetric matrices. The (pseudo)norm* $\| \mathbf{r} \| = \sqrt{(\mathrm{tr}\, \mathbf{r}^2)/2}$ *of these matrices is any imaginary positive number. The near-zero subset* $\mathfrak{so}(3)_0$ *is made by suppressing from* $\mathfrak{so}(3)$ *all the matrices with norm* $i\pi$. *See appendix A.14 for the geometrical interpretation of this subset. The logarithmic image of* SO(3) *is analyzed in appendix A.7.*

---

[47]For instance, the matrix $\mathbf{C} = \mathrm{diag}(-\alpha, -1/\alpha)$ belongs to SL(2), but $\mathbf{C}^\lambda = e^{\lambda i \pi}\, \mathrm{diag}(\alpha^\lambda, 1/\alpha^\lambda)$, is real only for integer values of λ. The matrix $\mathbf{c} = \log \mathbf{C} = \mathrm{diag}(i\pi + \log \alpha,\, i\pi - \log \alpha)$ belongs to ℓiSL(2), but $\lambda\, \mathbf{c}$ does not (in general).

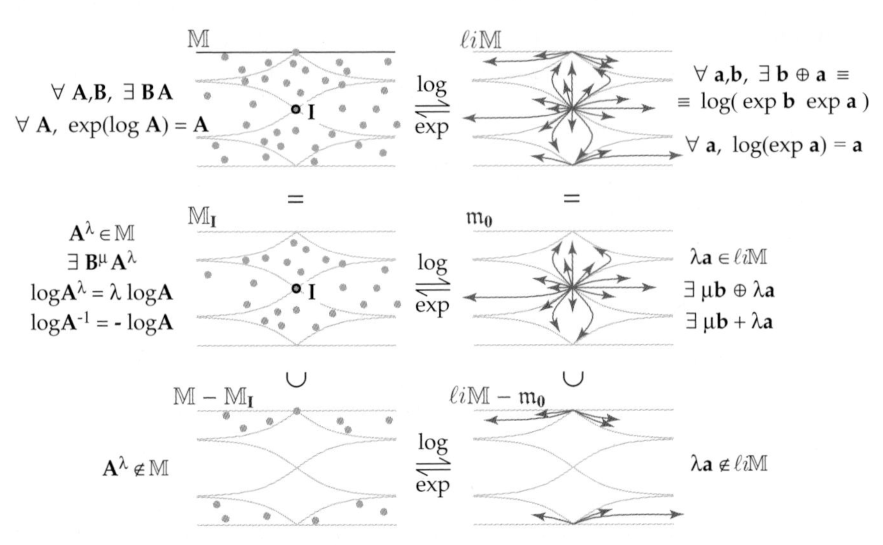

Fig. 1.8. *Top-left shows a schematic representation of the manifold attached to a Lie group (of matrices)* $\mathbb{M}$. *The elements (matrices) of the group, matrices* $\mathbf{A}, \mathbf{B}, \ldots$ *are represented as points. The group operation (matrix product) associates a point to any ordered pair of points. Via the log-exp duality, this multiplicative group is associated to its logarithmic image,* $\ell i\,\mathbb{M}$. *As shown later in the text, the elements of* $\ell i\,\mathbb{M}$ *are to be interpreted as the oriented geodesic (and autoparallel) segments of the manifold. The group operation here is the geometric sum* $\mathbf{b} \oplus \mathbf{a}$. *While the Lie group manifold can be separated into its near-identity subset* $\mathbb{M}_{\mathrm{I}}$ *and its complement,* $\mathbb{M} - \mathbb{M}_{\mathrm{I}}$, *the logarithmic image can be separated into the near-zero subset* $\mathfrak{m}_0$ *and its complement,* $\ell i\,\mathbb{M} - \mathfrak{m}_0$. *The elements of* $\ell i\,\mathbb{M} - \mathfrak{m}_0$ *can still be considered to be oriented geodesic segments on the manifold, but having their origin at a different point: the points of* $\mathbb{M} - \mathbb{M}_{\mathrm{I}}$ *cannot be geodesically connected to the origin* $\mathbf{I}$. *By definition, if* $\mathbf{A} \in \mathbb{M}_{\mathrm{I}}$, *then for any* $\lambda \in [-1,+1]$, $\mathbf{A}^{\lambda} \in \mathbb{M}_{\mathrm{I}}$. *Equivalently, if* $\mathbf{a} \in \mathfrak{m}_0$, *then for any* $\lambda \in [-1,+1]$, $\lambda\,\mathbf{a} \in \mathfrak{m}_0$. *The operation* $\mathbf{b} \oplus \mathbf{a}$ *induces the tangent operation* $\mathbf{b} + \mathbf{a}$, *and the linear combinations* $\mu\,\mathbf{b} + \lambda\,\mathbf{a}$ *of the matrices of* $\mathfrak{m}_0$ *generate* $\mathfrak{m}$, *the algebra of* $\mathbb{M}$ *(see figure 1.9). While the representations of this figure are only schematic for a general group, we shall see in section 1.4.6 that they are, in fact, quantitatively accurate for the Lie group* SL(2).

While the two examples above completely characterize the near-neutral subsets of SL(2) and SO(3) I don't know of any simple and complete characterization for $GL(n)$.

An operation which, to a real number and a member of a set, associates a member of the set is central in the definition of autovector space. Inside the near-identity subset and the near-zero subset, the two respective operations (1.149) and (1.150) are internal operations, and it is easy to see (demonstration outlined in appendix A.8) that they satisfy the axioms of a local autovector space (definitions 1.19 and 1.21). We thus arrive at the following properties.

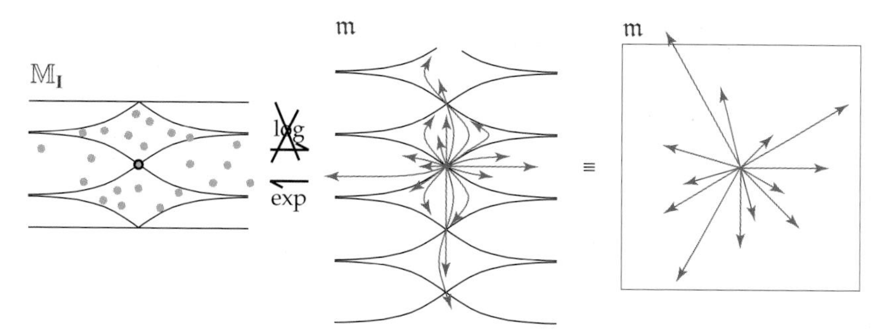

Fig. 1.9. *The algebra of* $\mathbb{M}$*, denoted* $\mathfrak{m}$ *is generated by the linear combinations* $\mathbf{c} = \mu\,\mathbf{b} + \lambda\,\mathbf{a}$ *of the elements of* $\mathfrak{m}_0$ *(see figure 1.8). The two images at the right of the figure suggest two possible representations of the algebra of a group. While the algebra is a linear space (representation on the right) we know that we can associate to any vector of a linear space an oriented geodesic segment on the manifold itself, this justifying the representation in the middle. The exponential function maps these vectors (or autovectors) into the near-identity subset of the Lie group manifold (representation on the left). Because of the periodic character of the matrix exponential function, this mapping is not invertible, i.e., we do not necessarily have* $\log(\exp\,\mathbf{a}) = \mathbf{a}$ *(an expression that is valid only if* $\mathbf{a}$ *belongs to the near-zero subset* $\mathfrak{m}_0$ *.*

Property 1.26 *Let* $\mathbb{M}$ *be a multiplicative group of matrices, and let* $\mathbb{M}_\mathrm{I}$ *be the near-identity subset of* $\mathbb{M}$*. With the two operations* $\{\mathbf{A},\mathbf{B}\} \mapsto \mathbf{B}\,\mathbf{A}$ *and* $\{\lambda,\mathbf{A}\} \mapsto \mathbf{A}^\lambda \equiv \exp(\lambda\,\log\mathbf{A})$*, the set* $\mathbb{M}_\mathrm{I}$ *is a (local) autovector space.*

Property 1.27 *Let* $\mathfrak{m}$ *be a matrix algebra, and let* $\mathfrak{m}_0$ *be the near-zero subset of* $\mathfrak{m}$*. With the two operations* $\{\mathbf{a},\mathbf{b}\} \mapsto \mathbf{b} \oplus \mathbf{a} \equiv \log(\exp\mathbf{b}\ \exp\mathbf{a})$ *and* $\{\lambda,\mathbf{a}\} \mapsto \lambda\,\mathbf{a}$*, the set* $\mathfrak{m}_0$ *is a (local) autovector space.*

Property 1.28 *The two autovector spaces in properties 1.26 and 1.27 are isomorphic, via the log-exp functions.*

All these different matrix groups are necessary if one wishes to associate to the group operation a geometric interpretation. Let $\mathbb{M}$ be a multiplicative group of matrices, and $\ell i\,\mathbb{M}$ its logarithmic image, with o-sum $\mathbf{b} \oplus \mathbf{a}$. Let $\mathbf{a}$ and $\mathbf{b}$ be two elements of $\mathfrak{m}_0$, the near-zero subset of $\ell i\,\mathbb{M}$, and let $\mathbf{c} = \mathbf{b} \oplus \mathbf{a}$. The $\mathbf{c}$ so defined is an element of $\ell i\,\mathbb{M}$, but not necessarily an element of $\mathfrak{m}_0$. If it belongs to $\mathfrak{m}_0$, then, as explained below (and demonstrated in appendix A.12), the operation $\mathbf{c} = \mathbf{b} \oplus \mathbf{a}$, is a sum of oriented segments (at the origin). This gives a precise sense to the locality property of the geometric sum: the three elements $\mathbf{a}$, $\mathbf{b}$ and $\mathbf{c} = \mathbf{b} \oplus \mathbf{a}$ must belong to the near-zero subset $\mathfrak{m}_0$.

1.4.5 The Geometry of GL(n)

Choosing an appropriate coordinate system always simplifies the study of a manifold. For some coordinates to be used over the Lie group manifold GL(n) a one-index notation, like x^α, is convenient, but for other coordinate systems it is better to use a double-index notation, like $x^\alpha{}_\beta$, to directly acknowledge the n^2 dimensionality of the manifold. Then, the coordinates defining a point can be considered organized as a matrix, $\mathbf{x} = \{x^\alpha{}_\beta\}$ as then, some of the coordinate manipulations to be found correspond to matrix multiplications.

The points of the Lie group manifold GL(n) are, by definition, the matrices of the set GL(n). The analysis of the parallel transport over the Lie group manifold is better done in the coordinate system defined as follows.

Definition 1.42 *The* exponential coordinates *of the point representing the matrix* $\mathbf{X} = \{X^\alpha{}_\beta\}$ *are the* $X^\alpha{}_\beta$ *themselves.*

It is clear that these coordinates cover the whole group manifold, as, by definition, the points of the manifold *are* the matrices of the multiplicative group.

We call this coordinate system 'exponential' to distinguish it from another possible (local) coordinate system, where the coordinates of a point are the $\mathbf{x} = \{x^\alpha{}_\beta\}$ defined as $\mathbf{x} = \log \mathbf{X}$. As shown in section A.12.6, these coordinates $x^\alpha{}_\beta$ are *autoparallel*, i.e., in fact, "locally linear". Calling the coordinates $X^\alpha{}_\beta$ exponential is justified because they are related through $\mathbf{X} = \exp \mathbf{x}$ to the locally linear coordinates $x^\alpha{}_\beta$.

Using a double index notation for the coordinates may be disturbing, and needs some training, but is is better to respect the intimate nature of GL(n) in our choice of coordinates. The components of all the tensors to be introduced below on the Lie group manifold are given in the natural basis associated to the coordinates $\{X^\alpha{}_\beta\}$. This implies, in particular, that the tensors have two times as many indices as when using coordinates with a single index. The squared distance element on the manifold, for instance, is written

$$ds^2 = g_\alpha{}^\beta{}_\mu{}^\nu \, dX^\alpha{}_\beta \, dX^\mu{}_\nu \quad , \tag{1.153}$$

this showing that the metric tensor has the components $g_\alpha{}^\beta{}_\mu{}^\nu$ instead of the usual $g_{\alpha\beta}$. Similarly, the torsion has components $T^\alpha{}_{\beta\mu}{}^\nu{}_\rho{}^\sigma$, instead of the usual $T^\alpha{}_{\beta\gamma}$.

The basic geometric properties of the Lie group manifold GL(n) (they are demonstrated in appendix A.12) are now listed.

(i) The connection of the manifold GL(n) at the point with coordinates $X^\alpha{}_\beta$ is (equation A.183)

$$\Gamma^\alpha{}_{\beta\mu}{}^\nu{}_\rho{}^\sigma = - \overline{X}^\sigma{}_\mu \, \delta^\alpha_\rho \, \delta^\nu_\beta \quad . \tag{1.154}$$

where a bar is used to denote the inverse of a matrix:

$$\overline{\mathbf{X}} \equiv \mathbf{X}^{-1} \quad ; \quad \overline{X}^\alpha{}_\beta \equiv (\mathbf{X}^{-1})^\alpha{}_\beta \quad . \tag{1.155}$$

(ii) The equation of the autoparallel line going from point $\mathbf{A} = \{A^\alpha{}_\beta\}$ to point $\mathbf{B} = \{B^\alpha{}_\beta\}$ is (equation A.186)

$$\mathbf{X}(\lambda) = \exp(\lambda \log(\mathbf{B}\,\mathbf{A}^{-1}))\,\mathbf{A} \quad ; \quad (0 \leq \lambda \leq 1) \quad . \tag{1.156}$$

(iii) On the natural basis at the origin $\mathbf{I}$ of the Lie group manifold, the components of the vector associated to the autoparallel line going from the origin $\mathbf{I}$ to point $\mathbf{A} = \{A^\alpha{}_\beta\}$ are (see equation (A.189)) the components $a^\alpha{}_\beta$ of the matrix

$$\mathbf{a} = \log \mathbf{A} \quad . \tag{1.157}$$

(iv) When taking two points $\mathbf{A}$ and $\mathbf{B}$ of the Lie group manifold GL(n) (i.e., two matrices of GL(n)) that are inside some neighborhood of the identity matrix $\mathbf{I}$, when considering the two oriented autoparallel segments going from the origin $\mathbf{I}$ to each of the two points, and making the geometric sum of the two segments (as defined in figure 1.3), one obtains the point (see (A.199))

$$\mathbf{C} = \mathbf{B}\,\mathbf{A} \quad . \tag{1.158}$$

This means that *when the geometric sum of two oriented autoparallel segments of the manifold* GL(n) *makes sense, it is the group operation* (see figure 1.10). Therefore, the general analytic expression for the o-sum

$$\mathbf{c} = \mathbf{b} \oplus \mathbf{a} = \log(\exp \mathbf{b}\ \exp \mathbf{a}) \quad , \tag{1.159}$$

an operation that is —by definition of the logarithmic image of multiplicative matrix group– always equivalent to the expression (1.158), can also be interpreted as the geometric sum of the two autovectors $\mathbf{a} = \log \mathbf{A}$ and $\mathbf{b} = \log \mathbf{B}$, producing the autovector $\mathbf{c} = \log \mathbf{C}$. The reader must remember that this interpretation of the group operation in terms of a geometric sum is only possible inside the region of the group around the origin (the corresponding subsets received a name in section 1.4.4).

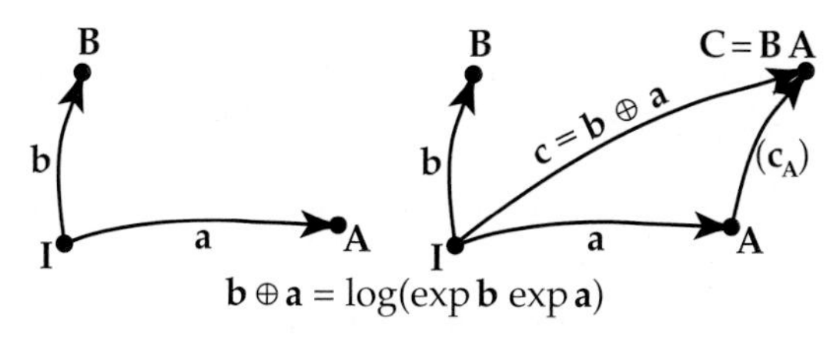

Fig. 1.10. *Recall of the geometric sum, as defined in figure 1.3. In a Lie group manifold, the points are the (multiplicative) matrices of the group, and the oriented autoparallel segments are the logarithms of these matrices. The geometric sum of the segments can be expressed as* $\mathbf{C} = \mathbf{B}\,\mathbf{A}$ *or, equivalently, as* $\mathbf{c} = \mathbf{b} \oplus \mathbf{a} = \log(\exp \mathbf{b}\ \exp \mathbf{a})$.

(v) The torsion of the manifold GL(n) at the point with coordinates $X^\alpha{}_\beta$ is (equation A.201)

$$T^\alpha{}_{\beta\mu}{}^\nu{}_\rho{}^\sigma = \overline{X}^\nu{}_\rho \, \delta^\sigma_\beta \, \delta^\alpha_\mu - \overline{X}^\sigma{}_\mu \, \delta^\alpha_\rho \, \delta^\nu_\beta \quad . \tag{1.160}$$

(vi) The Jacobi tensor of the Lie group manifold GL(n) identically vanishes (equation A.202):

$$\mathbf{J} = \mathbf{0} \quad . \tag{1.161}$$

(vii) The covariant derivative of the torsion of the Lie group manifold GL(n) identically vanishes (equation A.203)

$$\nabla\mathbf{T} = \mathbf{0} \quad . \tag{1.162}$$

(vii) The Riemann tensor of the Lie group manifold GL(n) identically vanishes (equation A.204)

$$\mathbf{R} = \mathbf{0} \quad . \tag{1.163}$$

Of course, the group operation being associative, the Anassociativity tensor (see equation (1.113)) also identically vanishes

$$\mathbf{A} = \mathbf{0} \quad . \tag{1.164}$$

(viii) The universal metric introduced in equation (1.31) (page 17) induces a metric over the Lie group manifold GL(n), whose components at the point with coordinates $X^\alpha{}_\beta$ are (equation A.206)

$$g_\alpha{}^\beta{}_\mu{}^\nu = \chi \, \overline{X}^\nu{}_\alpha \, \overline{X}^\beta{}_\mu + \frac{\psi - \chi}{n} \, \overline{X}^\beta{}_\alpha \, \overline{X}^\nu{}_\mu \quad , \tag{1.165}$$

the contravariant metric being

$$g^\alpha{}_\beta{}^\mu{}_\nu = \overline{\chi} \, X^\alpha{}_\nu \, X^\mu{}_\beta + \frac{\overline{\psi} - \overline{\chi}}{n} \, X^\alpha{}_\beta \, X^\mu{}_\nu \quad , \tag{1.166}$$

where $\overline{\chi} = 1/\chi$ and $\overline{\psi} = 1/\psi$.

(ix) The volume measure induced by this metric over the manifold is (see equation (A.210))

$$\sqrt{-\det \mathbf{g}} = \frac{(\psi \, \chi^{n^2-1})^{1/2}}{(\det \mathbf{X})^n} \quad . \tag{1.167}$$

Except for the specific constant factor, this corresponds to the well known *Haar measure* defined over (locally compact) Lie groups.

(x) The metric in equation (1.165) allows one to obtain an explicit expression for the squared distance between point $\mathbf{X} = \{X^\alpha{}_\beta\}$ and point $\mathbf{X}' = \{X'^\alpha{}_\beta\}$ (equation A.212):

$$D^2(\mathbf{X}',\mathbf{X}) \;=\; \|\,\mathbf{t}\,\|^2 \;\equiv\; \chi\,\mathrm{tr}\,\tilde{\mathbf{t}}^2 + \psi\,\mathrm{tr}\,\bar{\mathbf{t}}^2 \quad , \tag{1.168}$$

where

$$\mathbf{t} \;=\; \log(\mathbf{X}'\,\mathbf{X}^{-1}) \quad , \tag{1.169}$$

and where $\tilde{\mathbf{t}}$ and $\bar{\mathbf{t}}$ respectively denote the deviatoric and the isotropic parts of $\mathbf{t}$ (equations 1.34).

(xi) The covariant components of the torsion are defined as $T_\alpha{}^\beta{}_\mu{}^\nu{}_\rho{}^\sigma = g_\alpha{}^\beta{}_\epsilon{}^\pi\, T^\epsilon{}_{\pi\mu}{}^\nu{}_\rho{}^\sigma$, and this gives (equation A.214)

$$T_\alpha{}^\beta{}_\mu{}^\nu{}_\rho{}^\sigma \;=\; \chi\left(\overline{X}^\beta{}_\mu\,\overline{X}^\nu{}_\rho\,\overline{X}^\sigma{}_\alpha - \overline{X}^\beta{}_\rho\,\overline{X}^\nu{}_\alpha\,\overline{X}^\sigma{}_\mu\right) \quad . \tag{1.170}$$

One easily verifies the (anti)symmetries $T_\alpha{}^\beta{}_\mu{}^\nu{}_\rho{}^\sigma = -T_\mu{}^\nu{}_\alpha{}^\beta{}_\rho{}^\sigma = -T_\alpha{}^\beta{}_\rho{}^\sigma{}_\mu{}^\nu$, which demonstrate that the torsion of the Lie group manifold $\mathrm{GL}(n)$, endowed with the universal metric, is *totally antisymmetric*. Therefore, as explained in appendix A.11, geodesic lines and autoparallel lines coincide: when working with Lie group manifolds, the term 'autoparallel line' may be replaced by 'geodesic line'.

(xii) The Ricci of the universal metric is (equation A.217)

$$C_\alpha{}^\beta{}_\mu{}^\nu \;=\; \tfrac{1}{4}\,T^\rho{}_{\sigma\alpha}{}^\beta{}_\varphi{}^\phi\; T^\varphi{}_{\phi\mu}{}^\nu{}_\rho{}^\sigma \quad . \tag{1.171}$$

In fact, this expression corresponds, in our double index notation, to the usual definition of the Cartan metric of a Lie group (Goldberg, 1998): the so-called "Cartan metric" is the Ricci of the Lie group manifold $\mathrm{GL}(n)$ (up to a numerical factor).

For mode details on the geometry of $\mathrm{GL}(n)$, see appendix A.12.

1.4.6 Example: GL$^+$(2)

As already commented in the introduction to section 1.4, the geometry of a Lie group manifold may be quite complex. The manifold $\mathrm{GL}(n)$ —that because of Ado's theorem can be seen as containing all other Lie group manifolds— is made by the union of two unconnected manifolds. The submanifold $\mathrm{GL}^+(n)$, composed of all the matrices of $\mathrm{GL}(n)$ with positive determinant, is a group whose geometry we must understand (the other submanifold being essentially identical to this one, via the inversion of an axis).

The manifold $\mathrm{GL}^+(n)$ is connected, and simply connected. Yet the manifold is complex enough: it is not possible to join two arbitrarily chosen points by a geodesic line. In this section we shall understand how this may happen, thanks to a detailed analysis of the group $\mathrm{GL}^+(2)$.

In later sections of this chapter we will become interested in the notion of 'geotensor'. A geotensor essentially is a geodesic segment leaving the origin of a Lie group manifold. Therefore, the part of a Lie group manifold that is of interest to us is the part that is geodesically connected to the origin. Even this part of a Lie group manifold has a complex geometry, with light cones and two different sorts of geodesic lines, like the "temporal" and "spatial" lines of the relativistic space-time.

As these interesting properties are already present in $GL^+(2)$, it is important to explore the four-dimensional manifold $GL^+(2)$ here. In fact, as the four-dimensional manifold $GL^+(2)$ is a simple mixture of the three-dimensional group manifold SL(2) and the one-dimensional group of homotheties, $H^+(2)$, much of this section is, in fact, concerned with the three-dimensional SL(2) group manifold.

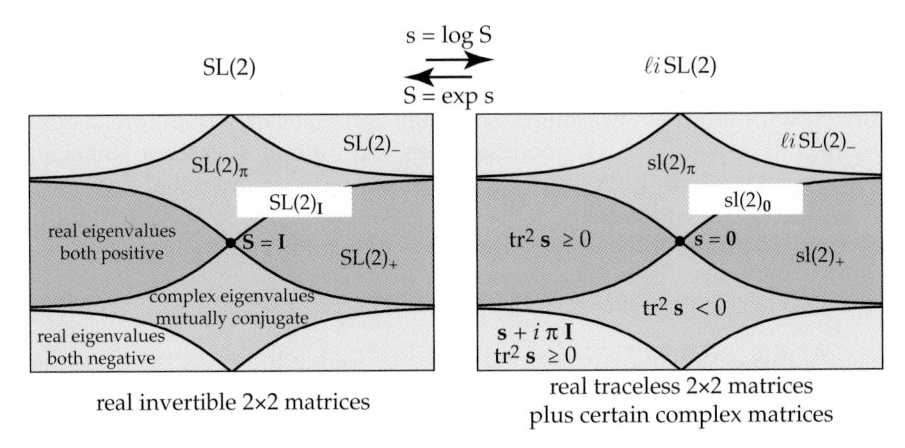

Fig. 1.11. *The sets appearing when considering the logarithm of* SL(2) *(see appendix A.6). In each of the two panels, the three sets represented correspond to the zones with a given level of gray. The meaning of the shapes attributed here to each subset will become clear when analyzing the metric properties of the space.*

1.4.6.1 Sets of Matrices

Let us start by studying the structure of the space $\ell i\, GL^+(2)$, i.e., the space of matrices that are the logarithm of the matrices in $GL^+(2)$. A matrix $\mathbf{G}$ in $GL^+(2)$ (i.e., a real 2×2 matrix with positive determinant) can always be written as

$$\mathbf{G} = \mathbf{H}\,\mathbf{S} \quad , \tag{1.172}$$

where $\mathbf{H}$ is a matrix of $H^+(2)$ (i.e., an isotropic matrix $H^\alpha{}_\beta = K\,\delta^\alpha_\beta$ with $K \geq 0$), and where $\mathbf{S}$ is a matrix of SL(2), (i.e., a real 2×2 matrix with unit determinant). As $\mathbf{H}\,\mathbf{S} = \mathbf{S}\,\mathbf{H}$, one has

$$\log \mathbf{G} = \log \mathbf{H} + \log \mathbf{S} \quad . \tag{1.173}$$

The characterization of the matrices $\mathbf{h} = \log \mathbf{H}$ is trivial: it is the set of all real isotropic matrices (i.e., the matrices with form $h^\alpha{}_\beta = k\, \delta^\alpha_\beta$, with k an arbitrary real number).

It remains, then, to characterize the sets SL(2) and $\ell i\,$SL(2) (the set of matrices that are the logarithm of the matrices in SL(2)). The basic results have been mentioned in section 1.4.2.4, and the details are in appendix A.6. Figure 1.11 presents the graphic correspondence between all these sets, using a representation inspired by the geodesic representations to be developed below (see, for instance, figure 1.13).

1.4.6.2 Exponential and Logarithm

Because of the Cayley-Hamilton theorem, any series of an $n \times n$ matrix can be reduced to a polynomial where the maximum power of the matrix is $n-1$. Then, any analytic function $\mathbf{f}(\mathbf{m})$ of a $2{\times}2$ matrix $\mathbf{m}$ must reduce to the form $\mathbf{f}(\mathbf{m}) = a\,\mathbf{I} + b\,\mathbf{m}$, where a and b are scalars depending on the invariants of $\mathbf{m}$ (and, of course, on the particular function $\mathbf{f}(\,\cdot\,)$ being considered). This, in particular, is true for the logarithm and for the exponential function. Let us find the corresponding expressions.

Property 1.29 *If* $\mathbf{s} \in \mathfrak{sl}(2)_0$ *, then* $\exp \mathbf{s} \in \mathrm{SL}(2)_\mathrm{I}$ *, and one has*

$$\exp \mathbf{s} \;=\; \frac{\sinh s}{s}\,\mathbf{s} + \cosh s\,\mathbf{I} \qquad ; \qquad s \;=\; \sqrt{\frac{\operatorname{tr} \mathbf{s}^2}{2}} \quad . \tag{1.174}$$

Reciprocally, if $\mathbf{S} \in \mathrm{SL}(2)_\mathrm{I}$ *, then* $\log \mathbf{S} \in \mathfrak{sl}(2)_0$ *, and one has*

$$\log \mathbf{S} \;=\; \frac{s}{\sinh s}\,(\,\mathbf{S} - \cosh s\,\mathbf{I}\,) \qquad ; \qquad \cosh s \;=\; \frac{\operatorname{tr} \mathbf{S}}{2} \quad . \tag{1.175}$$

The demonstration is given as a footnote.[48] Note that although the scalar s can be imaginary, both $\cosh s$ and $(\sinh s)/s$ are real, so $\mathbf{s} = \log \mathbf{S}$ and $\mathbf{S} = \exp \mathbf{s}$ given by these equations are real, as they should be.

Equation (1.174) is the equivalent for SL(2) of the Rodrigues' formula (equation (A.268), page 209), valid for SO(3).

[48]It follows from the Cayley-Hamilton theorem (see appendix A.4) that the square of a 2×2 traceless matrix is necessarily proportional to the identity, $\mathbf{s}^2 = s^2\,\mathbf{I}$, with $s = \sqrt{(\operatorname{tr} \mathbf{s}^2)/2}$. Then, for the even and odd powers of $\mathbf{s}$ one respectively has $\mathbf{s}^{2n} = s^{2n}\,\mathbf{I}$ and $\mathbf{s}^{2n+1} = s^{2n}\,\mathbf{s}$. The exponential of $\mathbf{s}$ is $\exp \mathbf{s} = \sum_{n=0}^\infty \frac{1}{n!}\,\mathbf{s}^n$. Separating the even from the odd powers, this gives $\exp \mathbf{s} = \sum_{n=0}^\infty \frac{1}{2n!}\,\mathbf{s}^{2n} + \sum_{n=0}^\infty \frac{1}{(2n+1)!}\,\mathbf{s}^{2n+1}$, i.e., $\exp \mathbf{s} = (\,\sum_{n=0}^\infty \frac{s^{2n}}{2n!}\,)\,\mathbf{I} + (\,\frac{1}{s}\sum_{n=0}^\infty \frac{s^{2n+1}}{(2n+1)!}\,)\,\mathbf{s}$. This is equation (1.174). Replacing $\mathbf{s}$ by $\log \mathbf{S}$ in this equation gives equation (1.175).

With these two equations at hand, it is easy to derive other properties. For instance, the power $\mathbf{G}^\lambda$ of a matrix $\mathbf{G} \in GL(n)_I$ is defined as $\mathbf{G}^\lambda = \exp(\lambda \log \mathbf{G})$. For $\mathbf{S} \in SL(2)_I$ one easily obtains

$$\mathbf{S}^\lambda = \frac{\sinh \lambda s}{\sinh s}\, \mathbf{S} + \left(\cosh \lambda s - \frac{\sinh \lambda s}{\sinh s} \cosh s\right) \mathbf{I} \quad , \tag{1.176}$$

where the scalar s is that given in (1.175). When λ is an integer, this gives the usual power of the matrix $\mathbf{S}$.

1.4.6.3 Geosum in SL(2)

The o-sum $\mathbf{g}_2 \oplus \mathbf{g}_1 \equiv \log(\exp \mathbf{g}_2 \ \exp \mathbf{g}_1)$ of two matrices of $\ell i\, GL(n)$ (the logarithmic image of $GL(n)$) is an operation that is always defined. We have seen that, if the two matrices are in the neighborhood of the origin, this analytic expression can be interpreted as a sum of autovectors. Let us work here in this situation.

To obtain the geosum $\mathbf{g}_2 \oplus \mathbf{g}_1 = \log(\exp \mathbf{g}_2 \ \exp \mathbf{g}_1)$ of two matrices of $\ell i\, GL(n)$ we can decompose them into trace and traceless parts ($\mathbf{g}_1 = \mathbf{h}_1 + \mathbf{s}_1$; $\mathbf{g}_2 = \mathbf{h}_2 + \mathbf{s}_2$), as, then,

$$\mathbf{g}_2 \oplus \mathbf{g}_1 = (\mathbf{h}_2 + \mathbf{h}_1) + (\mathbf{s}_2 \oplus \mathbf{s}_1) \quad . \tag{1.177}$$

The problem of expressing the geosum of matrices in $\ell i\, GL(n)$ is reduced to that of expressing the geosum of matrices in $\ell i\, SL(n)$. We can then limit our attention to the expression of the geosum of two matrices in the neighborhood of the origin of $\ell i\, SL(2)$.

The definition $\mathbf{s}_2 \oplus \mathbf{s}_1 = \log(\exp \mathbf{s}_2 \ \exp \mathbf{s}_1)$ easily leads to (using equations (1.175) and (1.174))

$$\boxed{\begin{aligned} \mathbf{s}_2 \oplus \mathbf{s}_1 = \frac{s}{\sinh s} \Big(& \frac{\sinh s_2}{s_2} \cosh s_1\, \mathbf{s}_2 + \cosh s_2 \frac{\sinh s_1}{s_1}\, \mathbf{s}_1 \\ & + \frac{1}{2} \frac{\sinh s_2}{s_2} \frac{\sinh s_1}{s_1} (\mathbf{s}_2\, \mathbf{s}_1 - \mathbf{s}_1\, \mathbf{s}_2) \Big) \quad , \end{aligned}} \tag{1.178}$$

where s_1 and s_2 are the respective norms[49] of $\mathbf{s}_1$ and $\mathbf{s}_2$, and where s is the scalar defined by[50]

$$\boxed{\cosh s = \cosh s_2 \cosh s_1 + \frac{1}{2} \frac{\sinh s_2}{s_2} \frac{\sinh s_1}{s_1} \operatorname{tr}(\mathbf{s}_2\, \mathbf{s}_1) \quad .} \tag{1.179}$$

The norm of $\mathbf{s} = \mathbf{s}_2 \oplus \mathbf{s}_1$ is s. A series expansion of expression (1.178) gives, of course, the BCH series

$$\mathbf{s}_2 \oplus \mathbf{s}_1 = (\mathbf{s}_2 + \mathbf{s}_1) + \tfrac{1}{2} (\mathbf{s}_2\, \mathbf{s}_1 - \mathbf{s}_1\, \mathbf{s}_2) + \ldots \quad . \tag{1.180}$$

[49] $\| \mathbf{s} \| = \sqrt{(\operatorname{tr} \mathbf{s}^2)/2}$.

[50] The sign of the scalar is irrelevant, as the equation (1.178) is symmetrical in $\pm s$.

1.4.6.4 Coordinates over the $GL^+(2)$ Manifold

We have seen that over the $GL(n)$ manifold, the components of a matrix can be used as coordinates. These coordinates are well adapted to analytic developments, but to understand the geometry of $GL(n)$ in some detail, other coordinate systems are preferable.

Here, we require a coordinate system that covers the four-dimensional manifold $GL^+(2)$. We use the parameters/coordinates $\{\kappa, e, \alpha, \varphi\}$ allowing one to express a matrix of $GL^+(2)$ as

$$\mathbf{M} = \exp\kappa \left[\cosh e \begin{pmatrix} \cos\alpha & -\sin\alpha \\ \sin\alpha & \cos\alpha \end{pmatrix} + \sinh e \begin{pmatrix} \sin\varphi & \cos\varphi \\ \cos\varphi & -\sin\varphi \end{pmatrix} \right] \quad . \tag{1.181}$$

The variable κ can be any real number, and the domains of variation of the other three coordinates are

$$0 \leq e < \infty \quad ; \quad -\pi < \varphi \leq \pi \quad ; \quad -\pi < \alpha \leq \pi \quad . \tag{1.182}$$

The formulas giving the parameters $\{\kappa, e, \alpha, \varphi\}$ as a function of the entries of the matrix $\mathbf{M}$ are given in appendix A.16 (where it is demonstrated that this coordinate system actually covers the whole of $GL^+(2)$). The inverse matrix is obtained by changing the sign of κ and α and by adding π to φ:

$$\mathbf{M}^{-1} = \exp{-\kappa} \left[\cosh e \begin{pmatrix} \cos\alpha & \sin\alpha \\ -\sin\alpha & \cos\alpha \end{pmatrix} - \sinh e \begin{pmatrix} \sin\varphi & \cos\varphi \\ \cos\varphi & -\sin\varphi \end{pmatrix} \right] \quad . \tag{1.183}$$

The logarithm $\mathbf{m} = \log\mathbf{M}$ is easily obtained decomposing the matrix as $\mathbf{M} = \mathbf{H}\,\mathbf{S}$, with $\mathbf{S}$ in $SL(2)$, and then using equation (1.175). One gets

$$\mathbf{m} = \kappa\,\mathbf{I} + \frac{\Delta}{\sinh\Delta} \left[\cosh e \begin{pmatrix} 0 & -\sin\alpha \\ \sin\alpha & 0 \end{pmatrix} + \sinh e \begin{pmatrix} \sin\varphi & \cos\varphi \\ \cos\varphi & -\sin\varphi \end{pmatrix} \right] , \tag{1.184}$$

where Δ is the scalar defined through

$$\cosh\Delta = \cosh e \, \cos\alpha \quad . \tag{1.185}$$

The eigenvalues of $\mathbf{m}$ are $\lambda_\pm = \kappa \pm \Delta$, and one has

$$\operatorname{tr}\mathbf{m} = 2\,\kappa \quad ; \quad \operatorname{tr}\mathbf{m}^2 = 2\,(\kappa^2 + \Delta^2) \quad . \tag{1.186}$$

The two expressions (1.184) and (1.185) present some singularities (where geodesics coming from the origin are undefined) that require evaluation of the proper limit. Along the axis $e = 0$ and on the plane $\alpha = 0$ one, respectively, has

$$\mathbf{m}(0,\alpha,\varphi) = \begin{pmatrix} 0 & -\alpha \\ \alpha & 0 \end{pmatrix} \quad ; \quad \mathbf{m}(e,0,\varphi) = e \begin{pmatrix} \sin\varphi & \cos\varphi \\ \cos\varphi & -\sin\varphi \end{pmatrix} \quad . \tag{1.187}$$

1.4.6.5 Metric

A matrix $\mathbf{M} \in GL^+(2)$ is represented by the four parameters/coordinates

$$\{x^0, x^1, x^2, x^3\} = \{\kappa, e, \alpha, \varphi\} \tag{1.188}$$

(see equation (1.181)). The components of the metric tensor at any point of $GL(n)$ were given in equation (1.165). Their expression for $GL^+(2)$ in the coordinates $\{\kappa, e, \alpha, \varphi\}$ can be obtained using equation (A.235) in appendix A.12. The partial derivatives $\Lambda^\alpha{}_{\beta i}$, defined in equations (A.224), are easily obtained, and the components of the metric tensor in these coordinates are then obtained using equation (A.227) (the inverse matrix $\mathbf{M}^{-1}$ is given in equation (1.183)). The metric so obtained (that —thanks to the coordinate choice— happens to be diagonal), is

$$\begin{pmatrix} g_{\kappa\kappa} & g_{\kappa e} & g_{\kappa\alpha} & g_{\kappa\varphi} \\ g_{e\kappa} & g_{ee} & g_{e\alpha} & g_{e\varphi} \\ g_{\alpha\kappa} & g_{\alpha e} & g_{\alpha\alpha} & g_{\alpha\varphi} \\ g_{\varphi\kappa} & g_{\varphi e} & g_{\alpha\varphi} & g_{\varphi\varphi} \end{pmatrix} = 2 \begin{pmatrix} \psi & 0 & 0 & 0 \\ 0 & \chi & 0 & 0 \\ 0 & 0 & -\chi \cosh^2 e & 0 \\ 0 & 0 & 0 & \chi \sinh^2 e \end{pmatrix} , \tag{1.189}$$

this giving to the expression $ds^2 = g_{ij}\, dx^i\, dx^j$ the form[51]

$$\boxed{ds^2 = 2\,\psi\, d\kappa^2 + 2\,\chi\, (\, de^2 - \cosh^2 e\, d\alpha^2 + \sinh^2 e\, d\varphi^2\,) \quad ,} \tag{1.190}$$

with the associated volume density

$$\sqrt{-\det \mathbf{g}} = 2\,\psi^{1/2}\,\chi^{3/2}\,\sinh 2e \tag{1.191}$$

This is the expression of the universal metric at any point of the Lie group manifold $GL^+(2)$.

From a metric point of view, we see that the four-dimensional manifold $GL^+(2)$ is, in fact, made up of an "orthogonal pile" (along the κ direction) of identical three-dimensional manifolds (described by the coordinates $\{e, \alpha, \varphi\}$). This, of course, corresponds to the decomposition of a matrix $\mathbf{G}$ in $GL^+(2)$ as the product of an isotropic matrix $\mathbf{H}$ by a matrix $\mathbf{S}$ in $SL(2)$: $\mathbf{G}(\kappa, e, \alpha, \varphi) = \mathbf{H}(\kappa)\, \mathbf{S}(e, \alpha, \varphi)$. The geodesic line from point $\{\kappa_1, e_1, \alpha_1, \varphi_1\}$ to point $\{\kappa_2, e_2, \alpha_2, \varphi_2\}$ simply corresponds to the line from κ_1 to κ_2 in the one-dimensional submanifold H(2) (endowed with the one-dimensional metric[52] $ds = d\kappa$) and, independently, to the line from point $\{e_1, \alpha_1, \varphi_1\}$ to point $\{e_2, \alpha_2, \varphi_2\}$ in the three-dimensional manifold SL(2) endowed with the three-dimensional metric[53]

---

[51] Choosing, for instance, $\psi = \chi = 1/2$, this simplifies to $ds^2 = d\kappa^2 + de^2 - \cosh^2 e\, d\alpha^2 + \sinh^2 e\, d\varphi^2$.

[52] The coordinate κ is a metric coordinate, and we can set $\psi = 1/2$.

[53] As the geodesic lines do not depend on the value of the parameter χ we can set $\chi = 1/2$.

$$ds^2 = de^2 - \cosh^2 e \, d\alpha^2 + \sinh^2 e \, d\varphi^2 \qquad (1.192)$$

This is why, when studying below the geodesic lines of the manifold $GL^+(2)$ we can limit ourselves to the study of those of $SL(2)$.

We may here remark that for small values of the coordinate e, the metric in SL(2) is

$$ds^2 \approx de^2 + e^2 \, d\varphi^2 - d\alpha^2 \quad . \qquad (1.193)$$

Locally (near the origin) the coordinates $\{e, \varphi, \alpha\}$ are cylindrical coordinates in a three-dimensional Minkowskian "space-time", the role of the time axis being played by the coordinate α. We can, therefore, anticipate the existence of the "light-cones" typical of the space-time geometry, cones that will be studied below in some detail.

1.4.6.6 Ricci

The Ricci of the metric can be obtained by direct evaluation from the metric just given (equation 1.189) or using the general expressions (1.189) and (1.190). One gets

$$\begin{pmatrix} C_{\kappa\kappa} & C_{\kappa e} & C_{\kappa\alpha} & C_{\kappa\varphi} \\ C_{e\kappa} & C_{ee} & C_{e\alpha} & C_{e\varphi} \\ C_{\alpha\kappa} & C_{\alpha e} & C_{\alpha\alpha} & C_{\alpha\varphi} \\ C_{\varphi\kappa} & C_{\varphi e} & C_{\alpha\varphi} & C_{\varphi\varphi} \end{pmatrix} = 2 \begin{pmatrix} 0 & 0 & 0 & 0 \\ 0 & 1 & 0 & 0 \\ 0 & 0 & -\cosh^2 e & 0 \\ 0 & 0 & 0 & \sinh^2 e \end{pmatrix} \quad . \qquad (1.194)$$

As already mentioned, it is this Ricci that corresponds to the so called Killing-Cartan "metric" in the literature.

1.4.6.7 Torsion

The torsion of the $GL^+(2)$ manifold can be obtained, for example, using equation (A.228) in appendix A.12. One gets

$$T_{ijk} = \frac{1}{\sqrt{\psi\chi}} \, \epsilon_{0ijk} \quad , \qquad (1.195)$$

where $\epsilon_{ijk\ell}$ is the Levi-Civita tensor of the space, i.e., the totally antisymmetric tensor defined by the condition $\epsilon_{0123} = \sqrt{-\det \mathbf{g}} = 2\,\psi^{1/2}\,\chi^{3/2}\,\sinh 2e$. In particular, all the components of the torsion T_{ijk} with an index 0 vanish.

As is the case for $GL(n)$, we see that the torsion of the manifold $GL^+(2)$ is totally antisymmetric. Therefore, autoparallel and geodesic lines coincide.

1.4.6.8 Geodesics

A line connecting two points of a manifold is called geodesic if it is the shortest of all the lines connecting the two points. It is well known that a geodesic line $x^\alpha = x^\alpha(s)$ satisfies the equation (see details in appendix A.11)

$$\frac{d^2 x^\alpha}{ds^2} + \{_{\beta\gamma}{}^\alpha\} \frac{dx^\beta}{ds} \frac{dx^\gamma}{ds} = 0 \ , \tag{1.196}$$

where $\{_{\beta\gamma}{}^\alpha\}$ is the Levi-Civita connection. In $GL^+(2)$, using the metric in equation (1.189), this gives the four equations

$$\begin{gathered} \frac{d^2\kappa}{ds^2} = 0 \ ; \quad \frac{d^2 e}{ds^2} + \sinh e \cosh e \left(\left(\frac{d\alpha}{ds}\right)^2 - \left(\frac{d\varphi}{ds}\right)^2 \right) = 0 \\ \frac{d^2\alpha}{ds^2} - 2 \tanh e \, \frac{de}{ds} \frac{d\alpha}{ds} = 0 \ ; \quad \frac{d^2\varphi}{ds^2} + 2 \operatorname{cotanh} e \, \frac{de}{ds} \frac{d\varphi}{ds} = 0 \ . \end{gathered} \tag{1.197}$$

Note that they do not depend on the two arbitrary constants ψ and χ that define the universal metric.

We have already seen that the only nontrivial aspect of the four-dimensional manifold $GL^+(2)$ comes from the three-dimensional manifold $SL(2)$. We can therefore forget the coordinate κ and concentrate on the final three equations in (1.197). We have seen that the three coordinates $\{e, \alpha, \varphi\}$ are cylindrical-like near the origin. This suggests the representation of the three-dimensional manifold SL(2) as in figure 1.12: the coordinate e is represented radially (and extends to infinity), the "vertical axis" corresponds to the coordinate α, and the "azimuthal variable" is φ (the "light-cones" represented in the figure are discussed below). As the variable α is cyclical, the surface at the top of the figure has to be imagined as glued to the surface at the bottom, so the two surfaces become a single one.

Once the representation is chosen, we can move to the calculation of the geodesics. Is it easy to see that all the geodesics passing though the origin ($e = 0$, $\alpha = 0$) are contained in a plane of constant φ. Therefore, it is sufficient to represent the geodesics in such a plane: the others are obtained by rotating the plane. The result of the numerical integration of the geodesic equations (1.197) is represented in figure 1.13, where, in addition to the geodesics passing through the origin, the geodesics passing though the anti-origin ($e = 0$, $\alpha = \pi$) have been represented.

To obtain an image of the whole geodesics of the space one should (besides interpolating between the represented geodesics) rotate the figure along the line $e = 0$, this corresponding to varying values of the coordinate φ. There is, in this space a light-cone, defined, as in relativistic space-time, by the geodesics with zero length: the surface represented in figure 1.12. The cone leaves the origin (the matrix $\mathbf{I}$), "goes to infinity" (in values of e), then comes back to close at the anti-origin (the matrix $-\mathbf{I}$).

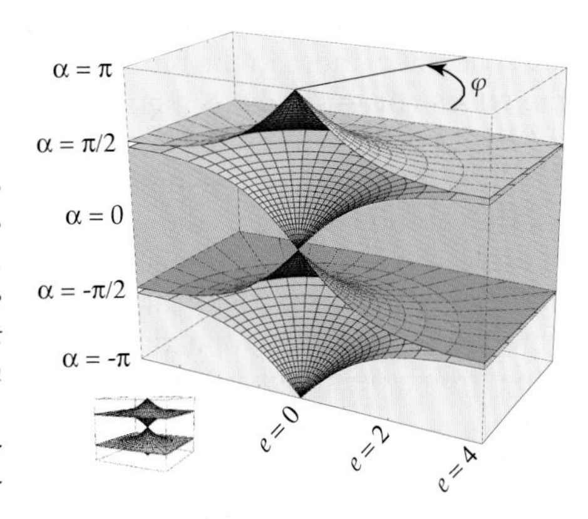

Fig. 1.12. *A representation of the three-dimensional Lie group manifold* SL(2)*, using the cylindrical-like coordinates* $\{e, \alpha, \varphi\}$ *defined by expression* (1.181)*. The light-like cones at the origin have been represented (see text). Unlike the light cone in a Minkowski space-time, the curvature of this space makes the cone close itself at the anti-origin point* $\mathcal{O}'$*, that because of the periodicities on* α*, can be reached from the origin* $\mathcal{O}$ *either with a positive (Euclidean) rotation (upwards) or with a negative (Euclidean) rotation (downwards).*

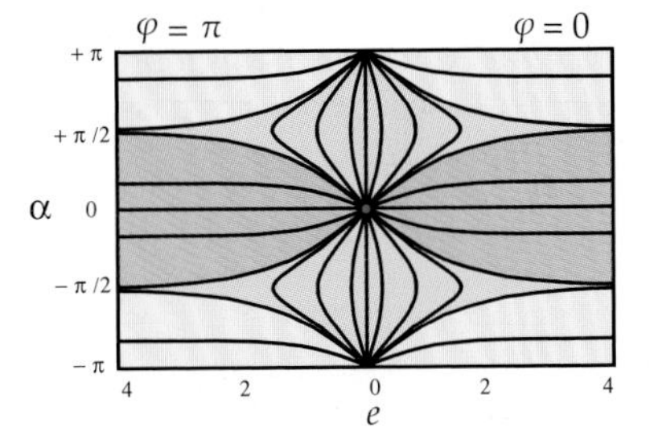

Fig. 1.13. *Geodesics in a section of* SL(2)*. As discussed in the text, the geodesics leaving the origin do not penetrate the yellow zone. This is the zone where the logarithm of the matrices in* SL(2) *takes complex values.*

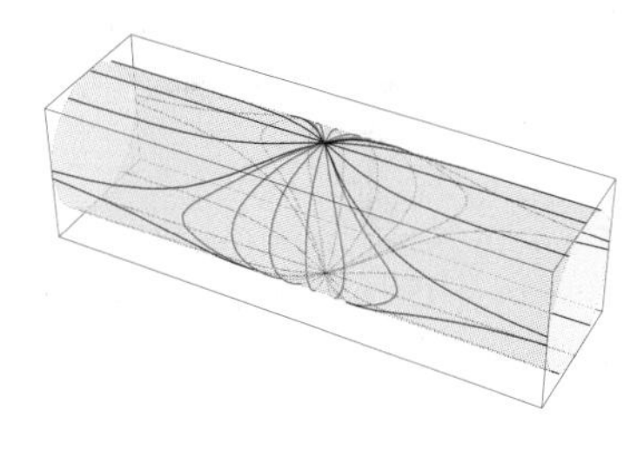

Fig. 1.14. *The same geodesics as in figure 1.13, but displayed here in a cylindrical representation. The axis of the cylinder corresponds to the coordinate* e*, the angular variable is* α*, and the whole cylinder corresponds to a fixed value of* φ*. This (metrically exact) representation better respects the topology of the two-dimensional submanifold defined by constant values of* φ*, but the visual extrapolation to the whole 3D manifold is not as easy as with the flat representation used in figure 1.13.*

As the line at the top of figure 1.13 has to be imagined as glued to the line at the bottom, there is an alternative representation of this two-dimensional surface, displayed in figure 1.14, that is topologically more correct for this 2D submanifold (but from which the extrapolation to the whole 3D manifold is less obvious).

To use a terminology reminiscent of that in used in relativity theory, the geodesics having a positive value of ds^2 are called *space-like* geodesics, those having a negative value of ds^2 (and, therefore, an imaginary value of the ds) are called *time-like* geodesics, and those having a vanishing ds^2 are called *light-like* geodesics. In figure 1.13, the geodesics in the green and yellow zones are space-like, those in the blue zones are time-like and the frontier between the zones corresponds to the zero length, light-like geodesics (that define the light-cone). In figure 1.14, the space-like geodesics are blue, the time-like geodesics are red, and the light-cone is not represented (but easy to locate).

We can now move to the geodesics that do not pass though the origin: figure 1.15 represents the geodesics passing through a point of the "vertical axis". They are identical to the geodesics passing through the origin (figure 1.13), excepted for a global vertical shift. The beam of geodesics radiating from a point outside the vertical axis is represented in figure 1.16.

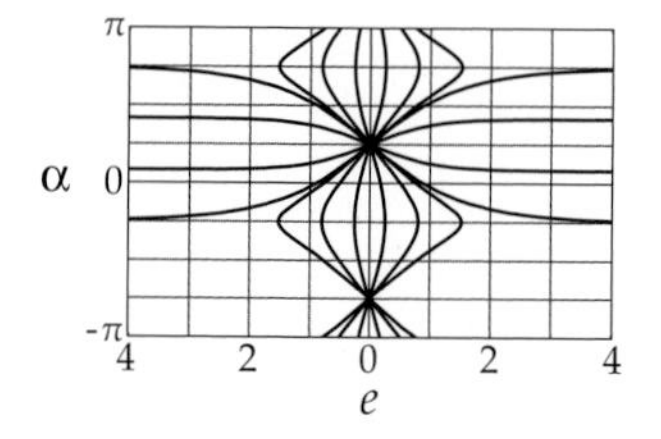

Fig. 1.15. *Some of the geodesics, generated by numerical integration of the differential system (1.197), that pass through the point* $(e, \alpha) = (0, \pi/4)$ *. Note that, in this representation, they look identical to the geodesics passing through the origin (figure 1.13), excepted for a global 'vertical shift'.*

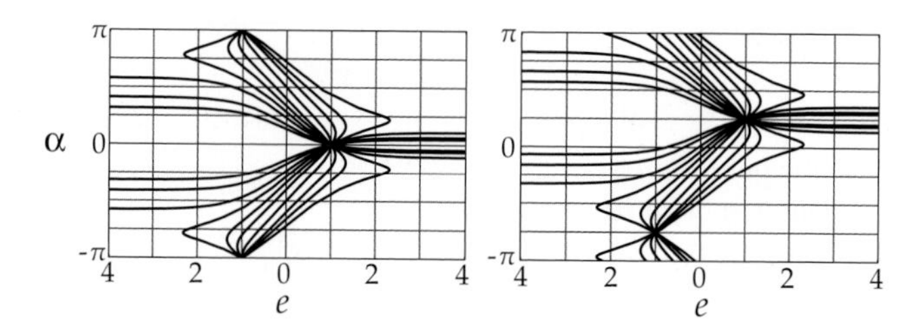

Fig. 1.16. *Some of the geodesics generated by numerical integration of the differential system (1.197). Here are displayed geodesics radiating from points that are not in the vertical axis of the representation. Note that these two figures are identical, except for a global vertical shift of the curves.*

Some authors have proposed qualitative representations of the SL(2) manifold, as, for instance, Segal (1995). The representation here proposed is quantitative.

The equation of the light-cones can be obtained by examination of the geodesic equations (1.197) or by simple considerations involving equation (1.185). One obtains the equation

$$\cosh e \, \cos \alpha = \pm 1 \quad . \tag{1.198}$$

The positive sign corresponds to the part of the light-cone leaving the origin, while the negative sign corresponds to the part of the light-cone converging to the point antipodal to the origin.

1.4.6.9 Pictorial Representation

By definition, each point of the Lie group manifold GL(2) corresponds to a matrix in the GL(2) set of matrices. As explained in appendix A.13, this set of matrices can be interpreted as the set of all possible vector bases in a two-dimensional linear space $\mathbb{E}_2$. Therefore a representation is possible, similar to those on previous pages, but where, at each point, a basis of $\mathbb{E}_2$ is represented.[54] Such a representation is proposed in figures 1.17 and 1.18.

The geodesic segment connecting any two points (i.e., any two bases) represents a linear transformation: that transforming one basis into the other. A segment connecting two points can be transported to the origin, so the set of transformations is, in fact, the set of geodesic segments radiating from the origin (or the anti-origin), a set represented in figure 1.13. The geometric sum of two such segments (examined below) then corresponds to the composition of two linear transformations.

It is easy to visually identify the transformation defined by any geodesic segment in figure 1.17. But one must keep in mind that no assumption has (yet) been made of a possible metric (scalar product) on the underlying space $\mathbb{E}_2$. Should the linear space $\mathbb{E}_2$ be endowed with an elliptic metric (i.e., should it correspond to an ordinary Euclidean space), then, the vertical axis in figure 1.17 corresponds to a rotation, and the horizontal axis to a 'deformation'. Alternatively, should the metric of the linear space $\mathbb{E}_2$ be hyperbolic (i.e., should it correspond to a Minkowskian space-time), then, it is the horizontal axis that corresponds to ("space-time") rotations and the vertical axis to ("space-time") deformations.

1.4.6.10 Other Coordinates

While the coordinates $\{e, \alpha, \varphi\}$ cover the whole manifold SL(2), we shall need, in chapter 4 (to represent the deformations of an elastic body), a coordinate system well adapted to the part of SL(2) that is geodesically connected to the origin. Keeping the coordinate φ, we can replace the two coordinates $\{e, \alpha\}$ by the two coordinates

$$\varepsilon = \frac{\Delta}{\sinh \Delta} \sinh e \quad ; \quad \theta = \frac{-\Delta}{\cosh \Delta} \cosh e \sin \alpha \quad , \tag{1.199}$$

where Δ is the parameter introduced in equation (1.185). Then, the matrix $\mathbf{m}$ in equation (1.184) becomes (taking $\kappa = 0$)

$$\mathbf{m} = \varepsilon \begin{pmatrix} \sin\varphi & \cos\varphi \\ \cos\varphi & -\sin\varphi \end{pmatrix} + \theta \begin{pmatrix} 0 & 1 \\ -1 & 0 \end{pmatrix} \tag{1.200}$$

[54]The presentation corresponds, in fact, to SL(2), which means that the homotheties have been excluded from the representation.

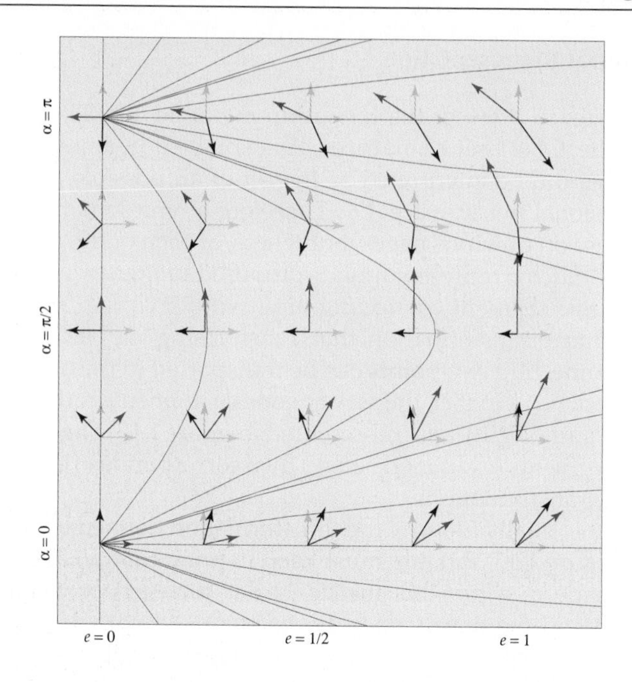

Fig. 1.17. *A 2D section of* SL(2) *, with* $\varphi = 0$ *. Each point corresponds to a basis of a 2D linear space, represented by the two arrows. See also the figures in chapter 4.*

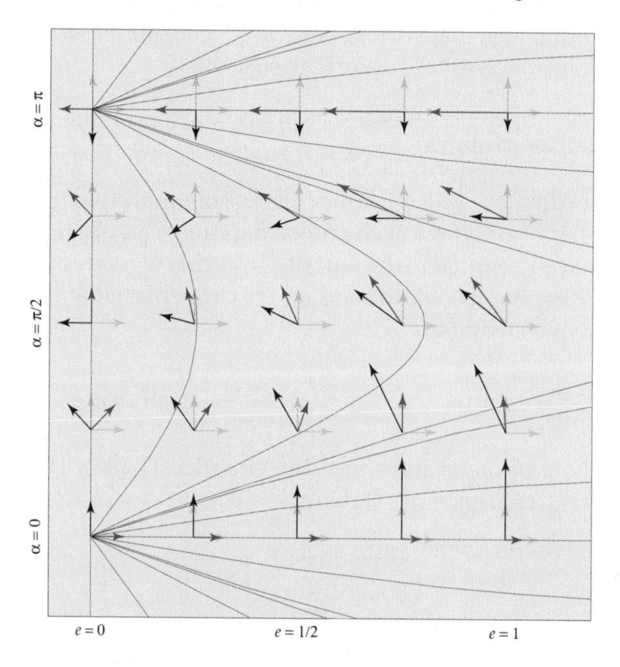

Fig. 1.18. *Same as figure 1.17, but for* $\varphi = \pi$ *.*

are then useful. The exponential of this matrix can be obtained using formula (1.174)

$$\mathbf{M} = \exp \mathbf{m} = \frac{\sinh s}{s}\, \mathbf{m} + \cosh s\, \mathbf{I} \quad ; \quad s = \sqrt{\varepsilon^2 - \theta^2} \ . \qquad (1.201)$$

When $\varepsilon^2 - \theta^2 < 0$, one should remember that $\sinh ix = \sin x$ and $\cosh ix = \cos x$. Here, ε takes any positive real value, and θ any real value.

The light-cone passing through the origin is now given by $\theta = \pm\varepsilon$, and the other light cone is at $\varepsilon = \infty$ and $\theta = \infty$. This coordinate change is represented in figure 1.19.[55]

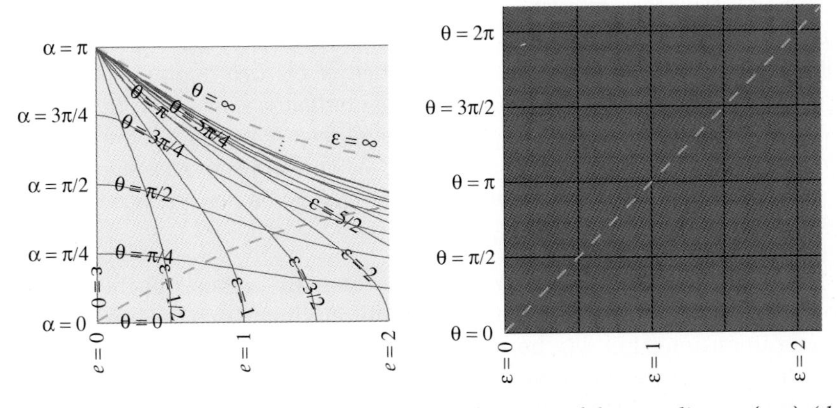

Fig. 1.19. *In the left, the coordinates* $\{\varepsilon, \theta\}$ *as a function of the coordinates* $\{e, \alpha\}$ *(the coordinate* φ *is the same). While the coordinates* $\{e, \alpha, \varphi\}$ *cover the whole of* SL(2)*, the coordinates* $\{\varepsilon, \theta, \varphi\}$ *cover the part of* SL(2) *that is geodesically connected to the origin. They are useful for the analysis of the deformation of a continuous medium (see chapter 4). When representing the part of* SL(2) *geodesically connected to the origin using the coordinates* $\{\varepsilon, \theta, \varphi\}$*, one obtains the representation at the right.*

1.5 Geotensors

The term autovector has been coined for the set of oriented autoparallel segments on a manifold that have a common origin. The Lie group manifolds are quite special manifolds: they are homogeneous and have an absolute notion of parallelism. Autoparallel lines and geodesic lines coincide. Thanks

[55] The expression of the metric (1.192) in the coordinates $\{\varepsilon, \theta, \varphi\}$ is $\{g_{ij}\} = \frac{2\chi}{\Lambda^2}\left(\begin{pmatrix} \varepsilon^2 & -\varepsilon\,\theta & 0 \\ -\varepsilon\,\theta & \theta^2 & 0 \\ 0 & 0 & 0 \end{pmatrix} + \sinh^2\!\Lambda \left(\begin{pmatrix} 0 & 0 & 0 \\ 0 & 0 & 0 \\ 0 & 0 & \varepsilon^2 \end{pmatrix} - \frac{1}{\Lambda^2}\begin{pmatrix} \theta^2 & -\varepsilon\,\theta & 0 \\ -\varepsilon\,\theta & \varepsilon^2 & 0 \\ 0 & 0 & 0 \end{pmatrix}\right)\right)$, with $\Lambda = \sqrt{\varepsilon^2 - \theta^2}$.

to Ado's theorem, we know that it is possible to represent the geodesic segments of the manifold as matrices. We shall see that in physical applications these matrices are, in fact, tensors. Almost.

In fact, although it is possible to define the ordinary sum of two such "tensors", say $\mathbf{t}_1 + \mathbf{t}_2$ it will generally not make much sense. But the geosum $\mathbf{t}_2 \oplus \mathbf{t}_1 = \log(\exp \mathbf{t}_2 \ \exp \mathbf{t}_1)$ is generally a fundamental operation.

Example 1.21 *In 3D Euclidean space, let* $\mathbf{R}$ *be a rotation operator,* $\mathbf{R}^* = \mathbf{R}^{-1}$. *Associated to this orthogonal tensor is the rotation pseudo-vector* $\boldsymbol{\rho}$, *whose direction is the rotation axis, and whose norm is the rotation angle. This pseudo-vector is the dual of an antisymmetric tensor* $\mathbf{r}$, $\rho_i = \frac{1}{2} \epsilon_{ijk} r^{jk}$. *This antisymmetric tensor* $\mathbf{r}$ *is the logarithm of the orthogonal tensor* $\mathbf{R}$: $\mathbf{r} = \log \mathbf{R}$ *(see details in appendix A.14). The composition of two rotations can be obtained as the product of the two orthogonal tensors that represent them:* $\mathbf{R} = \mathbf{R}_2 \, \mathbf{R}_1$. *If, instead, we are dealing with the two rotation 'vectors',* $\mathbf{r}_1$ *and* $\mathbf{r}_2$, *the composition of the two rotations is given by* $\mathbf{r} = \mathbf{r}_2 \oplus \mathbf{r}_1 = \log(\exp \mathbf{r}_2 \ \exp \mathbf{r}_1)$, *while the ordinary sum of the two rotation 'vectors',* $\mathbf{r}_2 + \mathbf{r}_1$, *has no special geometric meaning. It is only when the rotation 'vectors' are small that, as* $\mathbf{r}_2 \oplus \mathbf{r}_1 \approx \mathbf{r}_2 + \mathbf{r}_1$, *the ordinary sum makes approximate sense. The antisymmetric rotation tensors* $\mathbf{r}_1$ *and* $\mathbf{r}_2$ *do not belong to a (linear) tensor space. They are not tensors, but geotensors.*

In the physics of the continuum, one usually represents the physical space, or the physical space-time, by a manifold that may have three, four, or more dimensions. Let $\mathfrak{M}_n$ be such an n-dimensional manifold. It may have arbitrary curvature and torsion at all points.

Selecting any given point $\mathcal{P}$ of $\mathfrak{M}_n$ as an origin, the set of all oriented autoparallel segments (having $\mathcal{P}$ as origin) form an autovector space, with the geosum defined via the parallel transport as the basic operation. In the limit of small autovectors, this defines a linear (vector) tangent space, $\mathbb{E}_n$, the usual tangent linear space considered in standard tensor theory. This linear space $\mathbb{E}_n$ has a dual, $\mathbb{E}_n^*$, and one can build the standard tensorial product $\mathbb{E}_n \otimes \mathbb{E}_n^*$, a linear (tensor) space with dimension n^2. As $\mathbb{E}_n$ was built as a linear space tangent to the manifold $\mathfrak{M}_n$ at point $\mathcal{P}$, one can say, with language abuse (but with a clear meaning), that $\mathbb{E}_n \otimes \mathbb{E}_n^*$ is also tangent to $\mathfrak{M}_n$ at $\mathcal{P}$. When selecting a basis $\mathbf{e}_i$ for $\mathbb{E}_n$, the dual basis $\mathbf{e}^i$ provides a basis for $\mathbb{E}_n{}^*$, and the basis $\mathbf{e}_i \otimes \mathbf{e}^j$ for $\mathbb{E}_n \otimes \mathbb{E}_n^*$.

The linear (tensor) space $\mathbb{E}_n \otimes \mathbb{E}_n^*$ is not the only n^2-dimensional tangent space that can be contemplated at $\mathcal{P}$. For the group manifold associated to $GL(n)$ has also n^2 dimensions, and accepts $\mathbb{E}_n \otimes \mathbb{E}_n^*$ as tangent space at any of its points. The identification of the basis $\mathbf{e}_i \otimes \mathbf{e}^j$ mentioned above with the natural basis in $GL(n)$ (induced by the exponential coordinates), solidly attaches the Lie group manifold $GL(n)$ as a manifold that is also tangent to $\mathfrak{M}_n$ at point $\mathcal{P}$.

So the manifold $\mathfrak{M}_n$ has, at a point $\mathcal{P}$, many tangent spaces, and among them:

- the linear (vector) space $\mathbb{L}_n$, whose elements are ordinary vectors;
- the linear (tensor) space $\mathbb{L}_n^* \otimes \mathbb{L}_n$, whose elements are ordinary tensors;
- the Lie group manifold $GL(n)$, whose elements (not seen as the multiplicative matrices $\mathbf{A}, \mathbf{B} \ldots$, but as the o-additive matrices $\mathbf{a} = \log \mathbf{A}$, $\mathbf{b} = \mathbf{B} \ldots$) are geotensors (oriented geodesic segments on the Lie group manifold).

While tensors are linear objects, geotensors have curvature, but they belong to a space where curvature and torsion combine to give the absolute parallelism of a Lie group manifold.

Definition 1.43 *Let $\mathfrak{M}_n$ be an n-dimensional manifold and $\mathcal{P}$ one of its points around which the manifold accepts a linear tangent space $\mathbb{E}_n$. A* geotensor *at point $\mathcal{P}$ is an element of the associative autovector space (built on the Lie group manifold $GL(n)$) that is tangent at $\mathcal{P}$ to the tensor space $\mathbb{E}_n \otimes \mathbb{E}_n^*$.*

While conventional physics heavily relies on the notion of tensor, it is my opinion that it has so far missed the notion of geotensor. This, in fact, is the explanation of why in the usual tensor theories, logarithms and exponentials of tensors are absent (while they play a fundamental role in scalar theories): it is not that tensors repel the logarithm and exponential functions, it is only that, in general, the usual tensor theories are linear approximations[56] to more complete theories.

The main practical addition of the notion of geotensor to tensor theory is to complete the usual tensor operations with an extra operation: in addition to tensor expressions of the form

$$\mathbf{C} = \mathbf{B}\,\mathbf{A} \quad ; \quad \mathbf{B} = \mathbf{C}\,\mathbf{A}^{-1} \tag{1.202}$$

and of the form

$$\mathbf{t} = \mathbf{s} + \mathbf{r} \quad ; \quad \mathbf{s} = \mathbf{t} - \mathbf{r} \quad , \tag{1.203}$$

geotensor theories may also contain expressions of the form

$$\mathbf{t} = \mathbf{s} \oplus \mathbf{r} \quad ; \quad \mathbf{s} = \mathbf{t} \ominus \mathbf{r} \quad . \tag{1.204}$$

From an analytical point of view, it is sufficient to know that $\mathbf{s} \oplus \mathbf{r} = \log(\exp \mathbf{s}\ \exp \mathbf{r})$ and that $\mathbf{t} \ominus \mathbf{r} = \log(\exp \mathbf{t}\ (\exp \mathbf{r})^{-1})$, but it is important to understand that the operations $\mathbf{s} \oplus \mathbf{r}$ and $\mathbf{t} \ominus \mathbf{r}$ have a geometrical root as, respectively, a geometric sum and a geometric difference of oriented geodesic segments in a Lie group manifold.

---

[56] Often *falsely linear* approximations.

2 Tangent Autoparallel Mappings

> *... if the points [...] approach one another and meet, I say, the angle [...] contained between the chord and the tangent, will be diminished in infinitum, and ultimately will vanish.*
>
> Philosofiæ Naturalis Principia Mathematica, Isaac Newton, 1687

When considering a mapping between two manifolds, the notion of 'linear tangent mapping' (at a given point) makes perfect sense, whether the manifolds have a connection or not. When the two manifolds are connection manifolds, it is possible to introduce a more fundamental notion, that of 'autoparallel tangent mapping'. While the 'derivative' of a mapping is related to the linear tangent mapping, I introduce here the 'declinative' of a mapping, which is related to the autoparallel tangent mapping (and involves a transport to the origin of the considered manifolds). As an example, when considering a time-dependent rotation $\mathbf{R}(t)$, where $\mathbf{R}$ is an orthogonal matrix, the derivative is $\dot{\mathbf{R}} = d\mathbf{R}/dt$, while the declinative happens to be $\omega = \dot{\mathbf{R}}\,\mathbf{R}^{-1}$: the instantaneous rotation velocity is not the derivative $\dot{\mathbf{R}}$, but the declinative ω. As far as some of the so-called tensors in physics are, in fact, the geotensors introduced in the previous chapter, well written physical equations should contain declinatives, not derivatives.

Why we Need a New Concept

An equation like

$$v^i(\mathbf{a}) - v^i(\mathbf{a}_0) \;=\; K_\alpha{}^i\,(a^\alpha - a^\alpha_0) + \dots \tag{2.1}$$

or, equivalently,

$$\mathbf{v}(\mathbf{a}) - \mathbf{v}(\mathbf{a}_0) \;=\; \mathbf{K}\,(\mathbf{a} - \mathbf{a}_0) + \dots \tag{2.2}$$

will possibly suggest to every physicist an expansion of a vector function $\mathbf{a} \mapsto \mathbf{v}(\mathbf{a})$. The operator $\mathbf{K}$, with components $K_\alpha{}^i = \partial v^i / \partial a^\alpha$, defining the linear tangent mapping, is generally named the differential (or, sometimes, the derivative).

We now know that, in addition to vectors, we may have autovectors, that don't operate with the linear operations $+$ and $-$, but with geometric sums and differences. Expressions like those above will still make sense (as any autovector space has a linear tangent space) but will not be fundamental. Instead, we shall face developments like

$$\mathbf{v}(\mathbf{a}) \ominus \mathbf{v}(\mathbf{a}_0) \;=\; \mathfrak{D}\,(\mathbf{a} \ominus \mathbf{a}_0) + \dots \tag{2.3}$$

The operator $\mathfrak{D}$ is named the *declinative*, and it does not define a linear tangent mapping, but an 'autoparallel tangent mapping'.

When working with connection manifolds, the geometric sum and difference involve parallel transport on the manifolds. For a mapping involving a Lie group manifold, the declinative operator corresponds to transport of the differential operator from the point where it is evaluated to the origin of the Lie group. When considering that a Lie group manifold representing a physical transformation (say, the group SO(3), representing a rotation) is tangent to the physical space, with tangent point the origin of the group, we understand that transport to the origin implicit in the concept of declinative, is of fundamental importance.

For instance, when developing this notion, we find the following two results:

- The declinative of a time-dependent rotation $\mathbf{R}(t)$ gives the rotation velocity
$$\omega \equiv \mathfrak{D} = \dot{\mathbf{R}}\,\mathbf{R}^t \quad . \tag{2.4}$$
- The declinative of a mapping from a multiplicative matrix group (with matrices $A_1{}^\alpha{}_\beta$, $A_2{}^\alpha{}_\beta$, ...) into another multiplicative matrix group (with matrices $M_1{}^i{}_j$, $M_2{}^i{}_j$, ...) has the components (denoting $\overline{\mathbf{M}} \equiv \mathbf{M}^{-1}$)
$$\mathfrak{D}_i{}^{j\alpha}{}_\beta = A^j{}_s \, \frac{\partial M^\alpha{}_\sigma}{\partial A^i{}_s} \, \overline{M}^\sigma{}_\beta \quad . \tag{2.5}$$

Evaluation of the declinative produces different results because in each situation the metric of the space (and, thus, the connection) is different. One should realize that, in the case of a rotation $\mathbf{R}(t)$, spontaneously obtaining the rotation velocity $\omega(t) \equiv \mathfrak{D} = \dot{\mathbf{R}}\,\mathbf{R}^t$ as the declinative of the mapping $t \mapsto \mathbf{R}(t)$ is quite an interesting result: while the demonstration that the rotation velocity equals $\dot{\mathbf{R}}\,\mathbf{R}^t$ usually requires intricate developments, with the present theory we could just say "what can the rotation velocity be other than the declinative of $\mathbf{R}(t)$?"

Notation. As many different types of structures are considered in this chapter, let us start by reviewing the notation used. Linear spaces (i.e., vector spaces) are denoted $\{\mathsf{A}, \mathsf{B}, \mathsf{E}, \mathsf{F}, \dots\}$, and their vectors $\{\mathbf{a}, \mathbf{b}, \mathbf{u}, \mathbf{v}, \mathbf{b}+\mathbf{a}, \mathbf{b}-\mathbf{a}, \dots\}$. The dual of A is denoted A^*. Autovector spaces are denoted $\{\mathbb{A}, \mathbb{B}, \mathbb{E}, \mathbb{F}, \dots\}$, and their autovectors $\{\mathbf{a}, \mathbf{b}, \mathbf{u}, \mathbf{v}, \mathbf{b}\oplus\mathbf{a}, \mathbf{b}\ominus\mathbf{a}, \dots\}$. The linear space tangent to an autovector space $\mathbb{A}$ is denoted $\mathsf{A} = \mathsf{T}(\mathbb{A})$. Manifolds are denoted $\{\mathfrak{A}, \mathfrak{B}, \mathfrak{M}, \mathfrak{N}, \dots\}$, and their points $\{\mathcal{A}, \mathcal{B}, \mathcal{P}, \mathcal{Q}, \dots\}$. The autovector space associated with a manifold $\mathfrak{M}$ and a point $\mathcal{P}$ is denoted $\mathbb{A}(\mathfrak{M}, \mathcal{P})$. The autovector from point $\mathcal{P}$ to point $\mathcal{Q}$ is denoted $\mathbf{a}(\mathcal{Q}, \mathcal{P})$. A mapping from an autovector space $\mathbb{E}$ into an autovector space $\mathbb{F}$ is written $\mathbf{a} \in \mathbb{E} \mapsto \mathbf{v} \in \mathbb{F}$, with $\mathbf{v} = \mathbf{f}(\mathbf{a})$. Finally, a mapping from a manifold $\mathfrak{M}$ into a manifold $\mathfrak{N}$ is written $\mathcal{A} \in \mathfrak{M} \mapsto \mathcal{P} \in \mathfrak{N}$, with $\mathcal{P} = \varphi(\mathcal{A})$.

Metric coordinates and Jeffreys coordinates. A coordinate x over a metric one-dimensional manifold is a *metric coordinate* if the distance between two points, with respective coordinates x_1 and x_2, is $D = | x_2 - x_1 |$. The (oriented) length element is, therefore, $ds = dx$. A positive coordinate X over a metric one-dimensional manifold such that the distance between two points, with respective coordinates X_1 and X_2, is $D = | \log(X_2/X_1) |$, is called, all through this book, a *Jeffreys coordinate*. The (oriented) length element at point X is, therefore, $ds = dX/X$. As will be explained in chapter 3, these coordinates shall typically correspond to positive physical quantities, like a frequency. For the distance between two musical notes, with frequencies ν_1 and ν_2, is typically defined as $D = | \log(\nu_2/\nu_1) |$.

2.1 Declinative (Autovector Spaces)

When a mapping is considered between two linear spaces, its tangent linear mapping is introduced, which serves to define the 'differential' of the mapping. We are about to see that when a mapping is considered between two autovector spaces, this definition has to be generalized, this introducing the 'declinative' of the mapping.

The section starts by recalling the basic terminology associated with linear spaces.

2.1.1 Linear Spaces

Let A be a p-dimensional linear space over $\Re$, with vectors denoted $\mathbf{a}, \mathbf{b}, \dots$, let V be a q-dimensional linear space, with vectors denoted $\mathbf{v}, \mathbf{w}, \dots$ and let $\mathbf{a} \mapsto \mathbf{v} = \mathbf{L}(\mathbf{a})$ be a mapping from A into V. The mapping $\mathbf{L}$ is called *linear* if the properties

$$\mathbf{L}(\lambda\, \mathbf{a}) = \lambda\, \mathbf{L}(\mathbf{a}) \quad ; \quad \mathbf{L}(\mathbf{b} + \mathbf{a}) = \mathbf{L}(\mathbf{b}) + \mathbf{L}(\mathbf{a}) \tag{2.6}$$

hold for any vectors $\mathbf{a}$ and $\mathbf{b}$ of A and any real λ. It is common for a linear mapping to use as equivalent the two types of notation $\mathbf{L}(\mathbf{a})$ and $\mathbf{L}\,\mathbf{a}$.

The multiplication of a linear mapping by a real number and the sum of two linear mappings are defined by the conditions

$$(\lambda\, \mathbf{L})(\mathbf{a}) = \lambda\, \mathbf{L}(\mathbf{a}) \quad ; \quad (\mathbf{L}_1 + \mathbf{L}_2)(\mathbf{a}) = \mathbf{L}_1(\mathbf{a}) + \mathbf{L}_2(\mathbf{a}) \quad . \tag{2.7}$$

This endows the space of all linear mapping from A into V with a structure of linear space. There is a one-to-one correspondence between this space of linear mappings and the tensor space $\mathsf{V} \otimes \mathsf{A}^*$ (the tensor product of V times the dual of A).

Let $\{\mathbf{e}_\alpha\} = \{\mathbf{e}_1, \dots, \mathbf{e}_p\}$ be a basis of A and $\{\mathbf{e}_i\} = \{\mathbf{e}_1, \dots, \mathbf{e}_q\}$ be a basis of V. Then, to vectors $\mathbf{a}$ and $\mathbf{v}$ one can associate the components $\mathbf{a} = a^\alpha\, \mathbf{e}_\alpha$

and $\mathbf{v} = v^i \mathbf{e}_i$. Letting $\{\mathbf{e}^\alpha\}$ be the dual of the basis $\{\mathbf{e}_\alpha\}$, one can develop the tensor $\mathbf{L}$ on the basis $\mathbf{e}_i \otimes \mathbf{e}^\alpha$, writing $\mathbf{L} = L^i{}_\alpha \, \mathbf{e}_i \otimes \mathbf{e}^\alpha$. To obtain an explicit expression for the components $L^i{}_\alpha$, one can write the expression $\mathbf{v} = \mathbf{L}\,\mathbf{a}$ as $v^j \mathbf{e}_j = \mathbf{L}(a^\alpha \mathbf{e}_\alpha) = a^\alpha (\mathbf{L}\,\mathbf{e}_\alpha)$, from which $\langle \mathbf{e}^i , v^j \mathbf{e}_j \rangle = a^\alpha \langle \mathbf{e}^i , \mathbf{L}\,\mathbf{e}_\alpha \rangle$, i.e., $v^i = L^i{}_\alpha a^\alpha$, where

$$L^i{}_\alpha = \langle \mathbf{e}^i , \mathbf{L}\,\mathbf{e}_\alpha \rangle \quad , \tag{2.8}$$

and one then has the following equivalent notations:

$$\mathbf{v} = \mathbf{L}\,\mathbf{a} \qquad \Longleftrightarrow \qquad v^i = L^i{}_\alpha a^\alpha \quad . \tag{2.9}$$

We have, in particular, arrived at the following

Property 2.1 *The linear mappings between the linear (vector) space* A *and the linear (vector) space* V *are in one-to-one correspondence with the elements of the tensor space* $\mathsf{V} \otimes \mathsf{A}^*$.

Definition 2.1 Characteristic tensor. *The tensor* $\mathbf{L} \in \mathsf{V} \otimes \mathsf{A}^*$ *associated with a linear mapping —from a linear space* A *into a linear space* V*— is called the* characteristic tensor *of the mapping. The same symbol* $\mathbf{L}$ *is used to denote a linear mapping and its characteristic tensor.*

While $\mathbf{L}$ maps A into V, its *transpose*, denoted $\mathbf{L}^t$, maps V^* into A^*. It is defined by the condition that for any $\mathbf{a} \in \mathsf{A}$ and any $\hat{\mathbf{v}} \in \mathsf{V}^*$,

$$\langle \hat{\mathbf{v}} , \mathbf{L}\,\mathbf{a} \rangle_{\mathsf{V}} = \langle \mathbf{L}^t \hat{\mathbf{v}} , \mathbf{a} \rangle_{\mathsf{A}} \quad . \tag{2.10}$$

One easily obtains

$$(L^t)_\alpha{}^i = L^i{}_\alpha \quad . \tag{2.11}$$

This property means that as soon as the components of a linear operator are known on given bases, the components of the transpose operator are also known. In particular, while for any $\mathbf{a} \in \mathsf{A}$,

$$\mathbf{v} = \mathbf{L}\,\mathbf{a} \quad \Rightarrow \quad v^i = L^i{}_\alpha a^\alpha \quad , \tag{2.12}$$

one has, for any $\hat{\mathbf{v}} \in \mathsf{V}^*$,

$$\hat{\mathbf{a}} = \mathbf{L}^t \hat{\mathbf{v}} \quad \Rightarrow \quad \hat{a}_\alpha = L^i{}_\alpha \hat{v}_i \quad , \tag{2.13}$$

where the same "coefficients" $L^i{}_\alpha$ appear. One should remember this simple property, as the following pages contain some "jiggling" between linear operators and their transposes.

In definition 1.11 (page 16) we introduced the Frobenius norm of a tensor. This easily generalizes to the present situation, if the spaces under consideration are metric:

Definition 2.2 Frobenius norm of a linear mapping. *When the two linear (vector) spaces* A *and* V *are scalar product vector spaces, with respective metric tensors* $\mathbf{g_A}$ *and* $\mathbf{g_V}$, *the* Frobenius norm *of the linear mapping* $\mathbf{L}$ *is defined as*

$$\| \mathbf{L} \| = \sqrt{\operatorname{tr} \mathbf{L}\, \mathbf{L}^t} = \sqrt{\operatorname{tr} \mathbf{L}^t\, \mathbf{L}} = \sqrt{(\mathbf{g_V})_{ij}\, (\mathbf{g_A})^{\alpha\beta}\, L^i{}_\alpha\, L^j{}_\beta} \quad . \tag{2.14}$$

The Frobenius norm of a mapping $\mathbf{L}$ between two linear spaces bears a formal resemblance to the (pseudo) norm of a linear endomorphism $\mathbf{T}$ (see, for instance, equation (A.212), page 195) but they are fundamentally different: in equation (2.14) the components of $\mathbf{L}$ appear, while the definition of the pseudonorm of an endomorphism $\mathbf{T}$ concerns the components of $\mathbf{t} = \log \mathbf{T}$.

It is easy to generalize the above definition to define the Frobenius norm of a mapping that maps a tensor product of linear spaces into another tensor product of linear tensor spaces. For instance, in chapter 4 we introduce a mapping $\mathbf{L}$ with components $L_{a\alpha}{}^{Aij}$, where the indices $a, b \dots$, $\alpha, \beta \dots$, $A, B \dots$ and $i, j \dots$ "belong" to different linear spaces, with respective metric tensors γ_{ab}, $\Gamma_{\alpha\beta}$, G_{AB} and g_{ij}. The Frobenius norm of the mapping is then defined through $\| \mathbf{L} \|^2 = \gamma^{ab}\, \Gamma^{\alpha\beta}\, G_{AB}\, g_{ik}\, g_{j\ell}\, L_{a\alpha}{}^{Aij}\, L_{b\beta}{}^{Bk\ell}$.

We do not need to develop further the theory of linear spaces here, as some of the basic concepts appear in a moment within the more general context of autovector spaces. We may just recall here the basic property of the differential mapping associated to a mapping, a property that can be used as a definition:

Definition 2.3 Differential mapping. *Let* $\mathbf{a} \mapsto \mathbf{v} = \mathbf{v}(\mathbf{a})$ *a sufficiently regular mapping from a linear space* A *into a linear space* V. *The* differential mapping *at* $\mathbf{a}_0$, *denoted* $\mathbf{d}_0$, *is the linear mapping from* V^* *into* A^* *satisfying the expansion*

$$\mathbf{v}(\mathbf{a}) - \mathbf{v}(\mathbf{a}_0) = \mathbf{d}_0^t\, (\mathbf{a} - \mathbf{a}_0) + \dots \quad , \tag{2.15}$$

where the dots denote terms that are at least quadratic in $\mathbf{a} - \mathbf{a}_0$.

Note that, as $\mathbf{d}_0$ maps V^* into A^*, its transpose $\mathbf{d}_0^t$ maps A into V, so this expansion makes sense. The technicality of not calling differential operator the operator appearing in the expansion (2.15), but its transpose, allows us to obtain compact formulas below. It is important to understand that while the indices denoting the components of $\mathbf{d}_0^t$ are $(\mathbf{d}_0^t)^i{}_\alpha$, those of the differential $\mathbf{d}_0$ are $(\mathbf{d}_0)_\alpha{}^i$, in this order, and, according to equation (2.11), one has $(\mathbf{d}_0)_\alpha{}^i = (\mathbf{d}_0^t)^i{}_\alpha$.

2.1.2 Autovector Spaces

As in what follows both an autovector space and its linear tangent space are considered, let us recall the abstract way of understanding the relation between an autovector space and its tangent space: over a common set of elements there are two different sums defined, the o-sum and the related tangent operation, the (usual) commutative sum. An alternative, more visual interpretation, is to consider that the autovectors are oriented autoparallel segments on a (possibly) curved manifold (where an origin has been chosen, and with the o-sum defined geometrically through the parallel transport), and that the linear tangent space is the linear space tangent (in the usual geometrical sense) to the manifold at its origin. As these two points of view are consistent, one may switch between them, according to the problem at hand.

Consider a p-dimensional autovector space $\mathbb{A}$ and a q-dimensional autovector space $\mathbb{V}$. The autovectors of $\mathbb{A}$ are denoted $\mathbf{a}, \mathbf{b} \dots$, and the o-sum and o-difference in $\mathbb{A}$ are respectively denoted $\boxplus$ and $\boxminus$. The autovectors of $\mathbb{V}$ are denoted $\mathbf{u}, \mathbf{v}, \mathbf{w} \dots$, and the o-sum and o-difference in $\mathbb{V}$ are respectively denoted $\oplus$ and $\ominus$. Therefore, one can write

$$\begin{array}{lclcl} \mathbf{c} = \mathbf{b} \boxplus \mathbf{a} & \Leftrightarrow & \mathbf{b} = \mathbf{c} \boxminus \mathbf{a} & ; & (\mathbf{a}, \mathbf{b}, \dots \in \mathbb{A}) \\ \mathbf{w} = \mathbf{v} \oplus \mathbf{u} & \Leftrightarrow & \mathbf{v} = \mathbf{w} \ominus \mathbf{u} & ; & (\mathbf{u}, \mathbf{v}, \dots \in \mathbb{V}) \quad . \end{array} \tag{2.16}$$

We have seen that the o-sum and the o-difference operations in autovector space operations admit tangent operations, that are denoted $+$ and $-$, without distinction of the space where they are defined (as they are the usual sum and difference in linear spaces). Therefore one can write, respectively in $\mathbb{A}$ and in $\mathbb{V}$,

$$\begin{array}{lcl} \mathbf{b} \boxplus \mathbf{a} = \mathbf{b} + \mathbf{a} + \dots & ; & \mathbf{b} \boxminus \mathbf{a} = \mathbf{b} - \mathbf{a} + \dots \\ \mathbf{w} \oplus \mathbf{v} = \mathbf{w} + \mathbf{v} + \dots & ; & \mathbf{w} \ominus \mathbf{v} = \mathbf{w} - \mathbf{v} + \dots \quad . \end{array} \tag{2.17}$$

The autovectors of $\mathbb{A}$, when operated on with the operations $+$ and $-$ form a linear space, denoted $\mathsf{L}(\mathbb{A})$ and called the linear tangent space to $\mathbb{A}$. Similarly, the autovectors of $\mathbb{V}$, when operated on with the operations $+$ and $-$ form the linear tangent space $\mathsf{L}(\mathbb{V})$.

Let $\mathbf{L}$ be a linear mapping from $\mathsf{L}(\mathbb{V})^*$ into $\mathsf{L}(\mathbb{A})^*$ (so $\mathbf{L}^t$ maps $\mathsf{L}(\mathbb{A})$ into $\mathsf{L}(\mathbb{V})$). Such a linear mapping can be used to introduce an affine mapping $\mathbf{a} \mapsto \mathbf{v} = \mathbf{v}(\mathbf{a})$, a mapping from $\mathsf{L}(\mathbb{A})$ into $\mathsf{L}(\mathbb{V})$, that can be defined through the expression

$$\mathbf{v}(\mathbf{a}) - \mathbf{v}(\mathbf{a}_0) = \mathbf{L}^t (\mathbf{a} - \mathbf{a}_0) \quad . \tag{2.18}$$

Alternatively, a linear mapping $\mathfrak{L}$ from $\mathsf{L}(\mathbb{V})^*$ into $\mathsf{L}(\mathbb{A})^*$ can be used to define another sort of mapping, this time mapping the autovector space $\mathbb{A}$ (with its two operations $\boxplus$ and $\boxminus$) into the autovector space $\mathbb{V}$ (with its two operations $\oplus$ and $\ominus$). This is done via the relation

$$\boxed{\mathbf{v}(\mathbf{a}) \ominus \mathbf{v}(\mathbf{a}_0) = \mathfrak{L}^t (\mathbf{a} \boxminus \mathbf{a}_0)} \quad . \tag{2.19}$$

When bases are chosen in the linear tangent spaces $\mathsf{L}(\mathbb{A})$ and $\mathsf{L}(\mathbb{V})$, one can also write $[\mathbf{v}(\mathbf{a}) \ominus \mathbf{v}_0]^\alpha = \mathfrak{L}_i{}^\alpha [\mathbf{a} \boxminus \mathbf{a}_0]^i$. Note that equation (2.19) can be written, equivalently, $\mathbf{v}(\mathbf{a}) = \mathfrak{L}^t (\mathbf{a} \boxminus \mathbf{a}_0) \oplus \mathbf{v}(\mathbf{a}_0)$.

Definition 2.4 Autoparallel mapping. *A mapping* $\mathbf{a} \mapsto \mathbf{v} = \mathbf{v}(\mathbf{a})$ *from an autovector space* $\mathbb{A}$ *into an autovector space* $\mathbb{V}$ *is* autoparallel at $\mathbf{a}_0$ *if there is some* $\mathfrak{L} \in \mathsf{L}(\mathbb{V}) \otimes \mathsf{L}(\mathbb{A})^*$ *such that for any* $\mathbf{a} \in \mathbb{A}$ *expression* (2.19) *holds. The tensor* $\mathfrak{L}$ *is called the* characteristic tensor *of the autoparallel mapping.*

When considering a mapping $\mathbf{a} \mapsto \mathbf{v} = \mathbf{v}(\mathbf{a})$ from an autovector space $\mathbb{A}$ into an autovector space $\mathbb{V}$, it may be affine, if it has the form (2.18), or it may be autoparallel, if it has the form (2.19). The notions of affine and autoparallel mappings coexist, and are not equivalent (unless the autovector spaces are, in fact, linear spaces).

Writing the autoparallel mapping (2.19) for two autovectors $\mathbf{a}_1$ and $\mathbf{a}_2$ gives $\mathbf{v}(\mathbf{a}_1) = \mathfrak{L}^t\,(\,\mathbf{a}_1 \boxminus \mathbf{a}_0\,) \oplus \mathbf{v}(\mathbf{a}_0)$ and $\mathbf{v}(\mathbf{a}_2) = \mathfrak{L}^t\,(\,\mathbf{a}_2 \boxminus \mathbf{a}_0\,) \oplus \mathbf{v}(\mathbf{a}_0)$. Making the o-difference gives $\mathbf{v}(\mathbf{a}_2) \ominus \mathbf{v}(\mathbf{a}_1) = (\,\mathfrak{L}^t\,(\,\mathbf{a}_2 \boxminus \mathbf{a}_0\,) \oplus \mathbf{v}(\mathbf{a}_0)\,) \ominus (\,\mathfrak{L}^t\,(\,\mathbf{a}_1 \boxminus \mathbf{a}_0\,) \oplus \mathbf{v}(\mathbf{a}_0)\,)$. For a general autovector space, there is no simplification of this expression. If the autovector space $\mathbb{V}$ is associative (i.e., if it is, in fact, a Lie group), we can use the property in equation (1.52) to simplify this expression into

$$\mathbf{v}(\mathbf{a}_2) \ominus \mathbf{v}(\mathbf{a}_1) \;=\; \mathfrak{L}^t\,(\,\mathbf{a}_2 \boxminus \mathbf{a}_0\,) \ominus \mathfrak{L}^t\,(\,\mathbf{a}_1 \boxminus \mathbf{a}_0\,) \quad . \tag{2.20}$$

Therefore, we have the following

Property 2.2 *When considering a mapping from an autovector space* $\mathbb{A}$ *into an* associative *autovector space* $\mathbb{V}$, *a mapping* $\mathbf{a} \mapsto \mathbf{v} = \mathbf{v}(\mathbf{a})$ *that is autoparallel at* $\mathbf{a}_0$ *verifies the relation* (2.20), *for any* $\mathbf{a}_1$ *and* $\mathbf{a}_2$.

If the mapping is autoparallel at the origin, $\mathbf{a}_0 = \mathbf{0}$, then,

$$\mathbf{v}(\mathbf{a}_2) \ominus \mathbf{v}(\mathbf{a}_1) \;=\; \mathfrak{L}^t\,\mathbf{a}_2 \ominus \mathfrak{L}^t\,\mathbf{a}_1 \quad . \tag{2.21}$$

We shall make use of this equation when studying elastic media in chapter 4.

Definition 2.5 Tangent mappings. *Let* $\mathbf{f}$ *and* $\mathbf{g}$ *be two mappings from the autovector space* $\mathbb{A}$, *with operations* $\boxplus$ *and* $\boxminus$ *into the autovector space* $\mathbb{V}$, *with operations* $\oplus$ *and* $\ominus$. *The two mappings are* tangent *at* $\mathbf{a}_0$ *if for any* $\mathbf{a} \in \mathbb{A}$ *(see figure 2.1),*

$$\lim_{\lambda \to 0} \frac{1}{\lambda}\,(\,\mathbf{f}(\,\lambda\mathbf{a} \boxplus \mathbf{a}_0\,) \ominus \mathbf{g}(\,\lambda\mathbf{a} \boxplus \mathbf{a}_0\,)\,) \;=\; \mathbf{0} \quad . \tag{2.22}$$

Definition 2.6 Tangent autoparallel mapping. *Let* $\mathbf{f}(\,\cdot\,)$ *and* $\mathbf{F}(\,\cdot\,)$ *be two mappings from an autovector space* $\mathbb{A}$ *into an autovector space* $\mathbb{V}$. *We say that* $\mathbf{F}(\,\cdot\,)$ *is the* tangent autoparallel mapping *to* $\mathbf{f}(\,\cdot\,)$ *at* $\mathbf{a}_0$ *if the two mappings are tangent at* $\mathbf{a}_0$ *and if* $\mathbf{F}(\,\cdot\,)$ *is autoparallel at* $\mathbf{a}_0$.

Definition 2.7 Declinative of a mapping. *Let* $\mathbf{a} \mapsto \mathbf{v} = \mathbf{v}(\mathbf{a})$ *a mapping from an autovector space* $\mathbb{A}$, *with operations* $\boxplus$ *and* $\boxminus$, *into an autovector space* $\mathbb{V}$, *with operations* $\oplus$ *and* $\ominus$. *The* declinative *of* $\mathbf{v}(\,\cdot\,)$ *at* $\mathbf{a}_0$, *denoted* $\mathfrak{D}_0$, *is the characteristic tensor of the autoparallel mapping that is tangent to* $\mathbf{v}(\,\cdot\,)$ *at* $\mathbf{a}_0$.

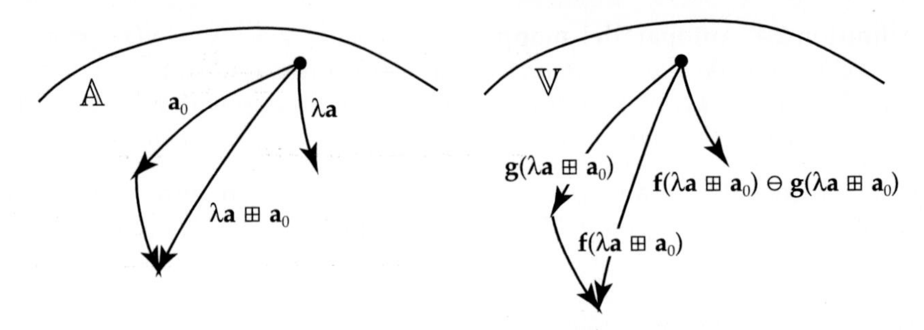

Fig. 2.1. *Two mappings* $\mathbf{f}(\cdot)$ *and* $\mathbf{g}(\cdot)$ *mapping an autovector space* $\mathbb{A}$, *with operations* $\{\boxplus,\boxminus\}$, *into another autovector space* $\mathbb{V}$, *with operations* $\{\oplus,\ominus\}$, *are tangent at* $\mathbf{a}_0$ *if for any* $\mathbf{a}$ *the limit* $\lim_{\lambda\to 0}\frac{1}{\lambda}(\,\mathbf{f}(\lambda\mathbf{a}\boxplus\mathbf{a}_0)\ominus\mathbf{g}(\lambda\mathbf{a}\boxplus\mathbf{a}_0)\,) = \mathbf{0}$ *(equation 2.22) holds.*

By definition, the declinative is an element of the tensor space $\mathsf{L}(\mathbb{A})^* \otimes \mathsf{L}(\mathbb{V})$. When some bases $\{\mathbf{e}_\alpha\}$ and $\{\mathbf{e}_i\}$ are chosen in $\mathsf{L}(\mathbb{A})$ and $\mathsf{L}(\mathbb{V})$, the components of $\mathfrak{D}_0$ are written $(\mathfrak{D}_0)_\alpha{}^i$ (note the order of the indices).

From the definition of tangent mappings follows

Property 2.3 *One has the expansion*

$$\boxed{\mathbf{v}(\mathbf{a})\ominus\mathbf{v}(\mathbf{a}_0) \;=\; \mathfrak{D}_0^t\,(\mathbf{a}\boxminus\mathbf{a}_0) + \dots \quad ,} \tag{2.23}$$

where the dots indicate terms that are, at least, second order in $(\mathbf{a}\boxminus\mathbf{a}_0)$.

The expansion is of course written in $\mathsf{L}(\mathbb{V})$. See figure 2.2 for a pictorial representation.

We know that to each autovector space operation is associated its tangent operation (see equation (1.66), page 24). Therefore, in addition to the expansion (2.23) we can introduce the expansion

$$\boxed{\mathbf{v}(\mathbf{a}) - \mathbf{v}(\mathbf{a}_0) \;=\; \mathbf{d}_0^t\,(\mathbf{a} - \mathbf{a}_0) + \dots \quad ,} \tag{2.24}$$

so we may set the following

Definition 2.8 Differential of a mapping. *Let* $\mathbf{a} \mapsto \mathbf{v} = \mathbf{v}(\mathbf{a})$ *be a mapping from an autovector space* $\mathbb{A}$ *into an autovector space* $\mathbb{V}$, *and let us denote, as usual, by* $+$ *and* $-$ *the associated tangent operations. The* differential *of* $\mathbf{v}(\cdot)$ *at* $\mathbf{a}_0$ *is the tensor* $\mathbf{d}_0$ *characteristic of the expansion 2.24.*

This definition is consistent with that made above for mappings between linear spaces (see equation 2.15).

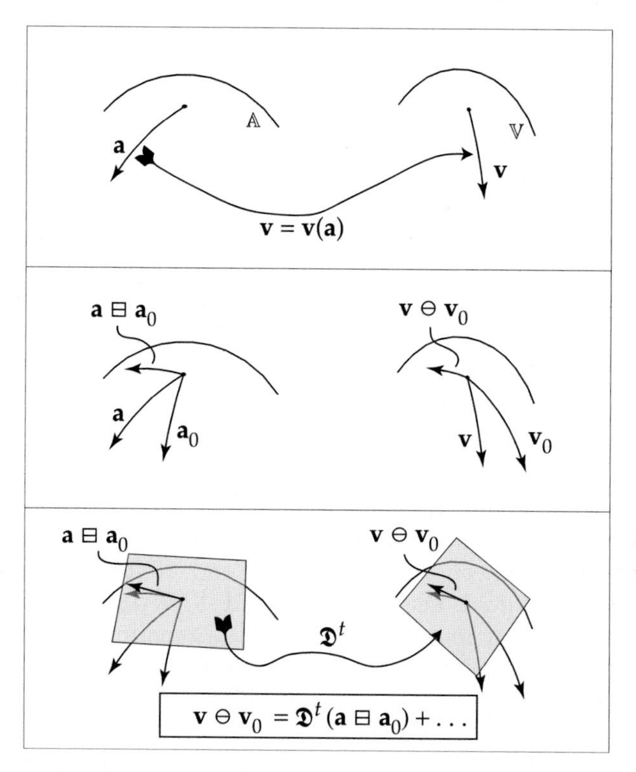

Fig. 2.2. *A mapping* $\mathbf{a} \mapsto \mathbf{v} = \mathbf{v}(\mathbf{a})$ *is considered that maps an autovector space* $\mathbb{A}$, *with operations* $\{\boxplus, \boxminus\}$, *into another autovector space* $\mathbb{V}$, *with operations* $\{\oplus, \ominus\}$. *The declinative* $\mathfrak{D}$ *of the mapping at* $\mathbf{a}_0$ *may be defined by the series development* $\mathbf{v} \ominus \mathbf{v}_0 = \mathfrak{D}^t\,(\mathbf{a} \boxminus \mathbf{a}_0) + \dots$.

2.2 Declinative (Connection Manifolds)

The goal of this chapter is to introduce the declinative of a mapping between two connection manifolds. We have, so far, defined the declinative of a mapping between two autovector spaces. But this is essentially enough, because once an origin is chosen on a connection manifold,[1] then, we can consider the set of all oriented autoparallel segments with the given origin, and use the connection on the manifold to define the sum and difference of segments. The manifold, then, has been transformed into an autovector space, and all the definitions made for autovector spaces apply. Let us develop this idea.

Let $\mathfrak{M}$ be a connection manifold, and $\mathcal{O}$ one particular point, named the *origin*. Let $\mathbf{a}(\mathcal{P}; \mathcal{O})$ denote the oriented autoparallel segment from the origin $\mathcal{O}$ to point $\mathcal{P}$. The sum (or geometric sum) of two such oriented autoparallel segments is defined as in section 1.3, this introducing the structure of a (local)

[1]For instance, when considering the Lie group manifold defined by a matrix group, the origin of the manifold is typically the identity matrix.

autovector space. An oriented autoparallel segment of the form $\mathbf{a}(\mathcal{P};\mathcal{O})$ is now called an autovector, and an expression like

$$\mathbf{a}(\mathcal{P}_3;\mathcal{O}) = \mathbf{a}(\mathcal{P}_2;\mathcal{O}) \oplus \mathbf{a}(\mathcal{P}_1;\mathcal{O}) \quad \Leftrightarrow \quad \mathbf{a}(\mathcal{P}_2;\mathcal{O}) = \mathbf{a}(\mathcal{P}_3;\mathcal{O}) \ominus \mathbf{a}(\mathcal{P}_1;\mathcal{O}) \tag{2.25}$$

makes sense, as makes sense, for a real λ inside some finite interval around zero, the expression

$$\mathbf{a}(\mathcal{P}_2;\mathcal{O}) = \lambda\, \mathbf{a}(\mathcal{P}_1;\mathcal{O}) \quad . \tag{2.26}$$

The autovector space associated to a connection manifold $\mathfrak{M}$ and origin $\mathcal{O}$ is denoted $\mathbb{A}(\mathfrak{M};\mathcal{O})$.

As we have seen in chapter 1, the limit

$$\mathbf{a}(\mathcal{P}_2;\mathcal{O}) + \mathbf{a}(\mathcal{P}_1;\mathcal{O}) \equiv \lim_{\lambda\to 0} \frac{1}{\lambda}(\,\lambda\, \mathbf{a}(\mathcal{P}_2;\mathcal{O}) \oplus \lambda\, \mathbf{a}(\mathcal{P}_1;\mathcal{O})\,) \tag{2.27}$$

defines an ordinary sum (i.e., a commutative and associative sum), this introducing the linear space tangent to $\mathbb{A}(\mathfrak{M};\mathcal{O})$, denoted $\mathbf{L}(\mathbb{A}(\mathfrak{M};\mathcal{O}))$.

Consider two connection manifolds $\mathfrak{M}$ and $\mathfrak{N}$. Let $\mathcal{O}_{\mathfrak{M}}$ and $\mathcal{O}_{\mathfrak{N}}$ the origins of each manifold. Let $\mathbb{A}(\mathfrak{M},\mathcal{O}_{\mathfrak{M}})$ and $\mathbb{V}(\mathfrak{N},\mathcal{O}_{\mathfrak{N}})$, be the associated autovector spaces. Any mapping $\mathcal{P} \mapsto \mathcal{Q} = \mathcal{Q}(\mathcal{P})$ mapping the points of $\mathfrak{M}$ into points of $\mathfrak{N}$ can be considered to be a mapping $\mathbf{a} \mapsto \mathbf{v} = \mathbf{v}(\mathbf{a})$ mapping autovectors of $\mathbb{A}(\mathfrak{M},\mathcal{O}_{\mathfrak{M}})$ into autovectors of $\mathbb{V}(\mathfrak{N},\mathcal{O}_{\mathfrak{N}})$, namely the mapping $\mathbf{a}(\mathcal{P},\mathcal{O}_{\mathfrak{M}}) \mapsto \mathbf{v}(\mathcal{Q},\mathcal{O}_{\mathfrak{N}}) = \mathbf{v}(\,\mathcal{Q}(\mathcal{P})\,,\,\mathcal{O}_{\mathfrak{N}}\,)$.

With this structure in mind, it is now easy to extend the basic definitions made in section 2.1.2 for autovector spaces into the corresponding definitions for connection manifolds.

Definition 2.9 Autoparallel mapping, characteristic tensor. *A mapping* $\mathcal{P} \mapsto \mathcal{Q} = \mathcal{Q}(\mathcal{P})$ *from the connection manifold* $\mathfrak{M}$ *with origin* $\mathcal{O}_{\mathfrak{M}}$ *into the connection manifold* $\mathfrak{N}$ *with origin* $\mathcal{O}_{\mathfrak{N}}$ *is* autoparallel *at point* $\mathcal{P}_0$ *if the mapping from the autovector space* $\mathbb{A}(\mathfrak{M},\mathcal{O}_{\mathfrak{M}})$ *into the autovector space* $\mathbb{V}(\mathfrak{N},\mathcal{O}_{\mathfrak{N}})$ *is autoparallel at* $\mathbf{a}(\mathcal{P}_0,\mathcal{O}_{\mathfrak{M}})$ *(in the sense of definition 2.4, page 85). The* characteristic tensor *of an affine mapping* $\mathcal{P} \mapsto \mathcal{Q} = \mathcal{Q}(\mathcal{P})$ *is the characteristic tensor of the associated affine autovector mapping.*

Definition 2.10 Geodesic mapping. *If the connection over the considered manifolds is the Levi-Civita connection (that results from a metric in each manifold), an autoparallel mapping is also called a* geodesic mapping.

Definition 2.11 Tangent mappings. *Two mappings from a connection manifold* $\mathfrak{M}$ *with origin* $\mathcal{O}_{\mathfrak{M}}$ *into a connection manifold* $\mathfrak{N}$ *with origin* $\mathcal{O}_{\mathfrak{N}}$ *are* tangent at point $\mathcal{P}_0 \in \mathfrak{M}$ *if the associated mappings from the autovector space* $\mathbb{A}(\mathfrak{M},\mathcal{O}_{\mathfrak{M}})$ *into the autovector space* $\mathbb{V}(\mathfrak{N},\mathcal{O}_{\mathfrak{N}})$ *are tangent at* $\mathbf{a}(\mathcal{P}_0,\mathcal{O}_{\mathfrak{M}})$ *(in the sense of definition 2.5, page 85).*

Definition 2.12 Declinative. *Let* $\mathcal{P} \mapsto \mathcal{Q} = \mathcal{Q}(\mathcal{P})$ *be a sufficiently smooth mapping from a connection manifold* $\mathfrak{M}$ *with origin* $\mathcal{O}_{\mathfrak{M}}$ *into a connection manifold* $\mathfrak{N}$ *with origin* $\mathcal{O}_{\mathfrak{N}}$*, and let* $\mathbf{a}(\mathcal{P}, \mathcal{O}_{\mathfrak{M}}) \mapsto \mathbf{v}(\mathcal{Q}, \mathcal{O}_{\mathfrak{N}}) = \mathbf{v}(\, \mathcal{Q}(\mathcal{P}) \,,\, \mathcal{O}_{\mathfrak{N}}\,)$ *be the associated mapping from the autovector space* $\mathbb{A}(\mathfrak{M}, \mathcal{O}_{\mathfrak{M}})$ *into the autovector space* $\mathbb{A}(\mathfrak{N}, \mathcal{O}_{\mathfrak{N}})$*, a mapping that, for short, we may denote as* $\mathbf{a} \mapsto \mathbf{v} = \mathbf{v}(\mathbf{a})$*. The* declinative *of the mapping* $\mathcal{Q}(\,\cdot\,)$ *at* $\mathcal{P}_0$*, denoted* $\mathfrak{D}_0$*, is the declinative of the mapping* $\mathbf{v}(\,\cdot\,)$ *at* $\mathbf{a}(\mathcal{P}_0, \mathcal{O}_{\mathfrak{M}})$ *(in the sense of definition 2.7).*

Therefore, denoting $\{\boxplus, \boxminus\}$ the geometric sum and difference in $\mathbb{A}(\mathfrak{M}; \mathcal{O}_{\mathfrak{M}})$, and $\{\oplus, \ominus\}$ those in $\mathbb{V}(\mathfrak{N}; \mathcal{O}_{\mathfrak{N}})$, the declinative $\mathfrak{D}_0$ allows one to write the expansion

$$\begin{aligned} &\mathbf{v}(\, \mathcal{Q}(\mathcal{P})\,;\, \mathcal{O}_{\mathfrak{N}}\,) \ominus \mathbf{v}(\, \mathcal{Q}(\mathcal{P}_0)\,;\, \mathcal{O}_{\mathfrak{N}}\,) = \\ &\qquad = \mathfrak{D}_0^t \left(\mathbf{a}(\, \mathcal{P}\,;\, \mathcal{O}_{\mathfrak{M}}\,) \boxminus \mathbf{a}(\, \mathcal{P}_0\,;\, \mathcal{O}_{\mathfrak{M}}\,) \right) + \dots \quad , \end{aligned} \tag{2.28}$$

where the dots represent terms that are at least quadratic in $\mathbf{a}(\mathcal{P}) \boxminus \mathbf{a}(\mathcal{P}_0)$. See a pictorial representation in figure 2.3. The series on the right of equation (2.28) is written in $\mathbf{L}(\, \mathbb{V}(\mathfrak{N}, \mathcal{O}_{\mathfrak{N}})\,)$.

The declinative $\mathfrak{D}_0$ defines a linear mapping that maps $\mathbf{L}(\, \mathbb{V}(\mathfrak{N}, \mathcal{O}_{\mathfrak{N}})\,)^*$ into $\mathbf{L}(\, \mathbb{A}(\mathfrak{M}, \mathcal{O}_{\mathfrak{M}})\,)^*$, i.e., to be more explicit, the declinative of the mapping $\mathcal{P} \mapsto \mathcal{Q} = \mathcal{Q}(\mathcal{P})$, evaluated at any point $\mathcal{P}_0$, is always a linear mapping that maps the dual of the linear space tangent to $\mathfrak{N}$ at its origin, $\mathcal{O}_{\mathfrak{N}}$, into the dual of the linear space tangent to $\mathfrak{M}$ at its origin, $\mathcal{O}_{\mathfrak{M}}$. This contrasts with the usual derivative at $\mathcal{P}_0$, that maps the dual of the linear space tangent to $\mathfrak{N}$ at $\mathcal{Q}(\mathcal{P}_0)$, into the dual of the linear space tangent to $\mathfrak{M}$ at $\mathcal{P}_0$. See figures 2.3 and 2.5.

Consideration of the tangent (i.e., linear) sum and difference associated to the geometric sum and difference on the two manifolds $\mathfrak{M}$ and $\mathfrak{N}$ allows us to introduce the following

Definition 2.13 Differential. *In the same context as in definition 2.12, the* differential *of the mapping* $\mathcal{P} \mapsto \mathcal{Q} = \mathcal{Q}(\mathcal{P})$ *at point* $\mathcal{P}_0$*, denoted* $\mathbf{d}_0$*, is the linear mapping that maps* $\mathbf{L}(\, \mathbb{V}(\mathfrak{N}; \mathcal{O}_{\mathfrak{N}})\,)^*$ *into* $\mathbf{L}(\, \mathbb{A}(\mathfrak{M}; \mathcal{O}_{\mathfrak{M}})\,)^*$ *defined by the expansion*

$$\begin{aligned} &\mathbf{v}(\, \mathcal{Q}(\mathcal{P})\,;\, \mathcal{O}_{\mathfrak{N}}\,) - \mathbf{v}(\, \mathcal{Q}(\mathcal{P}_0)\,;\, \mathcal{O}_{\mathfrak{N}}\,) = \\ &\qquad = \mathbf{d}_0^t \left(\mathbf{a}(\, \mathcal{P}\,;\, \mathcal{O}_{\mathfrak{M}}\,) - \mathbf{a}(\, \mathcal{P}_0\,;\, \mathcal{O}_{\mathfrak{M}}\,) \right) + \dots \quad , \end{aligned} \tag{2.29}$$

In addition to these two notions, declinative and differential, we shall also encounter the ordinary derivative. Rather than introducing this notion independently, let us use the work done so far. The two expressions (2.28) and (2.29) can be written when the origins $\mathcal{O}_{\mathfrak{M}}$ and $\mathcal{O}_{\mathfrak{N}}$ of the two manifolds $\mathfrak{M}$ and $\mathfrak{N}$ are moved to $\mathcal{P}_0$ and $\mathcal{Q}(\mathcal{P}_0)$ respectively. As $\mathbf{a}(\mathcal{O}_{\mathfrak{M}}; \mathcal{O}_{\mathfrak{M}}) = \mathbf{0}$, and $\mathbf{v}(\mathcal{O}_{\mathfrak{N}}; \mathcal{O}_{\mathfrak{N}}) = \mathbf{0}$, the two equations (2.28) and (2.29) then collapse into a single equation that we write

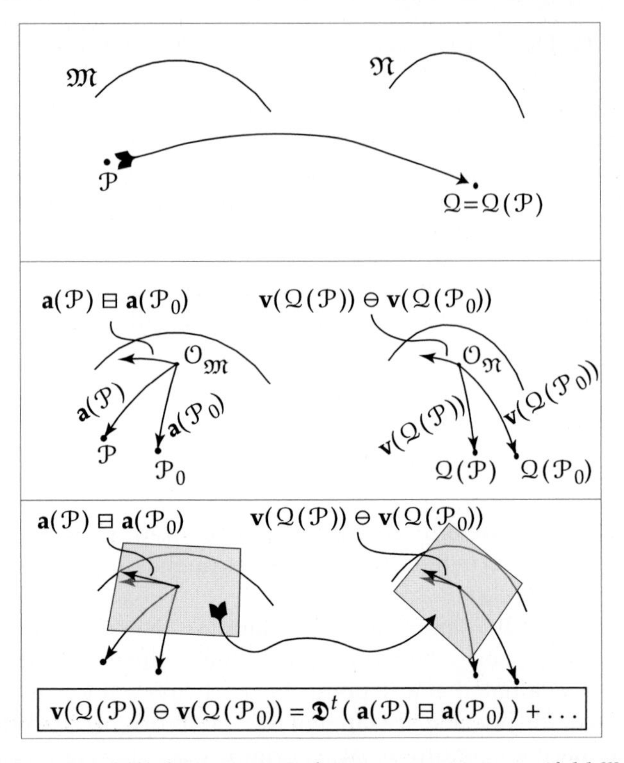

Fig. 2.3. *Let* $\mathcal{P} \mapsto \mathcal{Q} = \mathcal{Q}(\mathcal{P})$ *be a mapping from a connection manifold* $\mathfrak{M}$ *with origin* $\mathcal{O}_{\mathfrak{M}}$ *into a connection manifold* $\mathfrak{N}$ *with origin* $\mathcal{O}_{\mathfrak{N}}$ *. The derivative tensor at some point* $\mathcal{P}_0$ *, denoted* $\mathbf{D}$ *, defines a mapping between the (duals of the) linear spaces tangent to the manifolds at point* $\mathcal{P}_0$ *and* $\mathcal{Q}_0 = \mathcal{Q}(\mathcal{P}_0)$ *, respectively. The declinative tensor* $\mathfrak{D}$ *defines a mapping between the (duals of the) linear spaces tangent to the manifolds at their origin. Parallel transport of the derivative* $\mathbf{D}$ *from* $\mathcal{P}_0$ *to* $\mathcal{O}_{\mathfrak{M}}$ *, on one side, and from* $\mathcal{Q}_0 = \mathcal{Q}(\mathcal{P}_0)$ *to* $\mathcal{O}_{\mathfrak{N}}$ *, on the other side, gives the declinative* $\mathfrak{D}$ *.*

$$\boxed{\mathbf{v}(\,\mathcal{Q}(\mathcal{P})\,;\,\mathcal{Q}(\mathcal{P}_0)\,) \;=\; \mathbf{D}_0^t\,\mathbf{a}(\,\mathcal{P}\,;\,\mathcal{P}_0\,) + \dots \quad ,} \tag{2.30}$$

This leads to the following

Definition 2.14 Derivative. *Let* $\mathcal{P} \mapsto \mathcal{Q} = \mathcal{Q}(\mathcal{P})$ *be a mapping from a connection manifold* $\mathfrak{M}$ *into a connection manifold* $\mathfrak{N}$ *. The* derivative *of the mapping at point* $\mathcal{P}_0$ *, denoted* $\mathbf{D}_0$ *, equals the declinative (at the same point), provided that the origin* $\mathcal{O}_{\mathfrak{M}} = \mathcal{P}_0$ *is chosen on* $\mathfrak{M}$ *and the origin* $\mathcal{O}_{\mathfrak{N}} = \mathcal{Q}(\mathcal{P}_0)$ *is chosen on* $\mathfrak{N}$ *.*

Therefore, the derivative of the mapping $\mathcal{Q}(\,\cdot\,)$ at point $\mathcal{P}_0$ maps the dual of the linear space tangent to $\mathfrak{N}$ at $\mathcal{Q}(\mathcal{P}_0)$ into the dual of the linear space tangent to $\mathfrak{M}$ at $\mathcal{P}_0$. By definition, expression (2.30) holds.

We have just introduced three notions, the derivative, the differential, and the declinative of a mapping between manifolds. The following example, using results demonstrated later in this chapter, should allow us to understand in which sense they are different.

Example 2.1 *Consider a solid rotating around a fixed point of the Euclidean 3D space. Its* attitude *at time* t*, say* $\mathcal{A}(t)$*, is a point of the Lie group manifold* SO(3)*. When an origin is chosen on the manifold (i.e., a particular attitude), any other point (any other attitude) can be represented by a rotation (the rotation needed to transform one attitude into the other). Rotations can be represented by orthogonal matrices,* $\mathbf{R}\,\mathbf{R}^t = \mathbf{I}$*. The origin of the manifold is, then, the identity matrix* $\mathbf{I}$*, and the attitude of the solid at time* t *can be represented by the time-dependent orthogonal matrix* $\mathbf{R}(t)$*. Then, a time-dependent rotation is represented by a mapping* $t \mapsto \mathbf{R}(t)$*, mapping the points of the time axis*[2] *into points of* SO(3)*. We have seen in chapter 1 that the autovector of* SO(3) *connecting the origin* $\mathbf{I}$ *to the point* $\mathbf{R}$ *is the rotation vector (antisymmetric matrix)* $\mathbf{r} = \log \mathbf{R}$*. Then, as shown below,*

- *the* derivative *of the mapping* $t \mapsto \mathbf{R}(t)$ *is* $\dot{\mathbf{R}}$*;*
- *the* differential *of the mapping* $t \mapsto \mathbf{R}(t)$ *is* $\dot{\mathbf{r}}$*;*
- *the* declinative *of the mapping* $t \mapsto \mathbf{R}(t)$ *is the rotation velocity* $\boldsymbol{\omega} = \dot{\mathbf{R}}\,\mathbf{R}^t$*.*

To evaluate the derivative tensor, equation (2.30) has to be written in the limit $\mathcal{P} \to \mathcal{P}_0$, i.e., in the limit of vanishingly small autovectors $\mathbf{a}(\mathcal{P};\mathcal{P}_0)$ and $\mathbf{v}(\,\mathcal{Q}(\mathcal{P})\,;\,\mathcal{Q}(\mathcal{P}_0)\,)$. As any infinitesimal segment can be considered to be autoparallel, equation (2.30) is, in fact, independent of the particular connection that one may consider on $\mathfrak{M}$ or on $\mathfrak{N}$. Therefore one has

Property 2.4 *The derivative of a mapping is independent of the connection of the manifolds.*

In fact, it is defined whether the manifolds have a connection or not. This is not true for the differential and for the declinative, which are defined only for connection manifolds, and depend on the connections in an essential way.

The derivative is expressed by well known formulas. Choosing a coordinate system $\{x^\alpha\}$ over $\mathfrak{M}$ and a coordinate system $\{y^i\}$ over $\mathfrak{N}$, we can write a mapping $\mathcal{P} \mapsto \mathcal{Q} = \mathcal{Q}(\mathcal{P})$ as $\{x^1, x^2 \dots\} \mapsto y^i = y^i(x^1, x^2, \dots)$, or, for short,

$$x^\alpha \;\mapsto\; y^i = y^i(x^\alpha) \quad . \tag{2.31}$$

By definition of partial derivatives, we can write

$$dy^i \;=\; \frac{\partial y^i}{\partial x^\alpha}\, dx^\alpha \quad , \tag{2.32}$$

the partial derivatives being taken taken at point $\mathcal{P}_0$. Denoting by $\mathbf{x}_0$ the coordinates of the point $\mathcal{P}_0$ we can write the components of the derivative tensor as

[2] By "time axis", we understand here that we have a one-dimensional metric manifold, and that t is a metric coordinate along it.

$$(\mathbf{D}_0)_\alpha{}^i = \frac{\partial y^i}{\partial x^\alpha}(\mathbf{x}_0) \quad . \tag{2.33}$$

Should the two manifolds $\mathfrak{M}$ and $\mathfrak{N}$ be, in fact, metric manifolds, then, denoting $g_{\alpha\beta}$ the metric over $\mathfrak{M}$, and γ_{ij} that over $\mathfrak{N}$, the Frobenius norm the derivative tensor is (see definition 2.2, page 83) $\| \mathbf{D}(\mathbf{x}_0) \| = (\, \gamma_{ij}(\mathbf{y}(\mathbf{x}_0))\, g^{\alpha\beta}(\mathbf{x}_0)\, D^i{}_\alpha(\mathbf{x}_0)\, D^j{}_\beta(\mathbf{x}_0)\,)^{1/2}$, i.e., dropping the variable $\mathbf{x}_0$,

$$\| \mathbf{D} \| = \sqrt{\gamma_{ij}\, g^{\alpha\beta}\, D_\alpha{}^i\, D_\beta{}^i} \quad . \tag{2.34}$$

For general connection manifolds, there is no explicit expression for the differential or the declinative of a mapping, as the connections have to be explicitly used. It is only in Lie group manifolds, where the operation of parallel transport has an analytical expression, that an explicit formula for the declinative can be obtained, as shown in the next section.

2.3 Example: Mappings from Linear Spaces into Lie Groups

We will repeatedly encounter in this text mappings from a linear space into a Lie group manifold, as when, in elasticity theory, the strain —an oriented geodesic segment of $GL^+(3)$— depends on the stress (a bona fide tensor), or when the rotation of a body —an oriented geodesic segment of SO(3)— depends on time (a trivial one-dimensional linear space). Let us obtain, in this section, explicit expressions for an autoparallel mapping and for the declinative of a mapping valid in this context.

2.3.1 Autoparallel Mapping

Consider a mapping $\mathbf{a} \mapsto \mathbf{M}(\mathbf{a})$ from a linear space $\mathbf{A}$, with vectors $\mathbf{a}_1, \mathbf{a}_2 \ldots$, into a multiplicative matrix group $\mathfrak{G}$, with matrices $\mathbf{M}_1, \mathbf{M}_2 \ldots$. We have seen in chapter 1 that, the matrices can be identified with the points of the Lie group manifold, and that with the identity matrix $\mathbf{I}$ chosen as origin, the Lie group manifold defines an autovector space. In the linear space $\mathbf{A}$ we have the sum $\mathbf{a}_2 + \mathbf{a}_1$, while in the group $\mathfrak{G}$, the autovector from its origin to the point $\mathbf{M}$ is $\mathbf{m} = \log \mathbf{M}$, and we have the o-sum $\mathbf{m}_2 \oplus \mathbf{m}_1 = \log(\exp \mathbf{m}_2\ \exp \mathbf{m}_1)$.

The relation (2.19) defining an autoparallel mapping becomes, here,

$$\mathbf{m}(\mathbf{a}) \ominus \mathbf{m}(\mathbf{a}_0) = \mathfrak{L}^t\,(\,\mathbf{a} - \mathbf{a}_0\,) \quad , \tag{2.35}$$

where $\mathfrak{L}^t$ is a linear operator (mapping $\mathbf{A}$ into the linear space tangent to the group $\mathfrak{G}$ at its origin). In terms of the multiplicative matrices, i.e., in terms of the points of the group, this equation can equivalently be written $\log(\,\mathbf{M}(\mathbf{a})\,\mathbf{M}(\mathbf{a}_0)^{-1}\,) = \mathfrak{L}^t\,(\,\mathbf{a} - \mathbf{a}_0\,)$, i.e.,

$$\boxed{\mathbf{M}(\mathbf{a}) \;=\; \exp(\, \mathfrak{L}^t\, (\,\mathbf{a} - \mathbf{a}_0\,)\,)\, \mathbf{M}(\mathbf{a}_0) \quad .} \tag{2.36}$$

Each of the two equations (2.35) and (2.36) corresponds to the expression of a mapping from a linear space into a multiplicative matrix group that is autoparallel at $\mathbf{a}_0$. Choosing a basis $\{\mathbf{e}_i\}$ in $\mathbf{A}$ and the natural basis in the Lie group associated with the exponential coordinates, these two equations become, in terms of components,

$$(\,\mathbf{m}(\mathbf{a}) \ominus \mathbf{m}(\mathbf{a}_0)\,)^\alpha{}_\beta \;=\; \mathfrak{L}_i{}^\alpha{}_\beta\, (a^i - a^i_0) \quad , \tag{2.37}$$

and denoting $\exp(m^\alpha{}_\beta) \equiv (\exp \mathbf{m})^\alpha{}_\beta$,

$$M(\mathbf{a})^\alpha{}_\beta \;=\; \exp(\, \mathfrak{L}_i{}^\alpha{}_\sigma\, (a^i - a^i_0)\,)\, M(\mathbf{a}_0)^\sigma{}_\beta \quad . \tag{2.38}$$

Example 2.2 Elastic deformation (I). *The configuration (i.e., the "shape") of a homogeneous elastic body undergoing homogeneous elastic deformation can be represented (see chapter 4 for details) by a point in the submanifold of the Lie group manifold* $GL^+(3)$ *that is geodesically connected to the origin, i.e., by an invertible* 3×3 *matrix* $\mathbf{C}$ *with positive determinant and real logarithm. The reference configuration (origin of the Lie group manifold) is* $\mathbf{C} = \mathbf{I}$. *When passing from the reference configuration* $\mathbf{I}$ *to a configuration* $\mathbf{C}$, *the body experiences the* strain

$$\varepsilon \;=\; \log \mathbf{C} \quad . \tag{2.39}$$

The strain, being an oriented geodesic segment on $GL^+(3)$ *is a geotensor, in the sense of section 1.5. A medium is elastic (although, perhaps, not ideally elastic) when the configuration* $\mathbf{C}$ *depends only on the* stress σ *to which the body is submitted:* $\mathbf{C} = \mathbf{C}(\sigma)$. *The stress being a bona fide tensor, i.e., an element of a linear space (see chapter 4), the mapping* $\sigma \mapsto \mathbf{C} = \mathbf{C}(\sigma)$ *maps a linear space into a Lie group manifold. We shall say that an elastic medium is* ideally elastic *at* σ_0 *if the relation* $\mathbf{C}(\sigma)$ *is autoparallel at* σ_0, *i.e., if the relations 2.35 and 2.36 hold. This implies the existence of a tensor* $\mathbf{d}$ *(the compliance tensor) such that one has (equation 2.36)*

$$\mathbf{C}(\sigma) \;=\; \exp(\, \mathbf{d}\, (\,\sigma - \sigma_0\,)\,)\, \mathbf{C}(\sigma_0) \quad . \tag{2.40}$$

The stress σ_0 *is the pre-stress. Equivalently (equation 2.35),*

$$\varepsilon(\sigma) \ominus \varepsilon(\sigma_0) \;=\; \mathbf{d}\, (\sigma - \sigma_0) \quad . \tag{2.41}$$

Selecting a basis $\{\mathbf{e}_i\}$ *in the physical 3D space, and in* $GL^+(3)$ *the natural basis associated with the exponential coordinates in the group that are adapted*[3] *to the basis* $\{\mathbf{e}_i\}$, *the two equations (2.40) and (2.41) become, using the notation* $\exp(\varepsilon^i{}_j) \equiv (\exp \varepsilon)^i{}_j$,

[3]This is the usual practice, where the matrix $C^i{}_j$ and the tensor $\sigma^i{}_j$ have the same kind of indices.

$$C(\sigma)^i{}_j \;=\; \exp(\, d^i{}_{sk}{}^{\ell}\, (\sigma^k{}_\ell - {\sigma_0}^k{}_\ell)\,)\; C(\sigma_0)^s{}_j \tag{2.42}$$

and

$$(\, \varepsilon(\sigma) \ominus \varepsilon(\sigma_0)\,)^i{}_j \;=\; d^i{}_{jk}{}^{\ell}\, (\sigma^k{}_\ell - {\sigma_0}^k{}_\ell) \quad . \tag{2.43}$$

The simplest kind of ideally elastic media corresponds to the case where there is no prestress, $\boldsymbol{\sigma}_0 = \mathbf{0}$. *Then, taking as reference configuration the unstressed configuration* ($\mathbf{C}(\mathbf{0}) = \mathbf{I}$), *the autoparallel relation simplifies to*

$$\mathbf{C}(\boldsymbol{\sigma}) \;=\; \exp(\, \mathbf{d}\, \boldsymbol{\sigma}\,) \quad , \tag{2.44}$$

i.e.,

$$\boldsymbol{\varepsilon}(\boldsymbol{\sigma}) \;=\; \mathbf{d}\, \boldsymbol{\sigma} \quad . \tag{2.45}$$

Example 2.3 Solid rotation (I). *When a solid is freely rotating around a fixed point in 3D space, the rotation at some instant* t *may be represented by an orthogonal rotation matrix* $\mathbf{R}(t)$. *As a by product of the results presented in example 2.5, one obtains the expression of an autoparallel mapping,*

$$\mathbf{R}(t) \;=\; \exp(\, (t - t_0)\, \boldsymbol{\omega}\,)\, \mathbf{R}(t_0) \quad , \tag{2.46}$$

where $\boldsymbol{\omega}$ *is a fixed antisymmetric tensor. Physically, this corresponds to a solid rotating with constant rotation velocity.*

2.3.2 Declinative

Consider, as above, a mapping from a linear space A, with vectors $\mathbf{a}_0$, $\mathbf{a}, \dots$ into a multiplicative group $\mathfrak{G}$, with matrices $\mathbf{M}_0$, $\mathbf{M}, \dots$. Some basis is chosen in the linear space A, and the components of a vector $\mathbf{a}$ are denoted $\{a^i\}$. On the matrix group manifold, we choose the "entries" $\{M^\alpha{}_\beta\}$ of the matrix $\mathbf{M}$ as coordinates, as suggested in chapter 1. The geotensor associated to a point $\mathbf{M} \in \mathfrak{G}$ is $\mathbf{m} = \log \mathbf{M}$. Then, the considered mapping can equivalently[4] be represented as $\mathbf{a} \mapsto \mathbf{M}(\mathbf{a})$ or as $\mathbf{a} \mapsto \mathbf{m}(\mathbf{a})$. The declinative $\mathfrak{D}$ is defined (equation 2.23) through $\mathbf{m}(\mathbf{a}) \ominus \mathbf{m}(\mathbf{a}_0) = \mathfrak{D}^t_0\, (\mathbf{a} - \mathbf{a}_0) + \dots$, or, equivalently,

$$\log(\, \mathbf{M}(\mathbf{a})\, \mathbf{M}(\mathbf{a}_0)^{-1}\,) \;=\; \mathfrak{D}^t_0\, (\, \mathbf{a} - \mathbf{a}_0\,) + \dots \quad . \tag{2.47}$$

Using the notation abuse $\log A^a{}_b \equiv (\log \mathbf{A})^a{}_b$, we can successively write (denoting, as usual in this text, $\overline{\mathbf{M}} = \mathbf{M}^{-1}$)

$$\begin{aligned}
&\log[\, M(\mathbf{a})^\alpha{}_\sigma\, \overline{M}(\mathbf{a}_0)^\sigma{}_\beta\,] \\
&= \log[\, (M(\mathbf{a}_0)^\alpha{}_\sigma + (\partial M^\alpha{}_\sigma / \partial a^i)(\mathbf{a}_0)\, (a^i - a^i_0) + \dots)\, \overline{M}(\mathbf{a}_0)^\sigma{}_\beta\,] \\
&= \log[\, \delta^\alpha{}_\beta + (\partial M^\alpha{}_\sigma / \partial a^i)(\mathbf{a}_0)\, \overline{M}(\mathbf{a}_0)^\sigma{}_\beta\, (a^i - a^i_0) + \dots] \\
&= (\partial M^\alpha{}_\sigma / \partial a^i)(\mathbf{a}_0)\, \overline{M}(\mathbf{a}_0)^\sigma{}_\beta\, (a^i - a^i_0) + \dots
\end{aligned} \tag{2.48}$$

[4]This is equivalent because $\mathbf{m}$ belongs to the logarithmic image of the group $\mathfrak{G}$.

so we have

$$\log(M(\mathbf{a})^{\alpha}{}_{\sigma} \, \overline{M}(\mathbf{a}_0)^{\sigma}{}_{\beta}) = \frac{\partial M^{\alpha}{}_{\sigma}}{\partial a^i}(\mathbf{a}_0) \, \overline{M}(\mathbf{a}_0)^{\sigma}{}_{\beta} \, (a^i - a^i_0) + \dots \quad . \tag{2.49}$$

This is exactly equation (2.47), with

$$(\mathfrak{D}_0)_i{}^{\alpha}{}_{\beta} = (\mathbf{D}_0)_i{}^{\alpha}{}_{\sigma} \, (\mathbf{M}(\mathbf{a}_0)^{-1})^{\sigma}{}_{\beta} \quad , \tag{2.50}$$

where the $(\mathbf{D}_0)_i{}^{\alpha}{}_{\sigma}$, components of the derivative tensor (see equation 2.33), are the partial derivatives

$$(\mathbf{D}_0)_i{}^{\alpha}{}_{\sigma} = \frac{\partial M^{\alpha}{}_{\sigma}}{\partial a^i}(\mathbf{a}_0) \quad . \tag{2.51}$$

With an obvious meaning, equation (2.50) can be written

$$\mathfrak{D}_0 = \mathbf{D}_0 \, \mathbf{M}_0^{-1} \quad , \tag{2.52}$$

or, dropping the index zero, $\mathfrak{D} = \mathbf{D}\,\mathbf{M}^{-1}$. We have thus arrived at the following

Property 2.5 *The declinative of a mapping* $\mathbf{a} \mapsto \mathbf{M}(\mathbf{a})$ *mapping a linear space into a multiplicative group of matrices is*

$$\boxed{\mathfrak{D} = \mathbf{D}\,\mathbf{M}^{-1} \quad ,} \tag{2.53}$$

where $\mathbf{D}$ *is the (ordinary) derivative.*

It is demonstrated in the appendix (see equation (A.194), page 190), that parallel transport of a vector from a point $\mathbf{M}$ to the origin $\mathbf{I}$ is done by right-multiplication by $\mathbf{M}^{-1}$. We can, therefore, interpret equation (2.53) as providing the declinative by transportation of the derivative from point $\mathbf{M}$ to point $\mathbf{I}$ of the Lie group manifold. We only need to "transport the Greek indices": the Latin index corresponds to a linear space, and transportation is implicit.

Example 2.4 Elastic deformation (II). *Let us continue here developing example 2.2, where the elastic deformation of a solid is represented by a mapping* $\sigma \mapsto \mathbf{C}(\sigma)$ *from the (linear) stress space into the configuration space, the Lie group* $GL^+(3)$. *The declinative of the mapping* $\sigma \mapsto \mathbf{C}(\sigma)$ *is expressed by equation (2.53), so here we only need to care about the use of the indices, as the stress space has now two indices:* $\sigma = \{\sigma^i{}_j\}$. *Also, this situation is special, as the manifold* $GL^+(3)$ *is tangent to the physical 3D space (as explained in section 1.5), so the configuration matrices "have the same indices" as the stress:* $\mathbf{C} = \{C^i{}_j\}$. *The derivative at* σ_0 *of the mapping has components*

$$D_{0i}{}^{jk}{}_{\ell} = \frac{\partial C^k{}_{\ell}}{\partial \sigma^i{}_j}(\sigma_0) \quad , \tag{2.54}$$

and the components of the declinative $\mathfrak{D}_0$ *at* σ_0 *are (equation 2.53)*

$$\mathfrak{D}_{0i}{}^{jk}{}_{\ell} = D_{0i}{}^{jk}{}_{s}\, \overline{C}(\sigma_0)^{s}{}_{\ell} \quad . \tag{2.55}$$

By definition of declinative we have (equation 2.47)

$$\log(\, \mathbf{C}(\sigma)\, \mathbf{C}(\sigma_0)^{-1}\,) = \mathfrak{D}_0^{t}\, (\, \sigma - \sigma_0\,) + \dots \quad , \tag{2.56}$$

while expression (2.40)*, defining an ideally elastic medium, can be written,*

$$\log(\, \mathbf{C}(\sigma)\, \mathbf{C}(\sigma_0)^{-1}\,) = \mathbf{d}\, (\, \sigma - \sigma_0\,) \quad . \tag{2.57}$$

This shows that the declinative at σ_0 *of the mapping* $\sigma \mapsto \mathbf{C}(\sigma)$ *(for an arbitrary, nonideal, elastic medium) has to be interpreted as the compliance tensor at* σ_0*: for small stress changes around* σ_0*, the medium will behave as an ideally elastic medium with the compliance tensor* $\mathbf{d} = \mathfrak{D}_0$*.*

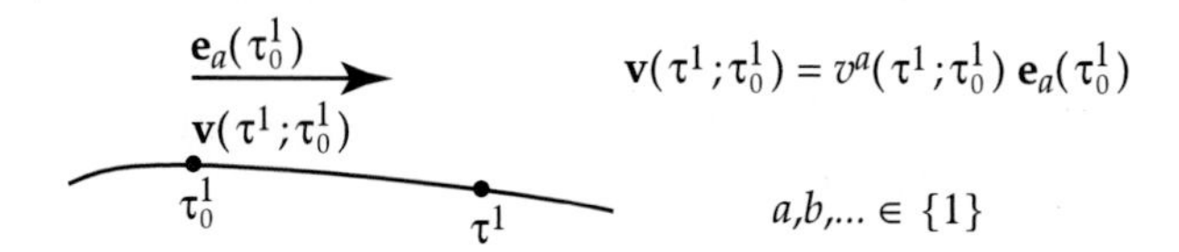

Fig. 2.4. The general definition of natural basis at a point of a manifold naturally applies to one-dimensional manifolds. Here a one-dimensional metric manifold is considered, endowed with a coordinate $\{\tau^a\} = \{\tau^1\}$. Here, the indices $\{a, b, \dots\}$ can only take the value 1. The length element at a point τ^1 is written as usual, $ds^2 = G_{ab}\, d\tau^a\, d\tau^b$. Let $\mathbf{v}(\tau^1;\tau_0^1)$ be the vector at point τ_0^1 associated with the segment $(\tau^1;\tau_0^1)$. Its norm, $\|\, \mathbf{v}(\tau^1;\tau_0^1)\, \|$, must equal the length of the interval, $\int_{\tau_0^1}^{\tau^1} d\ell\, \sqrt{G_{11}(\ell)}$. The natural basis at point τ_0^1 has the unique vector $\mathbf{e}_1(\tau_0^1)$, with norm $\|\, \mathbf{e}_1(\tau_0^1)\, \| = G_{11}(\tau_0^1)^{1/2}$. Writing $\mathbf{v}(\tau^1;\tau_0^1) = v^a(\tau^1;\tau_0^1)\, \mathbf{e}_a(\tau_0^1)$ defines the (unique) component of the vector $\mathbf{v}(\tau^1;\tau_0^1)$. The value of this component is $v^1(\tau^1;\tau_0^1) = (1/G_{11}(\tau_0^1)^{1/2}) \int_{\tau_0^1}^{\tau^1} d\ell\, \sqrt{G_{11}(\ell)}$. Should τ^1 be a Cartesian coordinate x (i.e., should one have $ds = dx$), then $v^1(x;x_0) = x - x_0$. Should τ^1 be a Jeffreys coordinate X (i.e., should one have $ds = dX/X$), then $v^1(X;X_0) = (1/X_0)\log(X/X_0)$.

Example 2.5 Solid rotation (II) (rotation velocity). *The rotation of a solid has already been mentioned in example 2.1. Let us here develop the theory (of the associated mapping). Consider a solid whose center of gravity is at a fixed point of a Euclidean 3D space, free to rotate, as time flows, around this fixed point. To every point* $\mathcal{T}$ *in the (one-dimensional) time space* $\mathfrak{T}$*, corresponds one point* $\mathcal{A}$ *in the space of attitudes* $\mathfrak{A}$*, and we can write*

$$\mathcal{T} \mapsto \mathcal{A} = \mathcal{A}(\mathcal{T}) \quad . \tag{2.58}$$

We wish to find an expression for the mapping $\mathcal{A}(\cdot)$ *that is autoparallel at some point* $\mathcal{T}_0$. *To characterize an instant (i.e., a point in time space* $\mathfrak{T}$ *) let us choose an arbitrary coordinate* $\{\tau^a\} = \{\tau^1\}$ *(the indices* $\{a, b, \dots\}$ *can only take the value* $\{1\}$*; see section 3.2.1 for details). The easiest way to characterize an attitude is to select one particular attitude* $\mathcal{A}_{\text{ref}}$, *once for all, and to represent any other attitude* $\mathcal{A}$ *by the rotation* $\mathcal{R}$ *transforming* $\mathcal{A}_{\text{ref}}$ *into* $\mathcal{A}$. *The mapping* (2.58) *can now be written as* $\tau^1 \mapsto \mathcal{R} = \mathcal{R}(\tau^1)$, *or if we characterize a rotation* $\mathcal{R}$ *(in the abstract sense) by the usual orthogonal tensor* $\mathbf{R}$,

$$\tau^1 \quad \mapsto \quad \mathbf{R} = \mathbf{R}(\tau^1) \quad . \tag{2.59}$$

To evaluate the declinative of this mapping we can either make a direct evaluation, or use the result in equation (2.53). *We shall take both ways, but let me first remember that in this text we are using explicit tensor notation even for one-dimensional manifolds. See figure 2.4 for the explicit introduction of the (unique) component of a vector belonging to (the linear space tangent to) a one-dimensional manifold. We start by expressing the derivative of the mapping (equation 2.33)*

$$D_a{}^i{}_j = dR^i{}_j \, / \, d\tau^a \quad . \tag{2.60}$$

Then the declinative is (equation 2.53) $\mathfrak{D}_a{}^i{}_j = D_a{}^i{}_s \overline{R}^s{}_j$, *but, as rotation matrices are orthogonal,*[5] *this is*

$$\mathfrak{D}_a{}^i{}_j = D_a{}^i{}_s \, R_j{}^s \quad , \tag{2.61}$$

or, in compact form,

$$\boldsymbol{\omega} \equiv \boldsymbol{\mathfrak{D}} = \mathbf{D}\,\mathbf{R}^t \quad , \tag{2.62}$$

where $\boldsymbol{\omega}$ *denotes the declinative, as it is going to be identified, in a moment, with the (instantaneous) rotation velocity. The tensor character of the index* a *in* $\omega_a{}^i{}_j$ *appears when changing in the time manifold the coordinate* τ^a *to another arbitrary coordinate* $\tau^{a'}$, *as the component* $\omega_a{}^i{}_j$ *would become*

$$\omega_{a'}{}^i{}_j = \frac{d\tau^a}{d\tau^{a'}} \, \omega_a{}^i{}_j \quad . \tag{2.63}$$

The norm of $\boldsymbol{\omega}$ *is, of course, an invariant, that we can express as follows. In the time axis there is a notion of distance between points, that corresponds to Newtonian time* t. *In the arbitrary coordinate* τ^1 *we shall have the relation* $dt^2 = G_{ab} \, d\tau^a \, d\tau^b$, *this introducing the one-dimensional metric* G_{ab} *(see section 3.2.1 for details). Denoting by* g_{ij} *the components of the metric of the physical space, in whatever coordinates we may use, the norm of* $\boldsymbol{\omega}$ *is*

$$\| \boldsymbol{\omega} \| = \sqrt{g_{ik} \, g^{j\ell} \, G^{ab} \, \omega_a{}^i{}_j \, \omega_b{}^k{}_\ell} \quad , \tag{2.64}$$

i.e., using a nonmanifestly covariant notation, $\| \boldsymbol{\omega} \| = (g_{ik} \, g^{j\ell} \, \omega_1{}^i{}_j \, \omega_1{}^k{}_\ell)^{1/2} / \sqrt{G_{11}}$. *The direct way of computing the declinative would start with equation* (2.23). *Introducing the geotensor* $\mathbf{r}(\tau^1) = \log \mathbf{R}(\tau^1)$ *this equation gives, here,*

---

[5]The condition $\mathbf{R}^{-1} = \mathbf{R}^t$ gives $\overline{R}^i{}_j = R_j{}^i$.

$$[\,\omega(\tau^1)\,]_1{}^i{}_j \;=\; \lim_{\tau'^1\to\tau^1} \frac{[\,\mathbf{r}(\tau'^1)\ominus\mathbf{r}(\tau^1)\,]^i{}_j}{\tau'^1-\tau^1} \;=\; \lim_{\tau'^1\to\tau^1} \frac{[\,\log(\,\mathbf{R}(\tau'^1)\,\mathbf{R}(\tau^1)^{-1}\,)\,]^i{}_j}{\tau'^1-\tau^1} \quad . \tag{2.65}$$

As we can successively write

$$\begin{aligned}\log[\,\mathbf{R}(\xi')\,\mathbf{R}(\xi)^{-1}\,] &= \log[\,(\,\mathbf{R}(\xi)+(d\mathbf{R}/d\xi)(\xi)\,(\xi'-\xi)+\dots)\,\mathbf{R}(\xi)^{-1}\,] \\ &= \log[\,\mathbf{I}+(d\mathbf{R}/d\xi)(\xi)\,\mathbf{R}(\xi)^{-1}\,(\xi'-\xi)+\dots)\,] \\ &= (d\mathbf{R}/d\xi)(\xi)\,\mathbf{R}(\xi)^{-1}\,(\xi'-\xi)+\dots\end{aligned} \tag{2.66}$$

we immediately obtain $\boldsymbol{\omega}_1 = (d\mathbf{R}/d\tau^1)\,\mathbf{R}^{-1}$ *as we should. There is only one situation where we can safely drop the index representing the variable in use on the one-dimensional manifold: when a metric coordinate is identified, an orientation is given to it, and it is agreed, once and for all, that only this oriented metric coordinate will be used. This is the case here, if we agree to always use Newtonian time* t, *oriented from past to future. We can then write* $D^i{}_j$ *instead of* $D_a{}^i{}_j$ *and* $\omega^i{}_j$ *instead of* $\omega_a{}^i{}_j$. *In addition, as traditionally done, we can use a dot to represent a (Newtonian) time derivative. Then, equation* (2.60) *becomes*

$$\mathbf{D} \;=\; \dot{\mathbf{R}} \quad , \tag{2.67}$$

and the declinative (equation 2.62) becomes

$$\boxed{\;\omega \;\equiv\; \mathfrak{D} \;=\; \dot{\mathbf{R}}\,\mathbf{R}^t \quad ,\;} \tag{2.68}$$

while equation (2.65) *can be written*

$$\omega \;=\; \lim_{\Delta t\to 0} \frac{\mathbf{r}(t+\Delta t)\ominus\mathbf{r}(t)}{\Delta t} \;=\; \lim_{\Delta t\to 0} \frac{\log(\,\mathbf{R}(t+\Delta t)\,\mathbf{R}(t)^{-1}\,)}{\Delta t} \quad . \tag{2.69}$$

This is the instantaneous rotation velocity.[6] *Note that the necessary antisymmetry of* $\boldsymbol{\omega}$ *comes from the condition of orthogonality satisfied by* $\mathbf{R}$.[7] *With the derivative and the declinative evaluated, we can now turn to the evaluation of the differential (that we can do directly using Newtonian time). The general expression* (2.24) *here gives*

$$\mathbf{d} \;=\; \dot{\mathbf{r}}(t) \;=\; \lim_{t'\to t} \frac{\mathbf{r}(t')-\mathbf{r}(t)}{t'-t} \quad . \tag{2.70}$$

We thus see that the differential of the mapping is the time derivative or the rotation "vector". The relation between this differential and the declinative can be found by using the particular expression for the operation $\ominus$ *in the group* SO(3) *(equation* (A.275), *page* 211*) and taking the limit. This gives*

[6]The demonstration that the instantaneous rotation velocity of a solid is, indeed, $\omega = \dot{\mathbf{R}}\,\mathbf{R}^t$ requires an intricate development. See Goldstein (1983) for the basic reference, or Baraff (2001) for a more recent demonstration (on-line).

[7]Taking the time derivative of the condition $\mathbf{R}\,\mathbf{R}^t = \mathbf{I}$ gives $\dot{\mathbf{R}}\,\mathbf{R}^t = -\mathbf{R}\,\dot{\mathbf{R}}^t = -(\dot{\mathbf{R}}\,\mathbf{R}^t)^t$.

$$\omega = \frac{\sin r}{r}\,\dot{\mathbf{r}} + \left(1 - \frac{\sin r}{r}\right)\frac{\dot{\mathbf{r}}\cdot\mathbf{r}}{r^2}\,\mathbf{r} - \frac{1-\cos r}{r^2}\,\dot{\mathbf{r}}\times\mathbf{r} \quad , \tag{2.71}$$

where r is the rotation angle (norm of the rotation "vector" $\mathbf{r}$). Figure 2.5 gives a pictorial representation of the relations between $\boldsymbol{\omega}$, $\dot{\mathbf{R}}$ and $\dot{\mathbf{r}}$.

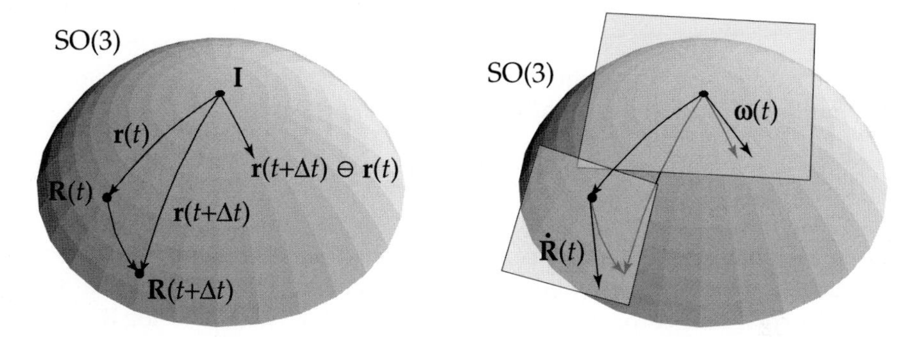

Fig. 2.5. *Relation between the rotation velocity $\omega(t)$ (the declinative) and $\dot{\mathbf{R}}(t)$. While the derivative $\dot{\mathbf{R}}(t)$ belongs to the linear tangent space at point $\mathbf{R}(t)$, the declinative $\omega(t)$ belongs to the linear tangent space at the origin of the Lie group SO(3) (the derivative $\dot{\mathbf{r}}(t)$ also belongs to this tangent space at the origin, but is different from $\omega(t)$).*

Example 2.6 *We shall see in chapter 4 that the configuration at (Newtonian) time t of an n-dimensional deforming body is represented by a matrix $\mathbf{C}(t) \in GL^+(n)$, the* strain *being*

$$\varepsilon(t) = \log \mathbf{C}(t) \quad . \tag{2.72}$$

The strain rate *is to be defined as the declinative of the mapping $t \mapsto \mathbf{C}(t)$:*

$$\nu(t) = \dot{\mathbf{C}}(t)\,\mathbf{C}^{-1}(t) \quad , \tag{2.73}$$

and this is different[8] *from $\dot{\varepsilon}(t)$. For instance, in an isochoric transformation of a 2D medium we have (equivalent to equation 2.71)*

$$\nu = \frac{\sinh 2\varepsilon}{2\varepsilon}\,\dot{\varepsilon} + \left(1 - \frac{\sinh 2\varepsilon}{2\varepsilon}\right)\frac{\mathrm{tr}\,(\dot{\varepsilon}\,\varepsilon)}{2\,\varepsilon^2}\,\varepsilon + \left(\frac{1-\cosh 2\varepsilon}{4\,\varepsilon^2}\right)(\dot{\varepsilon}\,\varepsilon - \varepsilon\,\dot{\varepsilon}) \quad , \tag{2.74}$$

where $\varepsilon = \sqrt{(\mathrm{tr}\varepsilon^2)/2}$.

[8]Excepted when the transformation is geodesic and passes through the origin of $GL^+(n)$.

2.3.3 Logarithmic Derivative?

It is perhaps the right place here, after example 2.5, to make a comment. The *logarithmic derivative* of a scalar function $f(t)$ (that takes positive values) has two common definitions,

$$\frac{1}{f}\frac{df}{dt} \quad ; \quad \frac{d\log f}{dt} \quad , \tag{2.75}$$

that are readily seen to be equivalent. For a matrix $\mathbf{M}(t)$ that is an element of a multiplicative group of matrices, the two expressions

$$\frac{d\mathbf{M}}{dt}\,\mathbf{M}^{-1} \quad ; \quad \frac{d\log\mathbf{M}}{dt} \quad , \tag{2.76}$$

are *not* equivalent.[9] For instance, in the context of the previous example, the first expression corresponds to the declinative, $\omega = \dot{\mathbf{R}}\,\mathbf{R}^{-1}$, while the second expression corresponds to the differential $\dot{\mathbf{r}}$, with $\mathbf{r} = \log\mathbf{R}$, and we have seen that $\dot{\mathbf{r}}$ is related to ω in a complex way (equation 2.71). To avoid confusion, we should not use the term 'logarithmic derivative': in one side we have the *declinative,* $\omega = \dot{\mathbf{R}}\,\mathbf{R}^{-1}$, and in the other side we have the *differential,* $\dot{\mathbf{r}}$.

2.4 Example: Mappings Between Lie Groups

This section is similar to section 2.3, but instead of considering mappings that map a linear space into a Lie group, we consider mappings that map a Lie group into another Lie group. The developments necessary here are similar to those in section 2.3, so I give the results only, leaving to the reader, as an exercise, the derivations.

2.4.1 Autoparallel Mapping

We consider here a mapping $\mathbf{A} \mapsto \mathbf{M}(\mathbf{A})$ mapping a multiplicative matrix group $\mathfrak{G}_1$, with matrices $\mathbf{A}_1$, $\mathbf{A}_2 \ldots$, into another multiplicative matrix group $\mathfrak{G}_2$, with matrices $\mathbf{M}_1$, $\mathbf{M}_2 \ldots$. We know that the matrices of a multiplicative group can be identified with the points of the Lie group manifold, and that with the identity matrix $\mathbf{I}$ chosen as origin, the Lie group manifold defines an autovector space. In the group $\mathfrak{G}_1$, the autovector going from its origin to the point $\mathbf{A}$ is $\mathbf{a} = \log\mathbf{A}$, and in the group $\mathfrak{G}_2$, the autovector from its origin to the point $\mathbf{M}$ is $\mathbf{m} = \log\mathbf{M}$. The o-sum

[9] In fact, it can be shown (J.M. Pozo, pers. commun.) that one has $(d\mathbf{M}/dt)\,\mathbf{M}^{-1} = \int_0^1 dt\,\mathbf{M}^t\,(d\log\mathbf{M}/dt)\,\mathbf{M}^{-t}$.

in each space is respectively given by $\mathbf{a}_2 \boxplus \mathbf{a}_1 = \log(\exp \mathbf{a}_2 \ \exp \mathbf{a}_1)$ and $\mathbf{m}_2 \oplus \mathbf{m}_1 = \log(\exp \mathbf{m}_2 \ \exp \mathbf{m}_1)$.

The expression of an autoparallel mapping in terms of geotensors is (equivalent to equation 2.35),

$$\mathbf{m}(\mathbf{a}) \ominus \mathbf{m}(\mathbf{a}_0) = \mathfrak{L}^t \, (\mathbf{a} \boxminus \mathbf{a}_0) \quad , \tag{2.77}$$

while in terms of the points in each of the two groups is (equivalent to equation 2.36)

$$\boxed{\mathbf{M}(\mathbf{A}) = \exp(\, \mathfrak{L}^t \, \log(\mathbf{A}\,\mathbf{A}_0^{-1})\,)\, \mathbf{M}(\mathbf{A}_0) \quad .} \tag{2.78}$$

In terms of components, the equivalent of equation (2.37) is

$$(\,\mathbf{m}(\mathbf{a}) \ominus \mathbf{m}(\mathbf{a}_0)\,)^\alpha{}_\beta = \mathfrak{L}_i{}^{j\alpha}{}_\beta \, (\,\mathbf{a} \ominus \mathbf{a}_0\,)^i{}_j \quad , \tag{2.79}$$

while the equivalent of equation (2.38) is (using the notation $\exp m^\alpha{}_\beta \equiv (\exp \mathbf{m})^\alpha{}_\beta$ and $\log A^i{}_j \equiv (\log \mathbf{A})^i{}_j$)

$$M(\mathbf{A})^\alpha{}_\beta = \exp[\, \mathfrak{L}_i{}^{j\alpha}{}_\sigma \, (\log[\, A^i{}_s \, \overline{A}_0{}^s{}_j \,])\,]\, M(\mathbf{A}_0)^\sigma{}_\beta \quad . \tag{2.80}$$

2.4.2 Declinative

The derivative at $\mathbf{A}_0$ of the mapping $\mathbf{A} \mapsto \mathbf{M}(\mathbf{A})$ is (equivalent to equation 2.51)

$$(\mathbf{D}_0)_i{}^{j\alpha}{}_\sigma = \frac{\partial M^\alpha{}_\sigma}{\partial A^i{}_j}(\mathbf{A}_0) \quad , \tag{2.81}$$

while the declinative of the mapping is[10] (equivalent to equation 2.53)

$$\boxed{(\mathfrak{D}_0)_i{}^{j\alpha}{}_\beta = (\mathbf{A}_0)^j{}_s \, (\mathbf{D}_0)_i{}^{s\alpha}{}_\sigma \, \overline{\mathbf{M}}(\mathbf{A}_0)^\sigma{}_\beta \quad .} \tag{2.82}$$

We could have arrived at this result by a different route, using explicitly the parallel transport in the two Lie group manifolds. The derivative $\mathbf{D}_0$ is the characteristic tensor of the linear tangent mapping at point $\mathbf{A}_0$. To pass from derivative to declinative, we must transport $\mathbf{D}_0$ from point $\mathbf{A}_0$ to the origin $\mathbf{I}$ in the manifold $\mathfrak{G}_1$, and from point $\mathbf{M}(\mathbf{A}_0)$ to the origin $\mathbf{I}$ in the manifold $\mathfrak{G}_2$. We have, thus, to transport, from one side, "the indices" ${}_i{}^j$, and, for the other side, "the indices" ${}^\alpha{}_\beta$. The indices ${}_i{}^j$ are those of a form, and to transport a form from point $\mathbf{A}_0$ to point $\mathbf{I}$ we use the formula (A.205), i.e., we multiply by the the matrix $\mathbf{A}_0$. The indices ${}^\alpha{}_\beta$ are those of a vector, and to transport a vector from point $\mathbf{M}(\mathbf{A}_0)$ to point $\mathbf{I}$ we use the formula (A.194), i.e., we multiply by the *inverse* of the matrix $\mathbf{M}(\mathbf{A}_0)$. When caring with the indices, this exactly gives equation 2.82.

---

[10] An expansion similar to that in equation (2.48) first gives $\log(\, M(\mathbf{A})^\alpha{}_\sigma \, \overline{M}(\mathbf{A}_0)^\sigma{}_\beta \,) = (\partial M^\alpha{}_\sigma \,/\, \partial A^i{}_j)(\mathbf{A}_0)\, \overline{M}(\mathbf{A}_0)^\sigma{}_\beta \, (A^i{}_j - A_0{}^i{}_j) + \dots$. Inserting here the expansion $\mathbf{A} - \mathbf{A}_0 = \log(\,\mathbf{A}\,\mathbf{A}_0^{-1}\,)\ \mathbf{A}_0 + \dots$ (that is found by developing the expression $\log(\,\mathbf{A}\,\mathbf{A}_0^{-1}\,) = \log(\,(\,\mathbf{A}_0 + (\mathbf{A} - \mathbf{A}_0) + \dots)\,\mathbf{A}_0^{-1}\,)$), directly produces the result.

2.5 Covariant Declinative

Tensor *fields* are quite basic objects in physics. One has a tensor field on a manifold when there is a tensor (or a vector) defined at every point of the manifold. When one has a vector at a point $\mathcal{P}$ of a manifold $\mathfrak{M}$, the vector belongs to $\mathbf{T}(\mathfrak{M},\mathcal{P})$, the linear space tangent to $\mathfrak{M}$ at $\mathcal{P}$. When one has a more general tensor, it belongs to one of the tensor spaces that can be built at the given point of the manifold by tensor products of $\mathbf{T}(\mathfrak{M},\mathcal{P})$ and its dual, $\mathbf{T}(\mathfrak{M},\mathcal{P})^*$. For instance, in a manifold $\mathfrak{M}$ with some coordinates $\{x^\alpha\}$ and the associated natural basis $\{\mathbf{e}_\alpha\}$ at each point, a tensor $t^\alpha{}_{\beta\gamma}$ at point $\mathcal{P}$ belongs to $\mathbf{T}(\mathfrak{M},\mathcal{P})\otimes\mathbf{T}(\mathfrak{M},\mathcal{P})^*\otimes\mathbf{T}(\mathfrak{M},\mathcal{P})^*$. It is for these objects that the *covariant derivative* has been introduced, that depends on the connection of the manifold. The definition of covariant derivative is recalled below.

When following the ideas proposed in this text, in addition to tensor fields, one finds geotensor fields. In this case, at each point $\mathcal{P}$ of $\mathfrak{M}$ we have an oriented geodesic segment of the Lie group $GL(n)$, that is tangent at point $\mathcal{P}$ to $\mathbf{T}(\mathfrak{M},\mathcal{P})\otimes\mathbf{T}(\mathfrak{M},\mathcal{P})^*$, this space then being interpreted as the algebra of the group. Given two geotensors $\mathbf{t}_1$ and $\mathbf{t}_2$ at a point of a manifold, we can make sense of the two sums $\mathbf{t}_2\oplus\mathbf{t}_1$ and $\mathbf{t}_2+\mathbf{t}_1$. The geometric sum $\oplus$ does not depend on the connection of the manifold $\mathfrak{M}$, but on the connection of $GL(n)$. The tangent operation $+$ is the ordinary (commutative) sum of the linear tangent space. When using the commutative sum $+$ we in fact consider the autovectors to be elements of a linear tangent space, and the covariant *derivative* of an autovector field is then defined as that of a tensor field. But when using the o-sum $\oplus$ we find the covariant *declinative* of the geotensor field.

Given a tensor field or a geotensor field $\mathbf{x}\mapsto\boldsymbol{\tau}(\mathbf{x})$, the tensor (or geotensor) obtained at point $\mathbf{x}_0$ by parallel transport of $\boldsymbol{\tau}(\mathbf{x})$ (from point $\mathbf{x}$ to point $\mathbf{x}_0$) is here denoted $\boldsymbol{\tau}(\mathbf{x}_0\|\mathbf{x})$.

2.5.1 Vector or Tensor Field

A tensor field is a mapping that to every point $\mathcal{P}$ of a finite-dimensional smooth manifold $\mathfrak{M}$ (with a connection) associates an element of the linear space $\mathbf{T}(\mathfrak{M},\mathcal{P})\otimes\mathbf{T}(\mathfrak{M},\mathcal{P})\otimes\ldots\mathbf{T}(\mathfrak{M},\mathcal{P})^*\otimes\mathbf{T}(\mathfrak{M},\mathcal{P})^*\otimes\ldots$

In what follows, let us assume that some coordinates $\{x^\alpha\}$ have been chosen over the manifold $\mathfrak{M}$. From now on, a point $\mathcal{P}$ of the manifold may be designated as $\mathbf{x}=\{x^\alpha\}$. A tensor $\mathbf{t}$ at some point of the manifold will have components $t^{\alpha\beta\ldots}{}_{\gamma\delta\ldots}$ on the local natural basis.

Let $\mathbf{x}\mapsto\mathbf{t}(\mathbf{x})$ be a tensor field. Using the connection of $\mathfrak{M}$ when transporting the tensor $\mathbf{t}(\mathbf{x}_a)$, defined at point $\mathbf{x}_a$, to some other point $\mathbf{x}_b$ gives $\mathbf{t}(\mathbf{x}_b\|\mathbf{x}_a)$, a tensor at point $\mathbf{x}_b$ (that, in general, is different from $\mathbf{t}(\mathbf{x}_b)$, the value of the tensor field at point $\mathbf{x}_b$).

The *covariant derivative* of the tensor field at a point $\mathbf{x}$, denoted $\nabla\mathbf{t}(\mathbf{x})$ is the tensor with components $(\ \nabla\mathbf{t}(\mathbf{x})\ )_\mu{}^{\alpha\beta\ldots}{}_{\gamma\delta\ldots}$ defined by the development (written at point $\mathbf{x}$)

$$(\mathbf{t}(\mathbf{x}\|\mathbf{x}+\delta\mathbf{x}) - \mathbf{t}(\mathbf{x}))^{\alpha\beta\ldots}{}_{\gamma\delta\ldots} = (\nabla\mathbf{t}(\mathbf{x}))_{\mu}{}^{\alpha\beta\ldots}{}_{\gamma\delta\ldots}\, \delta x^{\mu} + \ldots \quad , \tag{2.83}$$

where the dots denote terms that are at least quadratic in δx^{μ}. It is customary to use a notational abuse, writing $\nabla_{\mu}\, t^{\alpha\beta\ldots}{}_{\gamma\delta\ldots}$ instead of $(\nabla\mathbf{t})_{\mu}{}^{\alpha\beta\ldots}{}_{\gamma\delta\ldots}$.

It is well known that the covariant derivative can be written in terms of the partial derivatives and the connection as[11]

$$\begin{aligned} \nabla_{\mu}\, t^{\alpha\beta\ldots}{}_{\gamma\delta\ldots} &= \partial_{\mu}\, t^{\alpha\beta\ldots}{}_{\gamma\delta\ldots} \\ &+ \Gamma^{\alpha}{}_{\mu\sigma}\, t^{\sigma\beta\ldots}{}_{\gamma\delta\ldots} + \Gamma^{\beta}{}_{\mu\sigma}\, t^{\alpha\sigma\ldots}{}_{\gamma\delta\ldots} + \cdots \\ &- \Gamma^{\sigma}{}_{\mu\gamma}\, t^{\alpha\beta\ldots}{}_{\sigma\delta\ldots} - \Gamma^{\sigma}{}_{\mu\delta}\, t^{\alpha\beta\ldots}{}_{\gamma\sigma\ldots} - \cdots \quad . \end{aligned} \tag{2.84}$$

2.5.2 Field of Transformations

Assume now that at a given point $\mathbf{x}$ of the manifold, instead of having an "ordinary tensor", one has a geotensor, in the sense of section 1.5, i.e., an object with a natural operation that is not the ordinary sum, but the o-sum

$$\mathbf{t}_2 \oplus \mathbf{t}_1 = \log(\exp \mathbf{t}_2 \, \exp \mathbf{t}_1) \quad . \tag{2.85}$$

The typical example is when at every point of a 3D manifold (representing the physical space) there is a 3D rotation defined, that may be represented by the rotation geotensor $\mathbf{r}$.

The definition of declinative of a geotensor field is immediately suggested by equation (2.83). The *covariant declinative* of the geotensor field at a point $\mathbf{x}$, denoted $\mathfrak{D}(\mathbf{x})$ is the tensor with components $\mathfrak{D}(\mathbf{x})_{\mu}{}^{\alpha\beta\ldots}{}_{\gamma\delta\ldots}$ defined by the development (written at point $\mathbf{x}$)

$$\boxed{(\mathbf{t}(\mathbf{x}\|\mathbf{x}+\delta\mathbf{x}) \ominus \mathbf{t}(\mathbf{x}))^{\alpha\beta\ldots}{}_{\gamma\delta\ldots} = \mathfrak{D}(\mathbf{x})_{\mu}{}^{\alpha\beta\ldots}{}_{\gamma\delta\ldots}\, \delta x^{\mu} + \ldots \quad ,} \tag{2.86}$$

where the dots denote terms that are at least quadratic in δx^{μ}.

It is easy to see (the simplest way of demonstrating this is by using equation (2.91) below) that one has

$$\boxed{\mathfrak{D} = (\nabla \exp \mathbf{t})\, (\exp \mathbf{t})^{-1} \quad .} \tag{2.87}$$

---

[11] We may just outline here the elementary approach leading to the expression of the covariant derivative of a vector field. One successively has (using the notation in appendix A.9.1) $\mathbf{v}(\mathbf{x}\|\mathbf{x}+\delta\mathbf{x}) - \mathbf{v}(\mathbf{x}) = v^i(\mathbf{x}+\delta\mathbf{x})\, \mathbf{e}_i(\mathbf{x}\|\mathbf{x}+\delta\mathbf{x}) - v^i(\mathbf{x})\, \mathbf{e}_i(\mathbf{x}) = (v^i(\mathbf{x}) + \delta x^j\, (\partial_j v^i)(\mathbf{x}) + \ldots)\, (\mathbf{e}_i(\mathbf{x}) + \delta x^j\, \Gamma^k{}_{ji}(\mathbf{x})\, \mathbf{e}_k(\mathbf{x}) + \ldots) - v^i(\mathbf{x})\, \mathbf{e}_i(\mathbf{x}) = \delta x^j\, (\, (\partial_j v^i)(\mathbf{x}) + \Gamma^i{}_{jk}(\mathbf{x})\, v^k(\mathbf{x})\,)\, \mathbf{e}_i(\mathbf{x}) + \ldots$, i.e., $\mathbf{v}(\mathbf{x}\|\mathbf{x}+\delta\mathbf{x}) - \mathbf{v}(\mathbf{x}) = \delta x^j\, (\nabla_j v^i)(\mathbf{x})\, \mathbf{e}_i(\mathbf{x}) + \ldots$, where (dropping the indication of the point $\mathbf{x}$) $\nabla_j v^i = \partial_j v^i + \Gamma^i{}_{jk}\, v^k$.

Instead of the geotensor $\mathbf{t}$ one may wish to use the transformation[12] $\mathbf{T} = \exp \mathbf{t}$. The declinative of an arbitrary field of transformations $\mathbf{T}(\mathbf{x})$ is defined through (equivalent to equation (2.86)), $\log(\mathbf{T}(\mathbf{x}\|\mathbf{x} + \delta\mathbf{x})\, \mathbf{T}(\mathbf{x})^{-1}) = \mathfrak{D}(\mathbf{x})\, \delta\mathbf{x} + \dots$, or, equivalently,

$$\mathbf{T}(\mathbf{x}\|\mathbf{x} + \delta\mathbf{x}) = \exp(\mathfrak{D}(\mathbf{x})\, \delta\mathbf{x} + \dots)\, \mathbf{T}(\mathbf{x}) \quad . \tag{2.88}$$

A series development leads[13] to the expression

$$\mathbf{T}(\mathbf{x}\|\mathbf{x} + \delta\mathbf{x}) = \exp(\delta x^k\, (\nabla\mathbf{T})_k\, \mathbf{T}^{-1} + \dots)\, \mathbf{T}(\mathbf{x}) \quad , \tag{2.89}$$

where $\nabla\mathbf{T}$ is the covariant derivative

$$(\nabla\mathbf{T})_k{}^i{}_j \equiv \nabla_k T^i{}_j = \partial_k T^i{}_j + \Gamma^i{}_{ks}\, T^s{}_j - \Gamma^s{}_{kj}\, T^i{}_s \quad . \tag{2.90}$$

Comparison of the two equations (2.88) and (2.89) gives the declinative of a field of transformations in terms of its covariant derivative: $\mathfrak{D} = (\nabla\mathbf{T})\, \mathbf{T}^{-1}$. We have thus arrived at the following property (to be compared with property 2.5):

Property 2.6 *The declinative of a field of transformations* $\mathbf{T}(\mathbf{x})$ *is*

$$\boxed{\mathfrak{D} = (\nabla\mathbf{T})\, \mathbf{T}^{-1} \quad .} \tag{2.91}$$

where $(\nabla\mathbf{T})$ *is the usual covariant derivative.*

Using components, this gives

$$\mathfrak{D}_k{}^i{}_j = (\nabla_k T^i{}_s)\, \overline{T}^s{}_j \quad . \tag{2.92}$$

Example 2.7 *In the physical 3D space, let* $\mathbf{x} \mapsto \mathbf{R}(t)$ *be a field of rotations represented by the usual orthogonal rotation matrices. As* $\mathbf{R}^{-1} = \mathbf{R}^t$ *, equation* (2.91) *gives here*

$$\mathfrak{D} = (\nabla\mathbf{R})\, \mathbf{R}^t \quad , \tag{2.93}$$

i.e.,

$$\mathfrak{D}_k{}^i{}_j = (\nabla_k R^i{}_s)\, R_j{}^s \quad , \tag{2.94}$$

an equation to be compared with (2.61).

---

[12] Typically, $\mathbf{T}$ represents a field of rotations or a field of deformations.

[13] One has $\log(\mathbf{T}(\mathbf{x}\|\mathbf{x}+\delta\mathbf{x})\, \mathbf{T}(\mathbf{x})^{-1}) = \log(T^i{}_j(\mathbf{x}+\delta\mathbf{x})\, \mathbf{e}_i(\mathbf{x}\|\mathbf{x}+\delta\mathbf{s}) \otimes \mathbf{e}^j(\mathbf{x}\|\mathbf{x}+\delta\mathbf{x})\, \mathbf{T}(\mathbf{x})^{-1}) = \log([T^i{}_j + \delta x^k\, \partial_k T^i{}_j + \dots]\, [\mathbf{e}_i + \delta x^k\, \Gamma^\ell{}_{ki}\, \mathbf{e}_\ell + \dots] \otimes [\mathbf{e}^j - \delta x^k\, \Gamma^j{}_{ks}\, \mathbf{e}^s + \dots]\, \mathbf{T}(\mathbf{x})^{-1}) = \log([\, [T^i{}_j + \delta x^k (\partial_k T^i{}_j + \Gamma^i{}_{ks}\, T^s{}_j - \Gamma^s{}_{kj}\, T^i{}_s)\,]\, \mathbf{e}_i \otimes \mathbf{e}^j + \dots]\, \mathbf{T}^{-1}) = \log([\, \mathbf{T} + \delta x^k (\nabla\mathbf{T})_k{}^i{}_j\, \mathbf{e}_i \otimes \mathbf{e}^j + \dots]\, \mathbf{T}^{-1}) = \log(\mathbf{I} + \delta x^k (\nabla\mathbf{T})_k{}^i{}_j\, \mathbf{e}_i \otimes \mathbf{e}^j\, \mathbf{T}^{-1} + \dots) = \delta x^k\, (\nabla\mathbf{T})_k{}^i{}_j\, \mathbf{e}_i \otimes \mathbf{e}^j\, \mathbf{T}^{-1} + \cdots = \delta x^k\, (\nabla\mathbf{T})_k\, \mathbf{T}^{-1} + \dots$.

3 Quantities and Measurable Qualities

> *... alteration takes place in respect to certain qualities, and these qualities (I mean hot-cold, white-black, dry-moist, soft-hard, and so forth) are, all of them, differences characterizing the elements.*
>
> On Generation and Corruption, Aristotle, circa 350 B.C.

Temperature, inverse temperature, the cube of the temperature, or the logarithmic temperature are different quantities that can be used to quantify a 'measurable quality': the cold–hot quality. Similarly, the quality 'ideal elastic solid' may be quantified by the elastic compliance tensor $\mathbf{d} = \{d_{ijk\ell}\}$, its inverse, the elastic stiffness tensor $\mathbf{c} = \{c_{ijk\ell}\}$, etc. While the cold–hot quality can be modeled by a 1-D space, the quality 'ideal elastic medium' can be modeled by a 21-dimensional manifold (the number of degrees of freedom of the tensors used to characterize such a medium).

Within a given theoretical context, it is possible to define a unique distance in the 'quality spaces' so introduced. For instance, the distance between two linear elastic media can be defined, and it can be expressed as a function of the two stiffness tensors $\mathbf{c}_1$ and $\mathbf{c}_2$, or as a function of the two compliance tensors $\mathbf{d}_1$ and $\mathbf{d}_2$, and this expression has the same *form* when using the stiffnesses or the compliances.

Introduction

The properties of a physical system are represented by the values of *physical quantities*: temperature, electric field, stress, etc. Crudely speaking, a physical quantity is anything that can be measured. A physical quantity is defined by prescribing the experimental procedure that will measure it (Cook, 1994, discusses this point with clarity). To define a (useful) quantity, the physicist has in mind some context (do we represent our system by point particles or by a continuous medium? do we assume Galilean invariance or relativistic invariance?). She/he also has in mind some ideal circumstances, for instance the proportionality between the force applied to a particle and its acceleration —for small velocities and negligible friction— used to define the inertial mass.

Although current physical textbooks use the notion of 'physical quantity' as the base of physics, here I highlight a more fundamental concept: that of 'measurable physical quality'.

As a first example, an object may have the property of being cold or hot. We will talk about the cold–hot *quality*. The advent of thermodynamics has allowed us to quantify this quality, introducing the quantity 'temperature' T. But the same quality can be quantified by the inverse temperature[1] $\beta = 1/kT$, by the square of the temperature, $u = T^2$, its logarithm, $T^* = \log T/T_0$, the Celsius (or Fahrenheit) temperature t, etc. The quantities $T, \beta, u, T^*, t \dots$ are, in fact, different *coordinates* that can be used to describe the position of a point in the one-dimensional cold–hot quality manifold.

As a second example, an 'ideal elastic medium' may be defined by the condition of proportionality between the components σ_{ij} of the stress tensor σ and the components ε_{ij} of the strain tensor $\boldsymbol{\varepsilon}$. Writing this proportionality (Hooke's law) $\sigma_{ij} = c_{ij}{}^{k\ell}\, \varepsilon_{k\ell}$ quantifies the quality 'linear elastic medium' by the 21 independent components $c_{ij}{}^{k\ell}$ of the stiffness tensor $\mathbf{c}$. But it is also usual to write Hooke's law as $\varepsilon_{ij} = d_{ij}{}^{k\ell}\, \sigma_{k\ell}$, where the 21 independent components $d_{ij}{}^{k\ell}$ of the compliance tensor $\mathbf{d}$, inverse of the stiffness tensor, are used instead. The components of the tensors $\mathbf{c}$ or $\mathbf{d}$ may not be, in some circumstances, the best quantities to use, and their six eigenvalues and 15 orientation angles may be preferable. The 21 components of $\mathbf{c}$, the 21 components of $\mathbf{d}$, the 6 eigenvalues and 15 angles of $\mathbf{c}$ or of $\mathbf{d}$, or any other set of 21 values, related with the previous ones by a bijection, can be used to quantify the quality 'linear elastic medium'. These different sets of 21 quantities related by bijections can be seen as different *coordinates* over a 21-dimensional manifold, the *quality manifold* representing the property of a medium to be linearly elastic: each different linear elastic medium corresponds to a different point on the manifold, and can be referenced by the 21 values corresponding to the coordinates of the point, for whatever coordinate system we choose to use. As we shall see below, the 'stress space' and the 'strain space' are themselves examples of quality spaces.

This notion of 'physical measurable quality' would not be interesting if there was not an important fact: within a given theoretical context, it seems that we can always (uniquely) introduce a metric over a quality manifold, i.e., the distance between two points in a quality space can be defined with an absolute sense, independently of the coordinates (or quantities) used to represent the points. For instance, if two different 'ideal elastic media' $\mathcal{E}_1$ and $\mathcal{E}_2$ are characterized by the two compliances $\mathbf{d}_1$ and $\mathbf{d}_2$, or by the two stiffnesses $\mathbf{c}_1$ and $\mathbf{c}_2$, a sensible definition of distance between the two media is $D(\mathcal{E}_1, \mathcal{E}_2) = \| \log (\mathbf{d}_2 \cdot \mathbf{d}_1^{-1}) \| = \| \log (\mathbf{c}_2 \cdot \mathbf{c}_1^{-1}) \|$, i.e., the norm of the (tensorial) logarithm of "the ratio" of the two compliance tensors (or of the two stiffness tensors). That the expression of the distance is the same when using the compliance tensor or its inverse, the stiffness tensor, is one of the basic conditions defining the metric. I have no knowledge of a previous consideration of these metric structures over the physical measurable qualities.

As we are about to see, the metric is imposed by the invariances of the problem being investigated. In the most simple circumstances, the metric

[1] Here, k is the Boltzmann constant, $k = 1.380\,658\,\mathrm{J\,K^{-1}}$.

will be consistent with the usual definition of the norm in a vector or in a tensor space, but this only when we have bona fide elements of a linear space. For geotensors, of course, the metric will be that of the underlying Lie group. Most of the physical scalars are positive (mass, period, temperature. . .), and, typically, the distance is not related to the difference of values but, rather, to the logarithm of the ratio of values.

3.1 One-dimensional Quality Spaces

As physical quantities are going to be interpreted as coordinates over a quality manifold, we must start by recognizing the different kinds of quantities in common use. Let us examine one-dimensional quality spaces first.

3.1.1 Jeffreys (Positive) Scalars

We are here interested in scalar quantities such as 'the mass of a particle', 'the resistance of an electric wire', or 'the period of a repetitive phenomenon'. These scalars have some characteristics in common:

- they are positive, and span the whole range $(0, +\infty)$;
- one may indistinctly use the quantity or its inverse (conductance $C = 1/R$ instead of resistance $R = 1/C$, frequency $\nu = 1/T$ instead of period $T = 1/\nu$, readiness[2] $r = 1/m$ instead of mass $m = 1/r$, etc.);

As suggested above, these pairs of mutually inverse quantities can be seen as two possible coordinates over a given one-dimensional manifold. Many other coordinate systems can be imagined, as, for instance, any power of such a positive quantity, or the logarithm of the quantity.

As an example, let us place ourselves inside the theoretical framework of the typical Ohm's law for ordinary (macroscopic) electric wires. Ohm's law states that when imposing an electric potential U between the two extremities of an electric wire, the intensity I of the electric current established is[3] proportional to U. As the constant of proportionality depends on the particular electric wire under examination, this immediately suggests characterizing every wire by its *electric resistance* R or its *electric conductance* C defined respectively through the ratios

$$R = \frac{U}{I} \quad ; \quad C = \frac{I}{U} \quad . \tag{3.1}$$

[2]When the proportionality between the force $\mathbf{f}$ applied to a particle and the acceleration $\mathbf{a}$ of the particle is written $\mathbf{f} = m\,\mathbf{a}$, this defines the *mass* m. Writing, instead, $\mathbf{a} = r\,\mathbf{f}$ defines the *readiness* r. Readiness and mass are mutual inverses: $m\,r = 1$.

[3]Ordinary metallic wires satisfy this law well, for small values of U and I.

Resistance and conductance are mutually inverse quantities (one has $C = 1/R$ and $R = 1/C$), and may take, in principle, any positive value.[4] Consider, then, two electric wires, $\mathcal{W}_1$ and $\mathcal{W}_2$, with electric resistances R_1 and R_2, or, equivalently, with electric conductances C_1 and C_2. How should the *distance* $D(\mathcal{W}_1, \mathcal{W}_2)$ between the two wires be defined? It cannot be $D = | R_2 - R_1 |$ or $D = | C_2 - C_1 |$, as these two values are mutually inconsistent, and there is no argument—inside the theoretical framework of Ohm's law—that should allow us to prefer one to the other. In fact, if

- we wish the definition of distance to be additive (in a one-dimensional space, if a point [i.e., a wire] $\mathcal{W}_2$ is between points $\mathcal{W}_1$ and $\mathcal{W}_3$, then, $D(\mathcal{W}_1, \mathcal{W}_2) + D(\mathcal{W}_2, \mathcal{W}_3) = D(\mathcal{W}_1, \mathcal{W}_3)$), and if
- we assume that the coordinates $R = 1/C$ and $C = 1/R$ are such that a pair of wires with resistances (R_a, R_b) and conductances (C_b, C_a) is 'similar' to any pair of wires with resistances $(k\, R_a, k\, R_b)$, where k is any positive real number, or, equivalently, similar to any pair of wires with conductances $(k'\, C_b, k'\, C_a)$, where k' is any positive real number,

then, one easily sees that the distance is necessarily proportional to the expression

$$D(\mathcal{W}_1, \mathcal{W}_2) = \left| \log \frac{R_2}{R_1} \right| = \left| \log \frac{C_2}{C_1} \right| \quad . \tag{3.2}$$

A quantity having the properties just described is called, throughout this book, a *Jeffreys' quantity*, in honor of Sir Harold Jeffreys who, within the context of Probability Theory, was the first to analyze the properties of positive quantities (Jeffreys, 1939). As we will see in this book, the ubiquitous existence of Jeffreys quantities has profound implications in physics too.

Let R be a Jeffreys quantity, and C the inverse of R (so C is a Jeffreys quantity too). The infinitesimal distance associated to R and C is the absolute value of dR/R, which equals the absolute value of dC/C. In terms of the distance element

$$| ds_{\mathfrak{W}} | = \frac{|dR|}{R} = \frac{|dC|}{C} \quad . \tag{3.3}$$

By integration of this, one finds expression (3.2),

Further examples of positive (Jeffreys) quantities are the *temperature* of a normal medium $T = 1/k\beta$ and its inverse, the *thermodynamic parameter* $\beta = 1/kT$, where k is the is the Boltzmann constant; the *half-life* of radioactive nuclei $\tau = 1/\lambda$ and its inverse, the *disintegration rate* $\lambda = 1/\tau$; the *phase velocity* of a wave $c = 1/n$ and its inverse, the *phase slowness* $n = 1/c$, or the *wavelength* $\lambda = 2\pi/k$ and its inverse, the *wavenumber* $k = 2\pi/\lambda$; the *elastic incompressibility* $\kappa = 1/\gamma$ and its inverse, the *elastic compressibility* $\gamma = 1/\kappa$; or the *elastic shear modulus* $\mu = 1/\nu$ and its inverse $\nu = 1/\mu$; the *length* of

[4]Measuring exactly a zero resistance (infinite conductance) or a zero conductance (infinite resistance) is impossible if Ohm's law is valid.

an object $L = 1/S$ and its inverse, the *shortness* $S = 1/L$. There are plenty of other pairs of reciprocal parameters, like thermal conductivity–thermal resistivity; electric permittivity–electric impermittivity (inverse of electric permittivity); magnetic permeability–magnetic impermeability (inverse electric permeability); acoustic impedance–acoustic admittance.[5] There are also Jeffreys' parameters in other sciences, like in economics.[6]

So we can formally set the

Definition 3.1 Jeffreys quantity. *In a one-dimensional metric manifold, any coordinate X that gives to the distance element the form*

$$ds = k \frac{dX}{X} \quad , \tag{3.4}$$

where k is a real number, is called a Jeffreys coordinate. *Equivalently, a Jeffreys coordinate can be defined by the condition that the expression of the finite distance between the point of coordinate X_1 and the point of coordinate X_2 is*

$$D = k \left| \log \frac{X_2}{X_1} \right| \quad . \tag{3.5}$$

When the manifold represents a physical quality, such a coordinate is also called a Jeffreys quantity *(or Jeffreys* magnitude*).*

One has[7]

Property 3.1 *Let r be a real number, positive or negative. Any power $Y = X^r$ of a Jeffreys quantity X is a Jeffreys quantity.*

In particular,

Property 3.2 *The inverse of a Jeffreys quantity is a Jeffreys quantity.*

Let us explicitly introduce the following

Definition 3.2 Cartesian quantity. *In a one-dimensional metric manifold, any coordinate x that gives to the distance element the form*

$$ds = k\,dx \quad , \tag{3.6}$$

[5]This pair of quantities is one of the few pairs having a name: the term *immittance* designates any of the two quantities, impedance or admittance.

[6]As in the exchange rate of currency, where if α denotes the rate of US Dollars against Euros, $\beta = 1/\alpha$ denotes the rate of Euros against US Dollars, or as in the mileage–fuel consumption of cars, where if α denotes the number of miles per gallon (as measured in the US), $\beta = 1/\alpha$ is proportional to the number of liters per 100 km, as measured in Europe.

[7]Inserting $X = Y^{-r}$ in equation (3.4) it follows that $ds^2 = k\,r^2\,dY^2/Y^2$, which has the form (3.4).

where k *is a real number, is called a* Cartesian coordinate. *Equivalently, a Cartesian coordinate can be defined by the condition that the expression of the finite distance between the point of coordinate* x_1 *and the point of coordinate* x_2 *is*

$$D = k\,|\,x_2 - x_1\,| \quad . \tag{3.7}$$

When the manifold represents a physical quality, such a coordinate is also called a Cartesian quantity *(or* magnitude*).*

One then has the following

Property 3.3 *If a quantity* X *is Jeffreys, then, the quantity* $x = \log(X/X_0)$ *, where* X_0 *is any fixed value of* X *, is a Cartesian quantity.*

Clearly, the logarithm of a Jeffreys quantity takes any real value in the range $(-\infty, +\infty)$.

The symbol log stands for the natural, base e logarithms. When, instead, a logarithm in a base a is used, we write $\log_a$. While physicists may prefer to use natural logarithms to introduce a Cartesian quantity from a Jeffreys quantity, engineers may prefer the use of base 10 logarithms to find their usual 'decibel scales'. Musicians may prefer base 2 logarithms, as, then, the distance between two musical notes (as defined by their frequency or by their period) happens to correspond to the distance between notes expressed in 'octaves'.

For most of the positive parameters considered, the inverse of the parameter is usually also introduced, excepted for length. We have introduced above the notion of *shortness* of an object, as the inverse of its length, $S = 1/L$ (or the *thinness*, as the inverse of the thickness, etc.). One could name *delambre,* and denote d, the unit of shortness, in honor of Jean-Baptiste Joseph Delambre (Amiens, 1749–Paris, 1822), who measured with Pierre Méchain the length (or shortness?) of an Earth's meridian. The delambre is the inverse of the meter: $1\,\mathrm{d} = 1\,\mathrm{m}^{-1}$. A sheet of the paper of this book, for instance, has a *thinness* of about $9\,10^3$ delambres, which means that one needs to pile approximately 9 000 sheets of paper to make an object with a length of one meter (or with a shortness of one delambre).

When both quantities are in use, a Jeffreys quantity and its inverse, physicists often switch between one and the other, choosing to use the quantity that has a large number of units. For instance, when seismologists analyze acoustic waves, they will typically say that a wave has a period of 13 seconds (instead of a frequency 0.077 hertz), while another wave may have a frequency of 59 hertz (instead of a period of 0.017 seconds).

Many of the quantities used in physics are Jeffreys'. In fact, other types of quantities are often simply related to Jeffreys quantities. For instance, the 'Poisson's ratio' of an elastic medium is simply related to the eigenvalues of the stiffness tensor, which are Jeffreys (see section 3.1.4).

It should perhaps be mentioned here that the electric charge of a particle is *not* a Cartesian scalar. The electric charge of a particle is better understood inside the theory of electromagnetic continuous media, where the electric charge density appears as the temporal component of the 4D current density vector.[8]

There are not many Cartesian quantities in physics. Most of them are just the logarithms of Jeffreys' quantities, like the pH of an acid, the logarithm of the concentration (see details in section 3.1.4.2), the entropy of a thermodynamic system, the logarithm of the number of accessible states, etc.

3.1.2 Benford Effect

Many quantities in physics, geography, economics, biology, sociology, etc., take values that have a great tendency to start with the digit 1 or 2. Take, for instance, the list of the *States, Territories and Principal Islands of the World*, as given in the Times Atlas of the World (Times Books, 1983). The beginning of the list is shown in figure 3.1. In the three first numerical columns of the list, there are the surfaces (both in square kilometers and square miles) and populations of states, territories and islands. The statistic of the first digit is shown at the right of the figure: there is an obvious majority of ones, and the probability of the first digit being a 2, 3, 4, etc. decreases with increasing digit value. This observation dates back to Newcomb (1881), and is today known as the *Benford law* (Benford, 1938).

States, Territories, and Principal Islands of the World

Name	*Sq. km*	*Sq. miles*	*Population*
Afghanistan	636,267	245,664	15,551,358
Åland	1,505	581	22,000
Albania	28,748	11,097	2,590,600
Aleutian Islands	17,666	6,821	6,730
Algeria	2,381,745	919,354	18,250,000
American Samoa	197	76	30,600
Andorra	465	180	35,460
Angola	1,246,700	481,226	6,920,000
...	...	...	...

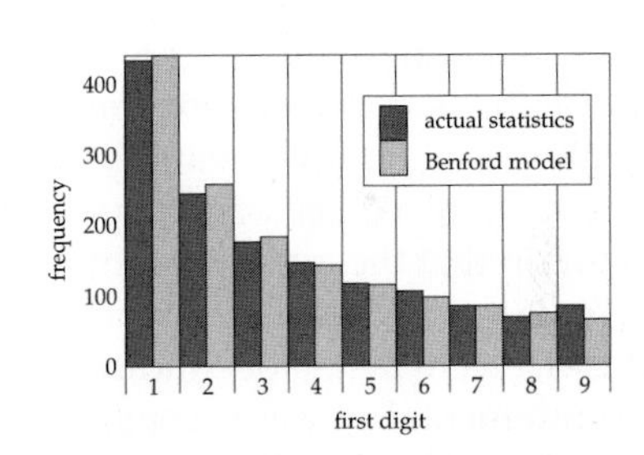

Fig. 3.1. *Left: the beginning of the list of the states, territories and principal islands of the World, in the Times Atlas of the World (Times Books, 1983), with the first digit of the surfaces (both in square kilometers and square miles) and populations highlighted. Right: statistics of the first digit (dark gray) and prediction from the Benford model (light gray).*

We can state the 'law' as follows.

Property 3.4 Benford effect. *Consider a Cartesian quantity x and a Jeffreys quantity*

[8]The component of a vector may take any value, and does not need to be positive (here, this allowing the classical interpretation of a positron as an electron "going backwards in time".

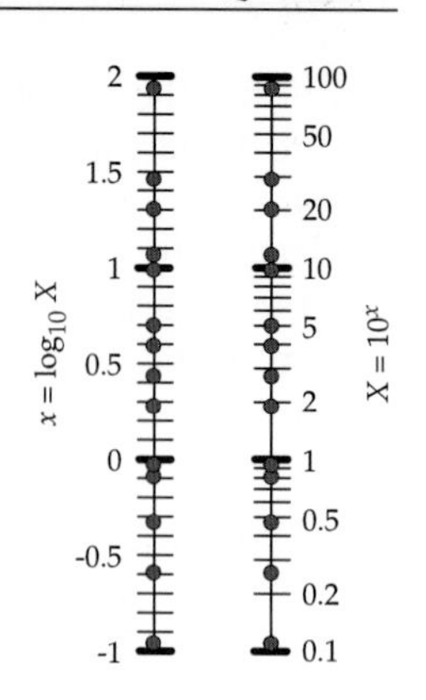

Fig. 3.2. *Generate points, uniformly at random, 'o'n the real axis" (left of the figure). The values $x_1, x_2 \dots$ will not have any special property, but the quantities $X_1 = 10^{x_1}, X_2 = 10^{x_2} \dots$ will present the Benford effect: as the figure suggests, the intervals 0.1–0.2, 1–2, 10–20, etc. are longer (so have greater probability of having points) than the intervals 0.2–0.3, 2–3, 20–30, etc., and so on. It is easy to see that the probability that the first digit of the coordinate X equals n is $p_n = \log_{10}(n+1)/n$ (Benford law). The same effect appears when, instead of base 10 logarithms, one uses natural logarithms, $X_1 = e^{x_1}, X_2 = e^{x_2} \dots$, or base 2 logarithms,, $X_1 = 2^{x_1}, X_2 = 2^{x_2} \dots$.*

$$X = b^x \quad , \tag{3.8}$$

where b is any positive base number (for instance, $b = 2$, $b = 10$, or $b = e = 2.71828\dots$). If values of x are generated uniformly at random, then the first digit of the values of X (that are all positive) has an uneven distribution. When using a base K system of numeration to represent the quantity X (typically, we write numbers in base 10, so $K = 10$), the probability that the first digit is n equals

$$p_n = \log_K(n+1)/n \quad . \tag{3.9}$$

The explanation of this effect is suggested in figure 3.2.

All Jeffreys quantities exhibit this effect, this meaning, in fact, that the logarithm of a Jeffreys quantity can be considered a 'Cartesian quantity'. That a table of values of a quantity exhibits the Benford effect is a strong suggestion that the given quantity may be a Jeffreys one.

This is the case for most of the quantities in physics: masses of elementary particles, etc. In fact, if one indiscriminately takes the first digits of a table of 263 fundamental physical constants, the Benford effect is conspicuous,[9] as demonstrated by the histogram in figure 3.3. This is a strong suggestion that most of the physical constants are Jeffreys quantities. It seems natural that this observation enters in the development of physical theories, as proposed in this text.

3.1.3 Power Laws

In the scientific literature, when one quantity is proportional to the power of another quantity, it is said that one has a power law. In biology, for instance, the metabolism rate of animals is proportional to the 3/4 power of their

[9]Negative values in the table, like the electric charge of the electron, should be excluded from the histogram, but they are not very numerous and do not change the statistics significantly.

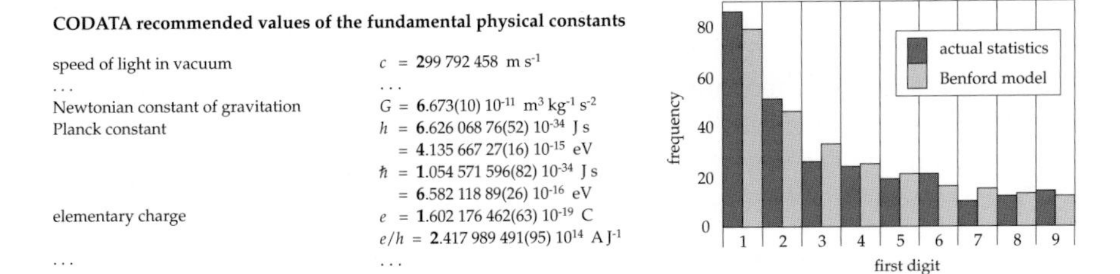

CODATA recommended values of the fundamental physical constants

speed of light in vacuum	c = 299 792 458 m s^{-1}
...	...
Newtonian constant of gravitation	G = 6.673(10) 10^{-11} m^3 kg^{-1} s^{-2}
Planck constant	h = 6.626 068 76(52) 10^{-34} J s
	= 4.135 667 27(16) 10^{-15} eV
	$\hbar$ = 1.054 571 596(82) 10^{-34} J s
	= 6.582 118 89(26) 10^{-16} eV
elementary charge	e = 1.602 176 462(63) 10^{-19} C
	e/h = 2.417 989 491(95) 10^{14} A J^{-1}
...	...

Fig. 3.3. *Left: the beginning of the table of Fundamental Physical Constants (1998 CODATA least-squares adjustment; Mohr and Taylor, 2001), with the first digit highlighted. Right: statistics of the first digit of the 263 physical constants in the table. The Benford effect is conspicuous.*

body mass, and this can be verified for body masses spanning many orders of magnitude. The quantities entering a power law are, typically, Jeffreys quantities.

That these power laws are so highlighted in biology or economics is probably because of their empirical character: in physics these laws are very common. For instance, Stefan's law states that the power radiated by a body is proportional to the 4th power of the absolute temperature. In fact, it is the hypothesis that power laws are ubiquitous, that gives sense to the dimensional analysis method (discovered by Fourier, 1822): physical relations between quantities can be guessed by just using dimensional arguments.

3.1.4 Ad Hoc Quantities

Many physical quantities have definitions that are justified only historically. As shown here below, this is the case for some of the coefficients used to define an elastic medium (like Poisson's ratio or Young's modulus), whereas the eigenvalues of the stiffness tensor should be used instead. As a second example, it is shown below how the usual definition of chemical concentration could be modified. There are many other ad hoc parameters, for instance the density parameter Ω in cosmological models (see Evrard and Coles, 1995). In each case, it is fundamental to recognize which is the Jeffreys' (or the Cartesian) parameter hidden behind the ad hoc parameter, and use it explicitly.

3.1.4.1 Elastic Poisson's Ratio

An ideal elastic medium $\mathcal{E}$ can be characterized by the stiffness tensor $\mathbf{c}$ or the compliance tensor $\mathbf{d} = \mathbf{c}^{-1}$. The distance between an ideal elastic medium $\mathcal{E}_1$, characterized by the stiffness $\mathbf{c}_1$ or the compliance $\mathbf{d}_1$ and an ideal elastic medium $\mathcal{E}_2$, characterized by the stiffness $\mathbf{c}_2$ or the compliance $\mathbf{d}_2$ is (see section 3.3)

$$D(\mathcal{E}_1,\mathcal{E}_2) = \| \log(\mathbf{c}_2\,\mathbf{c}_1^{-1}) \| = \| \log(\mathbf{d}_2\,\mathbf{d}_1^{-1}) \| \quad . \tag{3.10}$$

For an isotropic medium, the stiffness and the compliance tensor have two distinct eigenvalues. Let us, for instance, talk about stiffnesses, and denote χ and ψ the two eigenstiffnesses. These eigenstiffnesses are related to the common incompressibility modulus κ and shear modulus μ as

$$\chi = 3\kappa \quad ; \quad \psi = 2\mu \quad . \tag{3.11}$$

When computing the distance (as defined by equation (3.10)) between the elastic medium $\mathcal{E}_1 : (\chi_1,\psi_1)$ and the elastic medium $\mathcal{E}_2 : (\chi_2,\psi_2)$ one obtains

$$D(\mathcal{E}_1,\mathcal{E}_2) = \sqrt{\left(\log\frac{\chi_2}{\chi_1}\right)^2 + 5\left(\log\frac{\psi_2}{\psi_1}\right)^2} \quad . \tag{3.12}$$

The factors in this expression come from the fact that the eigenvalue χ has multiplicity one, while the eigenvalue ψ has multiplicity five.

Once the result is well understood in terms of the eigenstiffnesses, one may come back to the common incompressibility and shear moduli. The distance between the elastic medium $\mathcal{E}_1 : (\kappa_1,\mu_1)$ and the elastic medium $\mathcal{E}_2 : (\kappa_2,\mu_2)$ is immediately obtained by substituting parameter values in expression (3.12):

$$D(\mathcal{E}_1,\mathcal{E}_2) = \sqrt{\left(\log\frac{\kappa_2}{\kappa_1}\right)^2 + 5\left(\log\frac{\mu_2}{\mu_1}\right)^2} \quad . \tag{3.13}$$

Should one wish to use the logarithmic incompressibility modulus $\kappa^* = \log(\kappa/\kappa_0)$ and the logarithmic shear modulus $\mu^* = \log(\mu/\mu_0)$ (κ_0 and μ_0 being arbitrary constants), then,

$$D(\mathcal{E}_1,\mathcal{E}_2) = \sqrt{(\kappa_2^* - \kappa_1^*)^2 + 5\,(\mu_2^* - \mu_1^*)^2} \quad . \tag{3.14}$$

While the incompressibility modulus κ and the shear modulus μ are Jeffreys quantities, the logarithmic incompressibility κ^* and the logarithmic shear modulus μ^* are Cartesian quantities.

The distance element associated to this finite expression of distance clearly is

$$ds^2 = \left(\frac{d\kappa}{\kappa}\right)^2 + 5\left(\frac{d\mu}{\mu}\right)^2 = (d\kappa^*)^2 + 5\,(d\mu^*)^2 \quad . \tag{3.15}$$

In the Jeffreys coordinates $\{\kappa,\mu\}$ the components of the metric are

$$\begin{pmatrix} g_{\kappa\kappa} & g_{\kappa\mu} \\ g_{\mu\kappa} & g_{\mu\mu} \end{pmatrix} = \begin{pmatrix} 1/\kappa^2 & 0 \\ 0 & 5/\mu^2 \end{pmatrix} \quad , \tag{3.16}$$

while in the Cartesian coordinates $\{\kappa^*,\mu^*\}$ the metric matrix

$$\begin{pmatrix} g_{\kappa^*\kappa^*} & g_{\kappa^*\mu^*} \\ g_{\mu^*\kappa^*} & g_{\mu^*\mu^*} \end{pmatrix} = \begin{pmatrix} 1 & 0 \\ 0 & 5 \end{pmatrix} \quad . \tag{3.17}$$

Let us now express the distance element of the space of isotropic elastic media using as elastic parameters (i.e., as coordinates), two popular parameters, Young modulus Y and Poisson's ratio σ, that are related to the incompressibility and the shear modulus through

$$Y = \frac{9\,\kappa\,\mu}{3\,\kappa + \mu} \quad ; \quad \sigma = \frac{1}{2}\,\frac{3\,\kappa - 2\,\mu}{3\,\kappa + \mu} \quad , \tag{3.18}$$

or, reciprocally, $\kappa = Y/(3(1-2\,\sigma))$ and $\mu = Y/(2(1+\sigma))$. In these coordinates, the metric (3.15) then transforms[10] into

$$\begin{pmatrix} g_{YY} & g_{Y\sigma} \\ g_{\sigma Y} & g_{\sigma\sigma} \end{pmatrix} = \begin{pmatrix} \frac{6}{Y^2} & \frac{2}{Y(1-2\,\sigma)} - \frac{5}{Y(1+\sigma)} \\ \frac{2}{Y(1-2\,\sigma)} - \frac{5}{Y(1+\sigma)} & \frac{4}{(1-2\,\sigma)^2} + \frac{5}{(1+\sigma)^2} \end{pmatrix} \quad , \tag{3.19}$$

with associated surface element

$$dS_{Y\sigma}(Y,\sigma) = \sqrt{\det g}\; dY\,d\sigma = k\,\frac{dY\,d\sigma}{Y\,(1+\sigma)(1-2\,\sigma)} \quad , \tag{3.20}$$

where $k = 3\sqrt{5}$. To express the distance between the elastic medium $\mathcal{E}_1 = (Y_1, \sigma_1)$ and the elastic medium $\mathcal{E}_2 = (Y_2, \sigma_2)$, one could integrate the length element ds (associated to the metric in equation (3.19)) along the geodesic joining the points. It is much simpler to use the property that the distance is an invariant, and just rewrite expression (3.13) replacing the variables $\{\kappa, \mu\}$ by the variables $\{Y, \sigma\}$. This gives

$$D(\mathcal{E}_1, \mathcal{E}_2) = \sqrt{\left(\log \frac{Y_2\,(1-2\,\sigma_1)}{Y_1\,(1-2\,\sigma_2)}\right)^2 + 5\left(\log \frac{Y_2\,(1+\sigma_1)}{Y_1\,(1+\sigma_2)}\right)^2} \quad . \tag{3.21}$$

Although Poisson's ratio has historical interest, it is not a simple parameter, as shown by its theoretical bounds $-1 < \sigma < 1/2$, or the expression for the distance (3.21). In fact, the Poisson ratio σ depends only on the ratio κ/μ (incompressibility modulus over shear modulus), as we have

$$\frac{1+\sigma}{1-2\sigma} = \frac{3}{2}\,\frac{\kappa}{\mu} \, . \tag{3.22}$$

The ratio $J = \kappa/\mu$ of two independent[11] Jeffreys parameters being a Jeffreys parameter, we see that J, while depending only on σ, it is not an ad hoc parameter, as σ is. The only interest of σ is historical, and we should not use it any more.

[10] In a change of variables $x^i \rightleftharpoons x^I$, a metric g_{ij} changes to $g_{IJ} = \Lambda_I{}^i \Lambda_J{}^j g_{ij} = \frac{\partial x^i}{\partial x^I} \frac{\partial x^j}{\partial x^J} g_{ij}$.

[11] Independent in the sense of expression (3.16).

3.1.4.2 Concentration–Dilution

In a mixing of two substances, containing a mass m_a of the first substance, and a mass m_b of the second substance, one usually introduces the two concentrations

$$a = \frac{m_a}{m_a + m_b} \quad ; \quad b = \frac{m_b}{m_a + m_b} \quad , \tag{3.23}$$

and one has the relation

$$a + b = 1 \quad . \tag{3.24}$$

One has here a pair of quantities, that, like a pair of Jeffreys quantities, are reciprocal, but, here, it is not their product that equals one, it is their sum.

Let $\mathcal{P}_1$ be a point on the concentration–dilution manifold, that can either be represented by the concentration a_1 or the reciprocal concentration b_1, and let $\mathcal{P}_2$ be a second point, represented by the concentration a_2 or the reciprocal concentration b_2. As the expression $(a_2 - a_1)$ is easily seen to be identical to $-(b_2 - b_1)$, one may wish to introduce over the concentration–dilution manifold the distance

$$D(\mathcal{P}_1, \mathcal{P}_2) = | a_2 - a_1 | = | b_2 - b_1 | \quad . \tag{3.25}$$

It has the required properties: (i) it is additive, and (ii) its expression is formally identical using the concentration a or the reciprocal concentration b.

This simple definition of distance may be the correct one inside some theoretical context. For instance, when using the methods of chapter 4 to obtain simple physical laws (having invariant properties) it is this definition of distance that will automatically lead to Fick's law of diffusion.

In other theoretical contexts, a different definition of distance is necessary, typically, when a logarithmic notion like that of pH appears useful.

The chemical concentration of a solution is usually defined as

$$c = \frac{m_{\text{solute}}}{m_{\text{solute}} + m_{\text{solvent}}} \quad , \tag{3.26}$$

this introducing a quantity that varies between 0 and 1. One could, rather, define the quantity

$$\chi = \frac{m_{\text{solute}}}{m_{\text{solvent}}} \quad , \tag{3.27}$$

that we shall name *eigenconcentration*. It takes values in the range $(0, \infty)$, and it is obviously a Jeffreys quantity, its inverse having the interpretation of a *dilution*. The relationship between the concentration c and the eigenconcentration χ is

$$\chi = \frac{c}{1 - c} \quad ; \quad c = \frac{\chi}{\chi + 1} \quad . \tag{3.28}$$

For small concentrations, $c \approx \chi$. But, although for small concentrations c and χ tend to be identical, a logarithmic quantity (like the pH of an acid solution) should be defined as the logarithm of χ, not —as it is usually done— as the logarithm of c.

For a Jeffreys quantity like χ we have seen above that the natural definition of distance is $ds = |d\chi|/\chi$. This implies over the quantity $\chi^* = \log\chi$ the distance element $ds = |d\chi^*|$ and over the quantity c the distance element

$$ds = \frac{|dc|}{c\,(1-c)} . \tag{3.29}$$

It is, of course, possible to generalize this to the case where there are more than two chemical compounds, although the mathematics rapidly become complex (see some details in appendix A.19).

3.1.5 Quantities and Qualities

The examples above show that many different *quantities* can be used as *coordinates* for representing *points* in a one-dimensional *quality* space. Some of the coordinates are Jeffreys quantities, other are Cartesian quantities, and other are ad hoc. Present-day physical language emphasizes the use of quantities: one usually says "a temperature field", while we should say "a cold–hot field".

The following sections give some explicit examples of quality spaces.

3.1.6 Example: The Cold–hot Manifold

The *cold–hot manifold* can be imagined as an infinite one-dimensional space of points, with the infinite cold at one extremity and the infinite hot at the other extremity. This one-dimensional manifold (that, as we are about to see, may be endowed with a metric structure) shall be denoted by the symbol $\mathfrak{C}|\mathfrak{H}$.

The obvious coordinate that can be used to represent a point of the cold–hot quality manifold is the *thermodynamic temperature*[12] T. Another possible coordinate over the cold–hot space is the the thermodynamic parameter $\beta = 1/(k\,T)$, where k is Boltzmann's constant. And associated to these two coordinates we may introduce the logarithmic temperature $T^* = \log T/T_0$ (T_0 being an arbitrary, fixed temperature) and $\beta^* = \log \beta/\beta_0$ (β_0 being an arbitrary, fixed value of the thermodynamic parameter β).

When working inside a theoretical context where the temperature T and the thermodynamic parameter $\beta = 1/kT$ can be considered to be Jeffreys quantities (for, instance, in the context used by Fourier to derive his law of heat conduction), the distance between a point $\mathcal{A}_1$, characterized by the temperature T_1, or the thermodynamic parameter β_1, or the logarithmic

[12] In the International System of units, the temperature has its own physical dimension, like the quantities length, mass, time, electric current, matter quantity and luminous intensity.

temperature T_1^*, and the point $\mathcal{A}_2$, characterized by the temperature T_2, or the thermodynamic parameter β_2, or the logarithmic temperature T_2^*, is

$$D(\mathcal{A}_1, \mathcal{A}_2) = \left| \log \frac{T_2}{T_1} \right| = \left| \log \frac{\beta_2}{\beta_1} \right| = | T_2^* - T_1^* | \quad . \tag{3.30}$$

Equivalently, the distance element of the space is expressed as

$$| ds_{\mathfrak{C}|\mathfrak{H}} | = \frac{| dT |}{T} = \frac{| d\beta |}{\beta} = | dT^* | \quad . \tag{3.31}$$

Let us assume that we use an arbitrary coordinate λ^1, that can be one of the above, or some other one. It is assumed that the dependence of λ^1 on the other coordinates mentioned above is known. Therefore, the distance between points in the cold–hot space can also be expressed as a function of this arbitrary coordinate, and, in particular, the distance element. We write then

$$ds^2_{\mathfrak{C}|\mathfrak{H}} = \gamma_{\alpha\beta} \, d\lambda^\alpha \, d\lambda^\beta \quad ; \quad (\alpha, \beta, \ldots \in \{ 1 \}) \quad , \tag{3.32}$$

this defining the 1×1 metric tensor $\gamma_{\alpha\beta}$. We reserve the Greek indices $\{\alpha, \beta, \ldots\}$ to be used as tensor indices of the one-dimensional cold–hot manifold.

Should one use as coordinate λ the logarithmic temperature T^* then, $\sqrt{g_{T^*T^*}} = 1$. Should one use the temperature T, then, $\sqrt{g_{TT}} = 1/T$.

3.2 Space-Time

3.2.1 Time

The flowing of time is one of the most profoundly inscribed of human sensations. Two related notions have an innate sense: that of a time *instant* and that of a time *duration*.

While time, *per se*, is just a human perception, time durations are amenable to quantitative measure. In fact, it is difficult to find any good definition of time, excepted the obvious: *time (durations) is what clocks measure.* Time coordinates are defined by accumulating the durations realized by a clock (or a system of clocks). It is with the advent of Newtonian mechanics, the notion of an *ideal time* became clear (as a time for which the equations of Newtonian mechanics look simple).

This notion of *Newtonian time* remains inside Einstein's description of space-time: the existence of clocks measuring ideal time is a basic postulate of the general theory of relativity. In fact, the basic invariant of the theory, the "length" of a space-time trajectory, is, by definition, the (Newtonian) time measured by an ideal clock that describes the trajectory.[13]

[13]The basic difference between Newtonian and relativistic space-time is that while for Newton this time is the same for observers in the universe, in relativity, it is defined for individual clocks.

Any physical clock can only be an approximation to the ideal clock. The best clocks at present are atomic. In a *cesium fountain atomic clock*, cesium atoms are cooled (using laser beams), and are put in free-fall inside a cavity, where a microwave signal is tuned to different frequencies, until the frequency is found that maximizes the fluorescence of the atoms (because it excites "the transition between the two hyperfine levels of the ground state of the cesium 133 atom"). This frequency of 9 192 631 770 Hz is the frequency used to define the SI unit of *time duration*, the *second* (in reality, the SI unit of *frequency*, the *Hertz*.

Consider a one-dimensional manifold, the *time manifold* $\mathfrak{T}$, that has two possible orientations, from past to future and from future to past. The points of this manifold, $\mathcal{T}_1, \mathcal{T}_2 \dots$ are called (time) *instants*. A (perfect) clock can (in principle) be used to define a Newtonian time coordinate t over the the time manifold. The distance (duration) between two instants $\mathcal{T}_1$ and $\mathcal{T}_2$, with respective Newtonian time coordinates t_1 and t_2 as

$$\mathrm{dist}(\mathcal{T}_1, \mathcal{T}_2) \; = \; | \, t_2 - t_1 \, | \quad . \tag{3.33}$$

If instead of Newtonian time t one uses another arbitrary time coordinate τ^1, related to Newtonian time through $t = t(\tau^1)$, then,

$$\mathrm{dist}(\mathcal{T}_1, \mathcal{T}_2) \; = \; | \, t(\tau^1_2) - t(\tau^1_1) \, | \quad . \tag{3.34}$$

The duration element in the time manifold is then

$$ds_{\mathfrak{T}} \; = \; dt \; = \; \frac{dt}{d\tau^1} \, d\tau^1 \quad . \tag{3.35}$$

We introduce the 1×1 metric G_{ab} in the time manifold by writing

$$ds^2_{\mathfrak{T}} \; = \; G_{ab} \, d\tau^a \, d\tau^b \qquad ; \qquad (\, a, b, \dots \in \{ 1 \} \,) \quad , \tag{3.36}$$

reserving for the tensor notation related with $\mathfrak{T}$ the indices $\{a, b, \dots\}$. As $\mathfrak{T}$ is one-dimensional, these indices can only take the value 1. One has

$$\sqrt{G_{11}} \; = \; \left| \frac{dt}{d\tau} \right| \quad . \tag{3.37}$$

3.2.2 Space

The simplest and more fundamental example of measurable physical quality corresponds to the three-dimensional "physical space". All superior animals have developed the intuitive notion of physical space, know what the relative position of two points is, and have the notion of distance between points. Galilean physics considers that the space is an absolute notion, while in relativistic physics, only the space-time is absolute. The 3D physical space is the most basic example of a manifold. We denote it using the symbol $\mathfrak{E}$.

While in a flat (i.e., Euclidean) space, the notion of *relative position* of a point B with respect to a point A corresponds to the vector from A to B, in a curved space, it corresponds to the oriented geodesic from A to B (as there may be more than one geodesic joining two points, this notion may only make sense for any point B inside a finite neighborhood around point A). The *distance* between two points A and B is, by definition, the length of the geodesic joining the two points. In a flat space, this corresponds to the norm of the vector representing the relative position. To represent points in the space we use *coordinates*, and different observers may use different coordinate systems. The origin of the coordinates, or their orientation, may be different, or, while an observer may use Cartesian coordinates (if the space is Euclidean), another observer may use spherical coordinates, or any other coordinate system.

Because the distance between two points is a notion that makes sense independently from any choice of coordinates, it is possible to introduce the notion of *metric*: to each coordinate system $\{x^1, x^2, x^3\}$, it is possible to associate a *metric tensor* g_{ij} such that the squared distance element $ds^2_{\mathfrak{E}}$ can be written

$$ds^2_{\mathfrak{E}} = g_{ij}\, dx^i\, dx^j \quad ; \quad (\, i, j, \ldots \in \{1, 2, 3\}\,) \quad . \tag{3.38}$$

This metric may, of course, be nonflat (i.e., the associated Riemann may be non-zero).

The meter is presently defined as the length of the path travelled by light in vacuum during a time interval of 1/299 792 458 of a second. This definition fixes the speed of light in vacuum at exactly 299 792 458 m s^{-1}.

Example 3.1 Velocity "vector". *Let* $\{\tau^a\} = \{\tau^1\}$ *be one coordinate over the time space, not necessarily the Newtonian time, not necessarily oriented from past to future. This coordinate is related to the Newtonian notion of "distance" in the time space (in fact, of duration) as expressed in equation* (3.36). *Let* $\{x^i\}$; $i = 1, 2, 3$, *be a system of three coordinates on the space manifold, assumed to be a Riemannian (metric) manifold, not necessarily Euclidean. The distance element has been expressed in equation* (3.38). *The trajectory of a particle may be described by the functions*

$$\tau^1 \quad \mapsto \quad \{\, x^1(\tau^1)\, ,\, x^2(\tau^1)\, ,\, x^3(\tau^1)\,\} \quad . \tag{3.39}$$

The velocity *of the particle —at some point along the trajectory— is the derivative (tensor) of the mapping* (3.39). *As we have seen in the previous chapter, the components of the derivative (on the natural local bases associated to the coordinates* τ^a *and* x^i *being used) are*

$$v_a{}^i = \frac{\partial x^i}{\partial \tau^a} \quad . \tag{3.40}$$

The (Frobenius) norm of the velocity is $\| \mathbf{v} \| = \sqrt{G^{ab}\, g_{ij}\, v_a{}^i\, v_b{}^j}$ *, i.e.,*

$$\| \mathbf{v} \| = \frac{\sqrt{g_{ij}\, v_1{}^i\, v_1{}^j}}{\sqrt{G_{11}}} \quad . \tag{3.41}$$

Should we have agreed, once and for all, to use Newtonian time, $\tau^1 = t$, then $G_{11} = 1$. The Frobenius norm of the velocity tensor then equals the ordinary norm of the usual velocity vector.

3.2.3 Relativistic Space-Time

A point in space-time is called an *event*. One of the major postulates of special or general relativity, is the existence of a space-time metric, i.e., the possibility of defining the absolute length of a space-time line (corresponding to the proper time of a clock whose space-time trajectory is the given line). By absolute length it is meant that this length is measurable independently of any choice of space-time coordinates (which, in relativity, is equivalent to saying independently of any observer). In relativity, the (3D) physical space or the (1D) "time space" do not exist as individual entities.

Using the four space-time coordinates $\{x^0, x^1, x^2, x^3\}$ the squared line element is usually written as

$$ds^2 = g_{\alpha\beta}\, dx^\alpha\, dx^\beta \ , \tag{3.42}$$

where the four-dimensional metric $g_{\alpha\beta}$ has signature[14] $\{+,-,-,-\}$. For instance, in special relativity (where the space-time is flat), using Minkowski coordinates $\{t, x, y, z\}$

$$ds^2 = dt^2 - \frac{1}{c^2}\left(dx^2 + dy^2 + dz^2\right) , \tag{3.43}$$

but in general (curved) space-times, Minkowski coordinates do not exist. The *relative position* of space-time event B with respect to space-time event A (B being in the past or the future light-cone of A) is the oriented space-time geodesic from A to B. The *distance* between the two events A and B is the length of the space-time geodesic.

Example 3.2 *Assume that a space-time trajectory is parameterized by some parameter λ, as $x^\alpha = x^\alpha(\lambda)$. The velocity tensor* $\mathbf{U}$ *is, in terms of components,*

$$U_\lambda{}^\alpha = \frac{dx^\alpha}{d\lambda} \quad . \tag{3.44}$$

Writing the relations between the parameter λ and the proper time as $ds = \sqrt{\gamma_{\lambda\lambda}}\, |\, d\lambda\, |$, we obtain, for the (Frobenius) norm of the velocity tensor,

[14] Alternatively, the signature may be chosen $\{-,+,+,+\}$.

$$\| \mathbf{U} \| = \sqrt{\gamma^{\lambda\lambda}\, g_{\alpha\beta}\, U_\lambda{}^\alpha\, U_\lambda{}^\beta} \quad , \tag{3.45}$$

but, as $g_{\alpha\beta}\, U_\lambda{}^\alpha\, U_\lambda{}^\beta = g_{\alpha\beta}\,(dx^\alpha/d\lambda)\,(dx^\beta/d\lambda) = ds^2/d\lambda^2 = \gamma_{\lambda\lambda}$ *, we obtain*

$$\| \mathbf{U} \| = 1 \quad , \tag{3.46}$$

i.e.. the four-velocity tensor has unit norm.

Few physicists will doubt that there is *one* definition of distance in relativistic space-time. In fact, the distance we may need to introduce depends on the theoretical context. For instance, the coordinates of space-time events are measured using clocks and light rays, using, for instance Einstein's protocol. Every measurement has attached uncertainties, so the information we have on the actual coordinates of an event can be represented using a probability density on the space-time manifold. In the simplest simulation of an actual measurement, using imperfect clocks, one arrives at a Gaussian probability density,

$$f(t,x,y,z) = \frac{1}{(2\pi)^2\,\sigma^2}\,\exp\Big(-\frac{D^2}{2\,\sigma^2} \Big) \quad , \tag{3.47}$$

with (note the $\{+,+,+,+\}$ signature)

$$D^2 = (t-t_0)^2 + \frac{1}{c^2}\big((x-x_0)^2 + (y-y_0)^2 + (z-z_0)^2 \big) \quad . \tag{3.48}$$

This represents the information that the coordinates of the space-time event are approximately equal to (t_0, x_0, y_0, z_0) with uncertainties that are independent for each coordinate, and equal to σ . This elliptic distance is radically different from the hyperbolic distance in equation (3.43), yet we need to deal with this kind of elliptic distance in 4D space-time when developing the theory of space-time positioning.

3.3 Vectors and Tensors

For one-dimensional spaces, we have been through some long developments, in order to uncover the natural definition of distance. This distance is important because it gives to the one-dimensional space a structure of linear space, with an unambiguous definition of the sum of one-dimensional vectors.[15]

For bona fide vector spaces, there always is the ordinary sum of vectors, so we do not need special developments. For instance, if a particle is at some

[15]We have seen that, at a given point $\mathcal{P}_0$ of a metric one-dimensional manifold, a vector of the linear tangent space can be identified to an oriented segment going from $\mathcal{P}_0$ to some other point $\mathcal{P}$, the norm of the vector being equal to the distance between points.

point $\mathcal{P}$ of the physical space $\mathfrak{E}$, it may be submitted to some forces, $\mathbf{f}_1, \mathbf{f}_2 \dots$ that can be seen as vectors of the tangent linear space. The total force acting on the particle is the sum of forces $\mathbf{f} = \mathbf{f}_1 + \mathbf{f}_2 + \dots$.

Besides ordinary tensors, we may have the geotensors introduced in section 1.5. One example is the the strain tensor of the theory of finite deformation (extensively studied in section 4.3). As we have seen, geotensors are oriented geodesic segments on Lie group manifolds. The sum operation is always

$$\mathbf{t}_2 \oplus \mathbf{t}_1 = \log(\exp \mathbf{t}_2 \ \exp \mathbf{t}_1) \quad . \tag{3.49}$$

We may also mention here the spaces defined by positive tensors, like the elastic stiffness tensor $c_{ijk\ell}$ of elasticity theory. Let us examine this example in some detail here.

Example 3.3 The space of elastic media. *An ideal elastic medium can be characterized by the stiffness tensor* $\mathbf{c}$ *or the compliance tensor* $\mathbf{d} = \mathbf{c}^{-1}$. *Hooke's law relating strain* ε^{ij} *to (a sufficiently small) stress change,* $\Delta\sigma^{ij}$ *can be written, using the stiffness tensor as*

$$\Delta\sigma^{ij} = c^{ij}{}_{k\ell}\,\varepsilon^{k\ell} \quad , \tag{3.50}$$

or, equivalently, using the compliance tensor, as

$$\varepsilon^{ij} = d^{ij}{}_{k\ell}\,\Delta\sigma^{k\ell} \quad . \tag{3.51}$$

An elastic medium, say $\mathcal{E}$, *can equivalently be characterized by the stiffness tensor* $\mathbf{c}$ *or by the compliance tensor* $\mathbf{d}$. *Because of the symmetries of these tensors*[16] *an elastic medium is characterized by 21 quantities. Consider, then, an abstract, 21-dimensional manifold, where each point represents one different ideal elastic medium. As coordinates over this manifold we may choose 21 independent components of* $\mathbf{c}$, *or 21 independent components of* $\mathbf{d}$, *or the six eigenvalues and 15 angles defining one or the other of these two tensors, or any other set of 21 quantities related to these by a bijection. Each such set of 21 quantities defines a* coordinate system *over the* space of elastic media. *As mentioned in section 1.4 the spaces made by positive definite symmetric tensors are called 'symmetric spaces', and are submanifolds of Lie group manifolds. As such they are metric spaces with an unavoidable definition of distance between points. Let* $\mathcal{E}_1$ *be one elastic medium, characterized by the stiffness* $\mathbf{c}_1$ *or the compliance* $\mathbf{d}_1$, *and let* $\mathcal{E}_2$ *be a second elastic medium, characterized by the stiffness* $\mathbf{c}_2$ *or the compliance* $\mathbf{d}_2$. *The* distance *between the two elastic media, as inherited from the underlying Lie group manifold, is*

$$D(\mathcal{E}_1, \mathcal{E}_2) = \|\log(\mathbf{c}_2\,\mathbf{c}_1^{-1})\| = \|\log(\mathbf{d}_2\,\mathbf{d}_1^{-1})\| \quad , \tag{3.52}$$

where the norm of a fourth-rank tensor is defined as usual,

$$\|\psi\| = \sqrt{g_{ip}\,g_{jq}\,g^{kr}\,g^{\ell s}\,\psi^{ij}{}_{k\ell}\,\psi^{pq}{}_{rs}} \quad , \tag{3.53}$$

[16] For instance, $c_{ijk\ell} = c_{jik\ell} = c_{k\ell ij}$.

and where the logarithm of a tensor is as defined in chapter 1. The equality of the two expressions in equation (3.52) *results for the properties of the logarithm. In appendix A.23 the stiffness tensor of an isotropic elastic medium, its inverse and its logarithm are given. So introduced, the distance has two basic properties: (i) the expression of the distance is the same using the stiffness or its inverse, the compliance; (ii) the distance has an invariance of scale, i.e., the distance between the two media (characterized by)* $\mathbf{c}_1$ *and* $\mathbf{c}_2$ *is identical to that between the two media (characterized by)* $k\,\mathbf{c}_1$ *and* $k\,\mathbf{c}_2$*, where* k *is any positive real constant. This space of elastic media is one of the* quality spaces *highlighted in this text. We have seen in chapter 1 that Lie group manifolds have both curvature and torsion. The 21-dimensional manifold of elastic media being a submanifold (not a subgroup) of a Lie group manifold, it also has curvature and torsion.*

We have seen above that in the space-time of relativity, different theoretical developments may require the introduction of different definitions of distance, together with a fundamental one. This also happens in the theory of elastic media, where, together with the distance (3.52), one may introduce two other distances,

$$D_{\mathbf{c}}(\mathcal{E}_1,\mathcal{E}_2) \;=\; \|\, \mathbf{c}_2 - \mathbf{c}_1 \,\| \quad , \tag{3.54}$$

and

$$D_{\mathbf{d}}(\mathcal{E}_1,\mathcal{E}_2) \;=\; \|\, \mathbf{d}_2 - \mathbf{d}_1 \,\| \quad , \tag{3.55}$$

that appear when taking different "averages" of elastic media (Soize, 2001; Moakher, 2005).

4 Intrinsic Physical Theories

[...] the terms of an equation must have the same physical dimension. [...] Should this not hold, one would have committed some error in the calculation.

Théorie analytique de la chaleur, Joseph Fourier, 1822

Physical quantities (temperature, frequency...) can be seen as *coordinates* over manifolds representing *measurable qualities*. These quality spaces have a particular geometry (curvature, torsion, etc.). An acceptable physical theory has to be intrinsic: it has to be formulated independently on any particular choice of coordinates —i.e., independently of any particular choice of physical quantities for representing the physical qualities.— The theories so developed are tensorial in a stronger sense than the usual theories. These theories, in addition to ordinary tensors, may involve geotensors.

In this chapter two examples of intrinsic theories are given, the theory of heat transfer and the theory of ideal elastic media. The theories so obtained are quantitatively different from the commonly admitted theories. A prediction of the theory of elastic media is even absent from the standard theory (there are regions in the configuration space that cannot be reached through elastic deformation). The intrinsic theories here developed not only are mathematically correct; they better represent natural phenomena.

4.1 Intrinsic Laws in Physics

Physical laws are usually expressed as relations between quantities, but we have seen in chapter 3 that physical quantities can be interpreted as coordinates on (metric) quality spaces. In chapter 1 we found geotensors, intrinsic objects that belong to spaces with curvature and torsion.

Can we formulate physical laws without reference to particular coordinates, i.e., by using only the notion of physical quality, or of geotensor, and the (intrinsic) geometry of these spaces? The answer is yes, and the physical theories so obtained are not always equivalent to the standard ones.

Definition 4.1 *We shall say that a physical theory is* intrinsic *if it is formulated exclusively in terms of the geometrical properties (connection or metric) of the quality spaces involved.*

This, in particular, implies that physical theories depending in an essential way on the choice of physical quantities being used are not intrinsic.

This imposes that all quality spaces are to be treated tensorially, and not only, as is usually done, the physical space (in Newtonian physics) or space-time (in relativity). The equations so obtained "have more tensor indices" than standard equations: the reader may compare the equation (4.21) with the usual Fourier law (equation 4.29), or have a look at appendix A.20, where the equations describing the dynamics of a particle are briefly examined. Sometimes, the equations are equivalent (as is the case for the Newton's second law of dynamics), sometimes they are not (as is the case for the Fourier law).[1]

Besides the examples just mentioned, there is another domain where the invariance principle introduced above produces nontrivial results: when we face geotensors, as, for instance, when developing the theory of elastic deformation. There, it is the geometry of the Lie group manifolds[2] that introduces constraints that are not respected by the standard theories. The reader may, for instance, compare the elastic theory developed below with the standard theory.

4.2 Example: Law of Heat Conduction

Let us consider the law of heat conduction in the same context as that used by Fourier, i.e., typically in the ordinary heat conduction of ordinary metals (excluding, in particular, any quantization effect).

In the following pages, the theory is developed with strict adherence to the (extended) tensorial rule, but here we can anticipate the result using standard notation.

First, let us remember the standard (Fourier) law of heat conduction. In an ideal Fourier medium, the heat flux vector $\phi(\mathbf{x})$ at any point $\mathbf{x}$ is proportional (and opposed) to the gradient of the temperature field $T(\mathbf{x})$ inside the medium:

$$\phi(\mathbf{x}) = -k_F \, \mathbf{grad} \, T(\mathbf{x}) \quad . \tag{4.1}$$

Here, k_F is a constant (independent of both, temperature and spatial position) representing the particular medium being investigated. Instead, when using the intrinsic method developed here, it appears that the simplest law (in fact the *linear* law) relating heat flux to temperature gradient is

$$\phi(\mathbf{x}) = -k \, \frac{1}{T} \, \mathbf{grad} \, T(\mathbf{x}) \quad , \tag{4.2}$$

where k is a constant.

[1]Because we mention Fourier's work, it is interesting to note that the invariance principle used here bears some similarity with the condition that a physical equation must have homogeneous physical dimensions, a condition first explicitly stated by Fourier in 1822.

[2]Remember that a geotensor is an oriented geodesic (and autoparallel) segment of a Lie group manifold.

4.2.1 Quality Spaces of the Problem

Physical space. The *physical space* $\mathfrak{E}$ is modeled by a three-dimensional Riemannian (metric) manifold, not necessarily Euclidean, and described locally with a metric whose components on the natural basis associated to some coordinates $\{x^i\}$ are denoted g_{ij}, so the squared length element is

$$ds_{\mathfrak{E}}^2 = g_{ij}\, dx^i\, dx^j \quad ; \quad (\, i, j, \ldots \in \{1, 2, 3\}\,) \quad . \tag{4.3}$$

The distance has the physical dimension of a length.

Time. We work here in the physical context used by Fourier when establishing his law of heat conduction. Therefore, we assume Newtonian (nonrelativistic) physics, where time is "flowing" independently of space. The one-dimensional time manifold is denoted $\mathfrak{T}$, and an arbitrary coordinate $\{\tau^1\}$ is selected, that can be a Newtonian time t or can be any other coordinate (related to t through a bijection). For the tensor notation related with $\mathfrak{T}$ we shall use the indices $\{a, b, \ldots\}$, but as $\mathfrak{T}$ is one-dimensional, these indices can only take the value 1. We write the duration element as

$$ds_{\mathfrak{T}}^2 = G_{ab}\, d\tau^a\, d\tau^b \quad ; \quad (\, a, b, \ldots \in \{1\}\,) \quad , \tag{4.4}$$

this introducing the $1{\times}1$ metric tensor G_{ab}. As for Newtonian time, $ds_{\mathfrak{T}} = dt$, and as $dt = (dt/d\tau^1)\, d\tau^1$, the unique component of the metric tensor G_{ab} can be written

$$G_{11} = \left(\frac{dt}{d\tau^1}\right)^2 \quad . \tag{4.5}$$

Here, t is a Newtonian time, and τ^1 is the arbitrary coordinate being used on the time manifold to label instants.

Cold–hot space. The one-dimensional cold–hot manifold $\mathfrak{C}|\mathfrak{H}$ has been analyzed in section 3.1.6, where the distance between two points was expressed as

$$\operatorname{dist}(\mathcal{A}_1, \mathcal{A}_2) = \left| \log \frac{T_2}{T_1} \right| = \left| \log \frac{\beta_2}{\beta_1} \right| = \left| T_2^* - T_1^* \right| \quad . \tag{4.6}$$

Here, T is the absolute temperature, β is the thermodynamic parameter $\beta = 1/(\kappa T)$ (κ denoting here the Boltzmann's constant), and T^* is a logarithmic temperature $T^* = \log(T/T_0)$ (T_0 being an arbitrary constant value). The distance element was written

$$ds_{\mathfrak{C}|\mathfrak{H}}^2 = \gamma_{\alpha\beta}\, d\lambda^\alpha\, d\lambda^\beta \quad ; \quad (\, \alpha, \beta, \ldots \in \{1\}\,) \quad , \tag{4.7}$$

this introducing the 1×1 metric tensor $\gamma_{\alpha\beta}$. As explained in section 3.1.6, we reserve Greek indices $\{\alpha, \beta, \ldots\}$ for use as tensor indices of the one-dimensional cold–hot manifold. If using as coordinate λ^1 the temperature, the inverse temperature, or the logarithmic temperature,

$$\sqrt{\gamma_{11}} = 1/T \quad ; \quad \text{(if using temperature } T\text{)}$$
$$\sqrt{\gamma_{11}} = 1/\beta \quad ; \quad \text{(if using inverse temperature } \beta\text{)} \qquad (4.8)$$
$$\sqrt{\gamma_{11}} = 1 \quad ; \quad \text{(if using logarithmic temperature } T^*\text{)} \quad .$$

Thermal variation. When two thermodynamic reservoirs are put in contact, calories flow from the hot to the cold reservoir. Equivalently, frigories flow from the cold to the hot reservoir. While engineers working with heating systems tend to use the calorie quantity c, those working with cooling systems tend to use the frigorie quantity f. These two quantities —that may both take positive or negative values— are mutually opposite: $c = -f$. This immediately suggests that these two quantities are Cartesian coordinates in the "space of thermal variation" (that we may denote with the symbol $\mathfrak{H}$), endowed with the following definition: the distance between two points on the space of thermal variation is $D = |c_2 - c_1| = |f_2 - f_1|$, the associated distance element satisfying

$$|ds_{\mathfrak{H}}| = |dc| = |df| \quad . \qquad (4.9)$$

If instead of calories or frigories we choose to use an arbitrary coordinate $\{\kappa^A\} = \{\kappa^1\}$ over $\mathfrak{H}$, we write, using the standard notation,

$$ds^2_{\mathfrak{H}} = \Gamma_{AB}\, d\kappa^A\, d\kappa^B \quad ; \quad (A,B,\ldots \in \{1\}) \quad , \qquad (4.10)$$

this defining the 1×1 metric tensor Γ_{AB}. We reserve the upper-case indices $\{A,B,\ldots\}$ for use as tensor indices of the one-dimensional manifold $\mathfrak{H}$. Should one use as coordinate κ^1 the calorie or the frigorie, as usual, then $\Gamma_{11} = 1$. The reader may here note that while a (time) duration is a Jeffreys quantity, the (Newtonian) time coordinate is a Cartesian quantity. Both quantities are measured in seconds, but are quite different physically. The same happens here: while the total heat (i.e., energy) content of a thermodynamic system is a Jeffreys quantity, the thermal variation is a Cartesian quantity.

As there are many different symbols used in the four quality spaces, we need a table summarizing them:

quality manifold	coordinate(s)	distance element
physical space	$\{x^i\}$; $i,j,\ldots \in \{1,2,3\}$	$ds^2_{\mathfrak{E}} = g_{ij}\, dx^i\, dx^j$
time manifold	$\{\tau^a\}$; $a,b,\ldots \in \{1\}$	$ds^2_{\mathfrak{T}} = G_{ab}\, d\tau^a\, d\tau^b$
cold–hot manifold	$\{\lambda^\alpha\}$; $\alpha,\beta,\ldots \in \{1\}$	$ds^2_{\mathfrak{C}\vert\mathfrak{H}} = \gamma_{\alpha\beta}\, d\lambda^\alpha\, d\lambda^\beta$
thermal variation	$\{\kappa^A\}$; $A,B,\ldots \in \{1\}$	$ds^2_{\mathfrak{H}} = \Gamma_{AB}\, d\kappa^A\, d\kappa^B$

4.2.2 Thermal Flux

To measure the thermal flux, at a given point of the space, we consider a small surface element Δs_i. Then, we choose a small time vector whose (unique) component is $\Delta\tau^a$ (remember that we are not necessarily using Newtonian time). We are free to choose the orientation of this time vector (from past to future or from future to past) and its magnitude. Given a particular $\Delta\mathbf{s}$ and a particular $\Delta\boldsymbol{\tau}$ we can measure "how many frigories–calories" have crossed the surface, to obtain a vector in the thermal variation space, whose (unique) component is denoted $\Delta\kappa^A$. This vector indicates how many frigories–calories pass through the given surface element $\Delta\mathbf{s}$ during the given time lapse $\Delta\boldsymbol{\tau}$. Then, the *thermal flux* tensor, with components $\{\phi_a{}^{iA}\}$, is defined by the proportionality relation

$$\Delta\kappa^A = \phi_a{}^{iA}\,\Delta s_i\,\Delta\tau^a \quad . \tag{4.11}$$

The (Frobenius) norm of the thermal flux is $\|\,\boldsymbol{\phi}\,\| = \sqrt{G^{ab}\,\Gamma_{AB}\,g_{ij}\,\phi_a{}^{iA}\,\phi_b{}^{jB}}$, i.e., using noncovariant notation,

$$\|\,\boldsymbol{\phi}\,\| = \frac{\sqrt{\Gamma_{11}}}{\sqrt{G_{11}}}\,\sqrt{g_{ij}\,\phi_1{}^{i1}\,\phi_1{}^{j1}} \quad . \tag{4.12}$$

If we use calories c to measure the heat transfer, $\Gamma_{11} = 1$. If we use Newtonian time to measure time, $G_{11} = 1$. Then, $\|\,\boldsymbol{\phi}\,\| = \sqrt{g_{ij}\,\phi_1{}^{i1}\,\phi_1{}^{j1}}$.

4.2.3 Gradient of a Cold–Hot Field

A cold–hot field is a mapping that to any point $\mathcal{P}$ of the physical space $\mathfrak{E}$ associates a point $\mathcal{A}$ of the cold–hot manifold $\mathfrak{C}|\mathfrak{H}$.

The derivative of such a mapping may be called the *gradient of the cold–hot field*. When in the cold–hot space a coordinate λ^1 is used, and in the physical space a system of coordinates $\{x^i\}$ is used, a cold–hot field is described by a mapping

$$\{x^1, x^2, x^3\} \;\mapsto\; \lambda^\alpha(x^1, x^2, x^3) \quad ; \quad (\,\alpha = 1\,) \quad . \tag{4.13}$$

The derivative of the field (at a given point of space) is the (1×3) tensor $\mathbf{D}$ whose components (in the natural bases associated to the given coordinates) are

$$D^\alpha{}_i = \frac{\partial\lambda^\alpha}{\partial x^i} \quad . \tag{4.14}$$

The norm of the derivative is $\|\,\mathbf{D}\,\| = \sqrt{\gamma_{\alpha\beta}\,g^{ij}\,D^\alpha{}_i\,D^\beta{}_j}$, i.e., using noncovariant notation,

$$\|\,\mathbf{D}\,\| = \sqrt{\gamma_{11}}\,\sqrt{g^{ij}\,D^1{}_i\,D^1{}_j} \quad . \tag{4.15}$$

Should one use the temperature T as a coordinate over the cold–hot field, $\lambda^1 = T$, then $ds_{\mathfrak{C}|\mathfrak{H}} = dT/T$, $\sqrt{\gamma_{11}} = 1/T$, and

$$\| \mathbf{D} \| = \frac{1}{T} \sqrt{g^{ij} D^1{}_i D^1{}_j} \quad . \tag{4.16}$$

This is an invariant (we would obtain the same norm using inverse temperature or logarithmic temperature).

4.2.4 Linear Law of Heat Conduction

We shall say that a (heat) conduction medium is *linear* if the heat flux ϕ is proportional to the gradient of the cold–hot field $\mathbf{D}$. Using compact notation, this can be written $\phi = \mathbf{K} \cdot \mathbf{D}$, or, more explicitly, using the components of the tensors in the natural bases associated to the working coordinates (as introduced in the sections above),

$$\phi_a{}^{iA} = K_{a\alpha}{}^{ijA} D^\alpha{}_j \quad , \tag{4.17}$$

i.e.,

$$\boxed{\phi_a{}^{iA} = K_{a\alpha}{}^{ijA} \frac{\partial \lambda^\alpha}{\partial x^j} \quad .} \tag{4.18}$$

The $K_{a\alpha}{}^{ijA}$ are the components of the characteristic tensor of the linear mapping. Their sign is discussed below. The norm of this tensor is $\| \mathbf{K} \| = (G^{ab} \gamma^{\alpha\beta} \Gamma_{AB} \, g_{ik} \, g_{j\ell} \, K_{a\alpha}{}^{ijA} K_{b\beta}{}^{k\ell B})^{1/2}$, i.e., using noncovariant notation

$$\| \mathbf{K} \| = \frac{\sqrt{\Gamma_{11}}}{\sqrt{G_{11} \, \gamma_{11}}} \sqrt{g_{ik} \, g_{j\ell} \, K_{11}{}^{ij1} K_{11}{}^{k\ell 1}} \quad . \tag{4.19}$$

If the medium under investigation is isotropic, then there is a tensor $k_{a\alpha}{}^A$ such that

$$K_{a\alpha}{}^{ijA} = g^{ij} k_{a\alpha}{}^A \quad , \tag{4.20}$$

and the linear law of heat conduction simplifies to $\phi_{ai}{}^A = k_{a\alpha}{}^A D^\alpha{}_i$, i.e.,

$$\boxed{\phi_{ai}{}^A = k_{a\alpha}{}^A \frac{\partial \lambda^\alpha}{\partial x^i} \quad ,} \tag{4.21}$$

the norm of the tensor $\mathbf{k} = \{k_{a\alpha}{}^A\}$ being $\| \mathbf{k} \| = (G^{ab} \gamma^{\alpha\beta} \Gamma_{AB} \, k_{a\alpha}{}^A k_{b\beta}{}^B)^{1/2}$, i.e., using noncovariant notation,

$$k \equiv \| \mathbf{k} \| = \frac{\sqrt{\Gamma_{11}}}{\sqrt{G_{11} \, \gamma_{11}}} | k_{11}{}^1 | \quad . \tag{4.22}$$

It is the sign of the unique component, $k_{11}{}^1$, of the tensor $\mathbf{k}$ that determines in which sense calories (or frigories) flow. To match the behavior or natural media (or to match the conclusions of thermodynamic theories), we must supplement the definition of ideal conductive medium with a criterion for the sign of this unique component $k_{11}{}^1$ of $\mathbf{k}$:

- if a coordinate is chosen for time that runs from past to future (like the usual Newtonian time t with its usual orientation),
- if a coordinate is chosen for the cold–hot space that runs from cold to hot (like the absolute temperature T),
- and if "calories are counted positively" (and "frigories are counted negatively"),

then, ${k_{11}}^1$ is negative. Each change of choice of orientation in each of the three unidimensional quality spaces changes the sign of ${k_{11}}^1$.

Equation (4.21) can then be written, using the definition of k in equation (4.22),

$$ {\phi_{1i}}^1 \; = \; \pm\, k \; \frac{\sqrt{G_{11}\,\gamma_{11}}}{\sqrt{\Gamma_{11}}} \; \frac{\partial \lambda^1}{\partial x^i} \quad . \tag{4.23} $$

The parameter k, that may be a function of the space coordinates, characterizes the medium. As k is the norm of a tensor, it is a true (invariant) scalar, i.e., a quantity whose value is independent of the choice of coordinate λ^1 over the cold–hot space, the choice of coordinate τ^1 over the time space, and the choice of coordinate κ^1 over the space of thermal variations (and, of course, of the choice of coordinates $\{x^i\}$ over the physical space). The sign of the equation —which depends on the quantities being used— must correspond to the condition stated above.

To make the link with normal theory, let us particularize to the use of common quantities. When using calories to measure the thermal variation, $\kappa = c$, and Newtonian time to measure time variation, $\tau = t$, one simply has

$$ \sqrt{\Gamma_{11}} \; = \; 1 \qquad ; \qquad \sqrt{G_{11}} \; = \; 1 \quad . \tag{4.24} $$

In this situation, the components ${\phi_{1i}}^1$ are identical to the components of the ordinary heat flux tensor ϕ_i, and equation (4.23) particularizes to

$$ \phi_i \; = \; \pm\, k \, \sqrt{\gamma_{11}} \; \frac{\partial \lambda^1}{\partial x^i} \quad , \tag{4.25} $$

where, still, the coordinate λ^1 on the cold–hot manifold is arbitrary.

When using in the cold–hot field the (absolute) temperature $\lambda = T$, as we have seen when introducing the metric on the cold–hot manifold. Then, equation (4.25) particularizes to

$$ \phi_i \; = \; -k \, \frac{1}{T} \, \frac{\partial T}{\partial x^i} \quad , \tag{4.26} $$

where k is the parameter characterizing the linear medium under consideration. Should we choose instead of the temperature T the thermodynamic parameter $\beta = 1/(\kappa T)$, then, $\sqrt{\gamma_{\beta\beta}} \; = \; 1/\beta$, and we would obtain

$$ \phi_i \; = \; k \, \frac{1}{\beta} \, \frac{\partial \beta}{\partial x^i} \quad . \tag{4.27} $$

Putting these two equations together,

$$\phi_i = -k \frac{1}{T} \frac{\partial T}{\partial x^i} = k \frac{1}{\beta} \frac{\partial \beta}{\partial x^i} \quad . \tag{4.28}$$

The formal symmetry between these two expressions is an example of the invariance of the form of expressions that must be satisfied when changing one Jeffreys parameter by its inverse.[3] The usual Fourier's law

$$\phi_i = -k_F \frac{\partial T}{\partial x^i} = k_F \frac{1}{\beta^2} \frac{\partial \beta}{\partial x^i} \tag{4.29}$$

does not have this invariance of form.

We may now ask which of the two models, the law (4.28) or the Fourier law (4.29), best describes the thermal behavior of real bodies. The problem is that ordinary media have very complex mechanisms for heat conduction. An ideal medium should have the parameter k in equation (4.28) constant (or the parameter k_F in equation (4.29), if one believes in Fourier law). This is far from being the case, and it is in fact a quite difficult experimental task to tabulate the values of the conductivity "constant" as a function of temperature, especially at low temperatures. Let us consider here not a particular temperature range, where a particular mechanism may explain the heat transfer, but let us rather consider temperature ranges as large as possible, and let us ask the following question: "if a metal bar has at point x_1 the temperature T_1, and at point x_2 the temperature T_2, what is the variation of temperature outside the region between x_1 and x_2?"

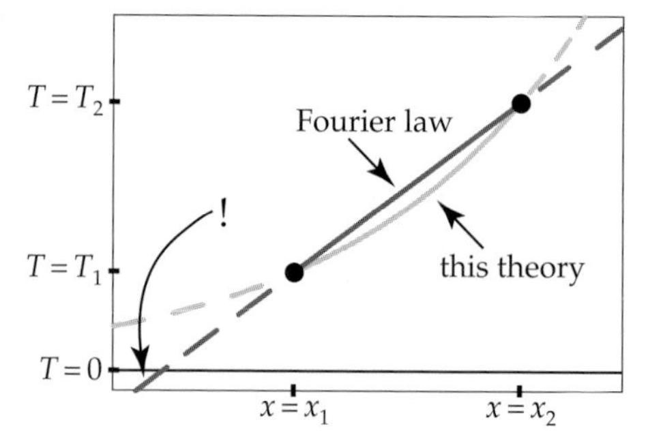

Fig. 4.1. *Assume that the temperature values T_1 and T_2 at two points x_1 and x_2 of a metallic bar in a stationary state are known. This figure shows the interpolation of the temperature values between the two points x_1 and x_2, and its extrapolation outside the points, as predicted by the law (4.27) and the Fourier law (4.29). While the Fourier law would predict negative temperatures, the model proposed here has the correct qualitative behavior.*

Figure 4.1 displays the prediction of the law (4.27) (with a constant value of k) and that of the Fourier law (4.29) (with a constant value of k_F): while the

[3] The change of sign here results from the breaking of tensoriality we have produced when choosing to use the pseudo-vector ϕ_i instead of the tensor $\phi_{ai}{}^A$.

Fourier law predicts a linear variation of temperature, the law (4.27) predicts an exponential variation.[4] It is clear that the Fourier prediction is qualitatively unacceptable, as it predicts negative temperatures (see figure 4.1). This suggests that an ideal heat conductor should be defined through the law (4.28), i.e., in fact, through equation (4.21) for an isotropic medium or equation (4.18) for a general medium.

4.3 Example: Ideal Elasticity

4.3.1 Introduction

Experiments suggest that there are bodies that have an elastic behavior: their shape (configuration) depends only on the efforts (tensions) being exerted on them (and not on the deformation history).

Simply put, an ideal elastic medium is defined by a proportionality between applied stress σ and obtained strain ε. Complications appear when trying to properly define the strain: the commonly accepted measures of strain (Lagrangian and Eulerian) do not conform to the geometric properties of the 'configuration space' (a space where each point corresponds to a shape of the medium). For the configuration space is a submanifold of the Lie group $GL^+(3)$, and the only possible measure of strain (as the geodesics of the space) is logarithmic.

It turns out that the general theory spontaneously contains 'micro-rotations' (in the sense of Cosserat[5]), this being intimately related to the existence of an antisymmetric part in the stress tensor. But at any stage of the theory, the micro-rotations may be assumed to vanish, and, still, the remaining theory, with symmetric stresses, differs from the usual theories.[6]

Although the possibility of a logarithmic definition of strain appears quite often in the literature, all authors tend to point out the difficulty (or even

[4]We use here the terms 'linear' and 'exponential' in the ordinary sense, that is at odds with the sense they have in this book. The relation $T(x) = T(x_0)\, \exp(\alpha\,(x-x_0))$ *is* the linear relation, as it can be written $\log(\, T(x)/T(x_0)\,) = \alpha\,(x-x_0)$. As $\log(\, T(x)/T(x_0)\,)$ is the expression of a distance in the cold–hot space, and $x - x_0$ is the expression of a distance in the physical space, the relation $T(x) = T(x_0)\, \exp(\alpha\,(x - x_0))$ just imposes the condition that the variations of distances in the cold–hot space are proportional to the variations of distances in the physical space.

[5]The brothers E. Cosserat and F. Cosserat published in 1909 their well known *Théorie des corps déformables* (Hermann, Paris), where a medium is not assumed to be composed of featureless points, but of small referentials. Their ancient notation makes reading the text a lengthy exercise.

[6]There is no general argument that the stress must be symmetric, provided that one allows for the existence of force moment density χ_{ij} acting from the outside into the medium, as one has $\sigma_{ij} - \sigma_{ji} = \chi_{ij}$. Arguments favoring the existence of an asymmetric stress are given by Nowacki (1986).

the impossibility) of reaching the goal (see some comments in section 4.3.8). Perhaps what has stopped many authors is the misinterpretation of the rotations appearing in the theory as macroscopic rotations, while they are to be interpreted, as we shall see, as micro-rotations. Then, of course, there also is today's lack of familiarity of many physicists with the logarithms of tensors.

Besides the connection of the work presented here with the Cosserattheory, and with Nowacki's*Theory of Asymmetric Elasticity* (1986), there are some connections with the works of Truesdell and Toupin (1960), Sedov (1973), Marsden and Hughes (1983), Ogden (1984), Ciarlet (1988), Kleinert (1989), Rougée (1997), and Garrigues (2002ab).

Although one could directly develop a theory valid for general heterogenous deformations, it is better, for pedagogical reasons, to split the problem in two, analyzing first (and mainly) the homogeneous deformations.

Here below, we assume a three-dimensional Euclidean space. With given coordinates $\{x^i\}$, the metric tensor (representing the Euclidean metric) has, at any point, components $g_{ij}(\mathbf{x})$ on the local basis at the given point. A continuous medium fills part of the space, and when a system of volume and surface forces acts on the medium, they create a stress field $\sigma^{ij}(\mathbf{x})$ at every point of it. It is assumed that the stress vanishes when there are no forces acting on the medium.

4.3.2 Configuration Space

Assume that a fixed *laboratory coordinate system* $\{x^i\}$, with metric tensor $g_{ij}(x^1, x^2, x^3)$, is given. The material point whose *current coordinates* are $\{x^i\}$ had some *initial coordinates* $\{X^i\}$. We can assume given any of the two equivalent functions

$$X^i = X^i(x^1, x^2, x^3) \quad ; \quad x^i = x^i(X^1, X^2, X^3) \quad . \tag{4.30}$$

One can then introduce the *displacement gradients*

$$S^i{}_j(x^1, x^2, x^3) = \frac{\partial X^i}{\partial x^j}(x^1, x^2, x^3) \tag{4.31}$$

and

$$T^i{}_j(x^1, x^2, x^3) = \frac{\partial x^i}{\partial X^j}(\, X^1(x^1, x^2, x^3) \, , \, X^2(x^1, x^2, x^3) \, , \, X^3(x^1, x^2, x^3) \,) \quad . \tag{4.32}$$

The displacement gradient $\mathbf{T}(x^1, x^2, x^3)$ can alternatively be computed as the inverse of $\mathbf{S}^{-1}(x^1, x^2, x^3)$:

$$\mathbf{T}(x^1, x^2, x^3) \equiv \mathbf{S}^{-1}(x^1, x^2, x^3) \quad . \tag{4.33}$$

In the absence of micro-rotations, the tensor field $T_i{}^j(x^1, x^2, x^3)$ has all necessary information on the "transformation" in the vicinity of every point.

The components of this tensor are defined on the natural basis associated (at each point) to the system of laboratory coordinates.

Much of what we are going to say would remain valid for a general transformation, where the field $T^i{}_j(x^1, x^2, x^3)$ may vary from point to point. But let us simplify the exposition by assuming, unless otherwise stated, that we have a homogeneous transformation (in an Euclidean space).

Example 4.1 *Consider the homogeneous transformation of a body in an Euclidean space, where a system of rectilinear coordinates is used. Then, the tensor* $\mathbf{T}$ *although a function of time, is constant in space (and its components* $T^i{}_j$ *only depend on time). The relation*

$$x^i = T^i{}_j X^j \quad , \tag{4.34}$$

then gives the final coordinates of a material point with initial coordinates X^i.

In the absence of micro-rotations ("symmetric elasticity"), the simplest way to express the stress-strain relation is to consider, at the "current time" when the evaluation is made, and at every point of the body, a polar decomposition of $\mathbf{T}$ (see appendix A.21.2) this defining a *macro-rotation* $\mathbf{R}$ and two symmetric positive definite tensors $\mathbf{E}$ and $\mathbf{F}$ such that one has

$$\mathbf{T} = \mathbf{R}\,\mathbf{E} = \mathbf{F}\,\mathbf{R} \quad . \tag{4.35}$$

The two symmetric tensors $\mathbf{E}$ and $\mathbf{F}$ are called *deformations*, they can be obtained as[7]

$$\mathbf{E} = (\mathbf{T}^*\,\mathbf{T})^{1/2} = (\mathbf{g}^{-1}\,\mathbf{T}^t\,\mathbf{g}\,\mathbf{T})^{1/2} \quad ; \quad \mathbf{F} = (\mathbf{T}\,\mathbf{T}^*)^{1/2} = (\mathbf{T}\,\mathbf{g}^{-1}\,\mathbf{T}^t\,\mathbf{g})^{1/2} \ , \tag{4.36}$$

and they are related via

$$\mathbf{F} = \mathbf{R}\,\mathbf{E}\,\mathbf{R}^{-1} \quad . \tag{4.37}$$

Expressions (4.35) can be interpreted as follows. The transformation $\mathbf{T}$ may have followed a complicated path between the initial time and the current time, but there are two simple ways that would give the current transformation: *(i)* applying first the deformation $\mathbf{E}$, then the rotation $\mathbf{R}$, or *(ii)* applying first the rotation $\mathbf{R}$, then the deformation $\mathbf{F}$. This interpretation suggests to name $\mathbf{E}$ the *unrotated deformation* and to name $\mathbf{F}$ the *rotated deformation*.

The stress-strain relation may be introduced using any of these two possible thought experiments, then verifying that they define the same stress. This stress is then taken, by definition, as the stress associated to the transformation $T^i{}_j$. These two experiments are considered in appendix A.24, and it is verified that they lead to the same state of stress.

In 3D elasticity, the space of transformations $\mathbf{T} = \{T^i{}_j\}$ can clearly be identified with $GL^+(3)$, but this space is not to be identified with the space

[7]Explicitly, $(\mathbf{E}^2)^i{}_j = g^{ik}\,T^\ell{}_k\,g_{\ell r}\,T^r{}_j$, and $(\mathbf{F}^2)^i{}_j = T^i{}_k\,g^{k\ell}\,T^r{}_\ell\,g_{rj}$.

of "configurations" of the body, for two reasons. First, there is not a one-to-one mapping[8] between the stress space and the space of the transformations $\mathbf{T}$. Second, the rotation $\mathbf{R}$ appearing in the polar decomposition $\mathbf{T} = \mathbf{R}\,\mathbf{E} = \mathbf{F}\,\mathbf{R}$ of a transformation is a macroscopic rotation, not related to the stress change, while a general theory of elasticity must be able to accommodate the possible existence of micro-rotations: in "micropolar media", the stress tensor needs not be symmetric, and each "molecule" may experience "micro-rotations". The surrounding molecules provide an elastic resistance to this micro-rotation, and this is the reason for the existence of an antisymmetric part of the stress (representing a force-moment density).

I suggest that the proper way to introduce the possibility of a micro-rotation into the theory is as follows (for heterogeneous transformations, see appendix A.25).

First, we get rid of the global body rotation, by just assuming $\mathbf{R} = \mathbf{I}$ in the equations above. In this case,

$$\mathbf{T} = \mathbf{E} = \mathbf{F} \quad , \tag{4.38}$$

and we choose to use the symbol $\mathbf{E}$ for this (symmetric) deformation.

Now, consider that part of $GL^+(3)$ that is geodesically connected to the origin of the group. This, in fact, is the set of matrices of $GL^+(3)$ whose logarithm is a real matrix. Let $\mathbf{C}$ be such a matrix, and let

$$\varepsilon = \log \mathbf{C} \tag{4.39}$$

be its logarithm (by hypothesis, it is a real matrix). The decomposition of ε into its symmetric part $\mathbf{e}$ and its antisymmetric part $\mathbf{s}$,

$$\mathbf{e} = \hat{\varepsilon} \equiv \tfrac{1}{2}\,(\varepsilon + \varepsilon^*) \quad ; \quad \mathbf{s} = \check{\varepsilon} \equiv \tfrac{1}{2}\,(\varepsilon - \varepsilon^*) \tag{4.40}$$

defines a (symmetric) deformation $\mathbf{E}$ and a rotation $\mathbf{S}$ (orthogonal tensor), respectively given by

$$\mathbf{E} = \exp \mathbf{e} \quad ; \quad \mathbf{S} = \exp \mathbf{s} \tag{4.41}$$

and, by definition, one has

$$\varepsilon = \mathbf{e} + \mathbf{s} \quad ; \quad \log \mathbf{C} = \log \mathbf{E} + \log \mathbf{S} \quad . \tag{4.42}$$

$\mathbf{E}$ corresponds to the (symmetric) deformation introduced above, and $\mathbf{S}$ corresponds to the micro-rotation (of the "molecules").

We shall see that this interpretation makes sense, as the simple proportionality between $\varepsilon = \log \mathbf{C}$ (that shall be interpreted as a strain) and the (possibly asymmetric) stress will provide a simple theory of elastic media. Therefore, we formally introduce the

[8]Compressing an isotropic body vertically and extending it horizontally, then rotating the body by 90 degrees, gives the same stress as extending the body vertically and compressing it horizontally, yet the two transformations $\{T^i{}_j\}$ are quite different.

Definition 4.2 Configuration space (of asymmetric elasticity). *In asymmetric elasticity, the configuration space* $\mathfrak{C}$ *is the subset of* $GL^+(3)$ *that is geodesically connected to the origin of the group, i.e., the set of matrices of* $GL^+(3)$ *whose logarithm is a real matrix.*[9]

If $\mathbf{C} \in \mathfrak{C}$, the decomposition made in equations (4.39)–(4.42) into a symmetric $\mathbf{E}$ and an orthogonal $\mathbf{S}$, corresponds to a (macroscopic) deformation and a micro-rotation.

Figure 4.2 suggests in which sense the micro-rotations of this theory coexist with, but are different from, the macroscopic (or "mesoscopic") rotations.

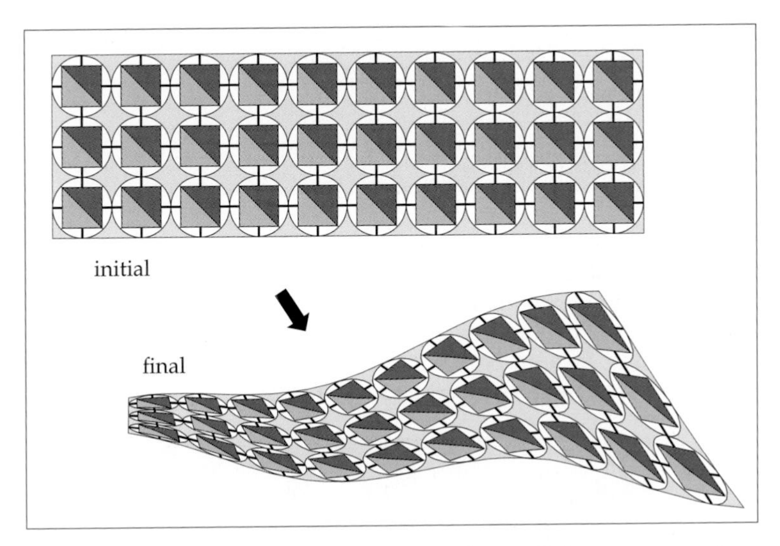

Fig. 4.2. *We consider here media made by "molecules" that may experience relative rotations. The different parts of the body may have macroscopic displacements and macroscopic rotations, and, in addition, there can be deformations and micro-rotations. In this two-dimensional sketch, besides the translations, one can observe some macroscopic rotations: at the rightmost part of the body, the macroscopic rotation has about 35 degrees, while at the leftmost part, it is quite small. There are also deformations, represented by small circles becoming small ellipses. Finally, there are micro-rotations, each molecule experiencing a rotation with respect to the neighboring molecules: the micro-rotations are zero at both, the left and the right part of the body, while they are of about 15 degrees in the middle (note that the black marks have lost their initial alignment there).*

Let us now see a series of sketches illustrating the configuration space in the case of 2D elasticity. Figures 4.3 and 4.4 show two similar sections of the configuration space. At each point of the configuration space a configuration

[9]In the terminology of chapter 1 (section 1.4.4), this set is the near identity subset $GL^+(3)_I$.

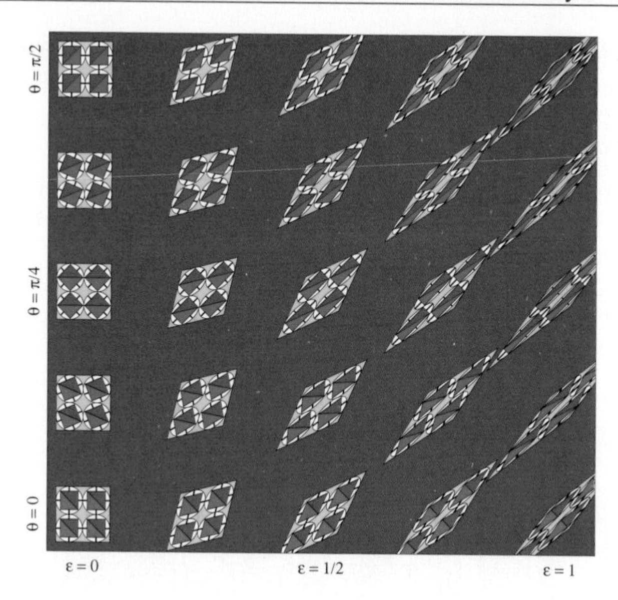

Fig. 4.3. *Section of* SL(2)*, interpreting each of the points as a configuration of a molecular body. The representation here corresponds to the representation at the right in figure 1.19, with $\varphi = 0$. Along the line $\theta = 0$ there are no micro-rotations. Along the other lines, there are both, a micro-rotation (i.e., a rotation of the molecules relatively to each other) and a shear. See text for details.*

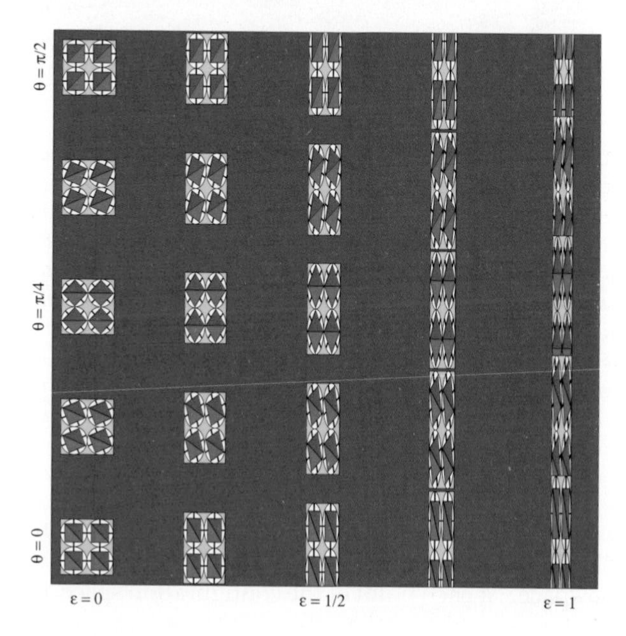

Fig. 4.4. *Same as figure 4.3, but for $\varphi = \pi/2$.*

of a molecular body is suggested. These two sections are represented using the coordinates $\{\varepsilon, \theta, \varphi\}$, and correspond to the representation of the configuration space suggested at the right of figure 1.19. Figure 4.3 is for $\varphi = 0$, and figure 4.4 is for $\varphi = \pi/2$.

Figure 4.5 represents the symmetric submanifold of the configuration space (only for isochoric transformations). There are no micro-rotations there, and this is the usual configuration space of the theory of symmetric elasticity (excepted that, here, this configuration space is identified with the symmetric subspace of SL(2)). Some of the geodesics of this manifold are represented in figure 4.6.

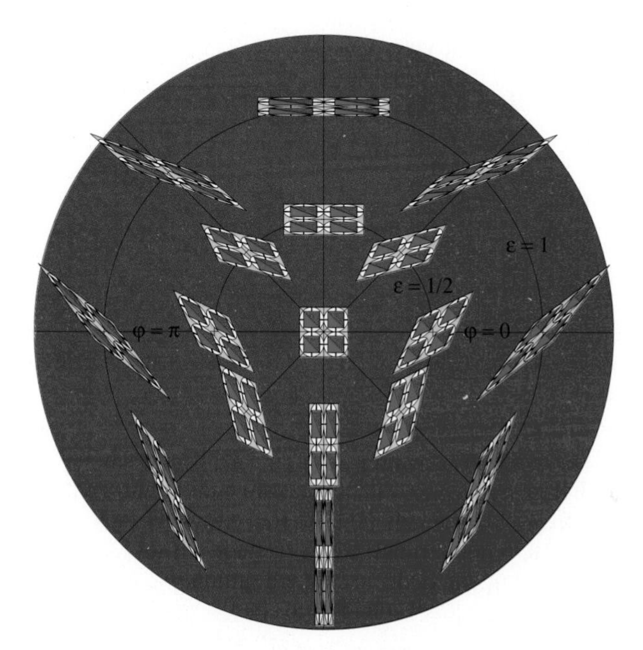

Fig. 4.5. *Representation of the symmetric (and isochoric) configurations of the configuration space (no micro-rotations). This two-dimensional space corresponds to the section $\theta = \alpha = 0$ of the three-dimensional SL(2) Lie group manifold, represented in figures 1.12 and 1.13 and in figures 1.17 and 1.18. Some of the geodesics of this 2D manifold are represented in figure 4.6.*

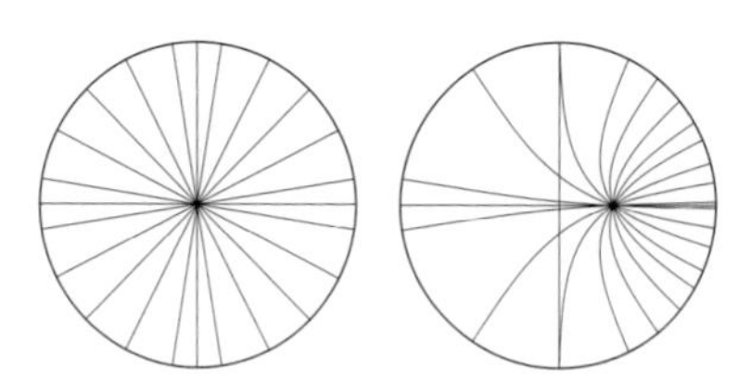

Fig. 4.6. *At the left, some of the geodesics leaving the origin in the configuration space of figure 4.5. At the right, some geodesics leaving a point that is not the origin.*

As an example of the type of expressions produced by this theory, let us ask the following question: which is the strain ε_{21} experienced by a body when it transforms from configuration $\mathbf{C}_1 = \exp \varepsilon_1$ to configuration $\mathbf{C}_2 = \exp \varepsilon_2$? A simple evaluation shows[10] that the unrotated strain (there may also be a macro-rotation involved) is

$$\varepsilon_{21} = \hat{\varepsilon}_{21} + \check{\varepsilon}_{21} \quad , \tag{4.43}$$

where the symmetric and the antisymmetric parts of the strain are given by[11]

$$\hat{\varepsilon}_{21} = \tfrac{1}{2}((\text{-}\,\hat{\varepsilon}_1) \oplus (2\,\hat{\varepsilon}_1) \oplus (\text{-}\,\hat{\varepsilon}_1)) \quad ; \quad \check{\varepsilon}_{21} = \check{\varepsilon}_2 \ominus \check{\varepsilon}_2 \quad . \tag{4.44}$$

A series expansion of $\hat{\varepsilon}_{21}$ gives

$$\hat{\varepsilon}_{21} = (\varepsilon_2 - \varepsilon_1) - \tfrac{1}{6}\,(\mathbf{e} + \mathbf{e}^*) + \dots \quad , \tag{4.45}$$

where $\mathbf{e} = \varepsilon_2^2\,\varepsilon_1 + \varepsilon_2\,\varepsilon_1^2 - \frac{1}{2}\,\varepsilon_1\,\varepsilon_2\,\varepsilon_1 - 2\,\varepsilon_2\,\varepsilon_1\,\varepsilon_2$ (there are no second order terms in this expansion).

4.3.3 Stress Space

As discovered by Cauchy, the 'state of tensions' at any point inside a continuous medium is not to be described by a system of vectors, but a 'two-index tensor': the *stress tensor* (in fact, this is the very origin of the name 'tensor' used today with a more general meaning).

As we are not going to assume any particular symmetry for the stress, the space of all possible states of stress at the considered point inside a continuous medium, is a nine-dimensional linear space. We are familiar with the usual basis $\{\mathbf{e}^i \otimes \mathbf{e}^j\}$ that is induced in such a space of tensors by a choice of basis $\{\mathbf{e}_i\}$ in the underlying 3D physical space. Then, any stress tensor can be written as

$$\sigma = \sigma_{ij}\,\mathbf{e}^i \otimes \mathbf{e}^j \quad . \tag{4.46}$$

It is immaterial whether we consider the covariant or the contravariant components of the stress, as we shall always assume here that the underlying space has a metric whose components (in the given basis) are g_{ij}.

---

[10]The configuration $\mathbf{C}_1$ corresponds to a symmetric deformation $\mathbf{E}_1$ (from the reference configuration) and to a micro-rotation $\mathbf{S}_1$, and one has $\varepsilon_1 = \log \mathbf{C}_1 = \log \mathbf{E}_1 + \log \mathbf{S}_1$, with a similar set of equations for $\mathbf{C}_2$. Moving in the configuration space from point $\mathbf{C}_1$ to point $\mathbf{C}_2$, produces the transformation $\mathbf{T}_{21} = \mathbf{E}_2\,\mathbf{E}_1^{-1}$ and the micro-rotation $\mathbf{S}_{21} = \mathbf{S}_2\,\mathbf{S}_1^{-1}$ The transformation $\mathbf{T}_{21}$ has a polar decomposition $\mathbf{T}_{21} = \mathbf{R}_{21}\,\mathbf{E}_{21}$, but as we are evaluating the unrotated strain, we disregard the macro-rotation $\mathbf{R}_{21}$, and evaluate $\varepsilon_{21} = \log \mathbf{E}_{21} + \log \mathbf{S}_{21}$.

[11]Using the geometric sum $\mathbf{t}_1 \oplus \mathbf{t}_2 \equiv \log((\exp \mathbf{t}_2)\,(\exp \mathbf{t}_1))$ and the geometric difference $\mathbf{t}_1 \ominus \mathbf{t}_2 \equiv \log((\exp \mathbf{t}_2)\,(\exp \mathbf{t}_1)^{-1})$, introduced in chapter 1.

At each point of a general continuous medium, the actions of the exterior world are described by a *force density* φ_i and a *moment-force density*[12] χ_{ij} (Truesdell and Toupin, 1960). The medium reacts by developing a *stress* σ_{ij} and a *moment-stress* $m_{ij}{}^k$. When considering a virtual surface inside the medium, with unit normal n_i, the efforts exerted by one side of the surface on the other side correspond to some *tractions* τ_i and some *moment-tractions* μ_{ij}, that are related to the stress and the moment-stress as

$$\tau_i = \sigma_i{}^j n_j \quad ; \quad \mu_{ij} = m_{ij}{}^k n_k \quad . \tag{4.47}$$

Writing the conditions of static equilibrium (total force and total moment-force must vanish) one easily arrives to the *conditions of static equilibrium*

$$\varphi_i + \nabla_j \sigma_i{}^j = 0 \quad ; \quad \chi_{ij} + \nabla_k m_{ij}{}^k = \sigma_{ij} - \sigma_{ji} \quad . \tag{4.48}$$

The analysis of a medium that can sustain a moment-stress is outside the scope of this text, so we assume $m_{ij}{}^k = 0$. Form this, it follows that the moment-traction is also zero: $\mu_{ij} = 0$. The conditions of static equilibrium then simplify to

$$\varphi_i + \nabla_j \sigma_i{}^j = 0 \quad ; \quad \chi_{ij} = \sigma_{ij} - \sigma_{ji} \quad . \tag{4.49}$$

We do not assume that the stress is necessarily symmetric; the equation at the right shows that this is only possible if a moment force density is applied to the body from the exterior.

The stress σ_{ij} is "generated" by the force density φ_i and the moment-force density χ_{ij}, plus, perhaps, the traction τ_i at the boundary of the medium. As these forces satisfy a principle of superposition (the resultant of a system of forces is the vector sum of the forces), it is natural to become interested in the linear space structure of the stress space, with the ordinary sum of tensors and the usual multiplication by a scalar as fundamental operations:

$$(\sigma_2 + \sigma_1)_{ij} = (\sigma_2)_{ij} + (\sigma_1)_{ij} \quad ; \quad (\lambda\, \sigma)_{ij} = \lambda\, (\sigma)_{ij} \quad . \tag{4.50}$$

While the strain is a geotensor, with an associated 'sum' that is not the ordinary sum, the stress is a bona-fide tensor: *the stress space is a linear space.* It is a normed space, the norm of any element $\boldsymbol{\sigma} = \{\sigma_i{}^j\}$ of the space being

$$\| \boldsymbol{\sigma} \| = \sqrt{g^{ik}\, g_{j\ell}\, \sigma_i{}^j\, \sigma_k{}^\ell} = \sqrt{\sigma_{ij}\, \sigma^{ij}} \quad . \tag{4.51}$$

Definition 4.3 Stress space. *In asymmetric elasticity, the stress space* $\mathfrak{S}$ *is the set of all (real) stress tensors, not necessarily symmetric.*[13] *It is a linear space.*

---

[12]For a comment on the representation of moments using antisymmetric tensors, see footnote 50, page 244.

[13]There are two different conventions of sign for the stress tensor in the literature: while in mechanics, it is common to take tensile stresses as positive, in geophysics it

4.3.4 Hooke's Law

It is assumed that there is a special configuration that corresponds to the unstressed state. Then, this special configuration is taken as the origin in the configuration space, i.e., the origin for the autovectors in the space $GL^+(3)$. Figure 4.7 proposes a schematic representation of both, the stress space and the configuration space.

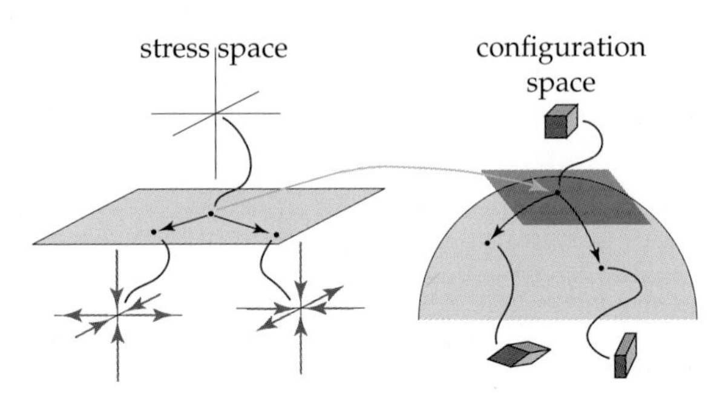

Fig. 4.7. *While the stress space is a linear space, the configuration space is a submanifold of the Lie group manifold $GL^+(3)$. The strain is a geotensor, i.e., an oriented geodesic segment over the configuration space. An ideal elastic medium corresponds, by definition, to a geodesic mapping from the stress space into the configuration space.*

We have just introduced the stress space $\mathfrak{S}$, a nine-dimensional linear space. The configuration space $\mathfrak{C}$, also nine-dimensional, is the part of $GL^+(3)$ that is geodesically connected to the origin of the group. It is a metric space, with the natural metric existing in Lie group manifolds. It is not a flat space.

Let $\mathcal{C}$ represent a point in the configuration space $\mathfrak{C}$, and $\mathcal{S}$ a point in the stress space $\mathfrak{S}$.

Definition 4.4 Elastic medium. *A medium is* elastic *if the configuration $\mathcal{C}$ depends only*[14] *on the stress $\mathcal{S}$,*

$$\mathcal{S} \mapsto \mathcal{C} = \mathcal{C}(\mathcal{S}) \quad , \tag{4.52}$$

with each stress corresponding one, and only one, configuration.[15]

is common to take compressive stresses as positive (see, for instance, Malvern, 1969). Here, we skip this complication by just choosing the mechanical convention, i.e., by counting tensile stresses as positive.

[14]And not on other variables, like the stress rate, or the deformation history.

[15]But, as we shall see, there are configurations that are not associated to any state of stress.

Representing the points of the configuration space by the matrices $C^i{}_j$ introduced above, and the elements of the stress space by the stress $\sigma_i{}^j$, we can write (4.52) more explicitly as

$$\sigma \quad \mapsto \quad \mathbf{C} = \mathbf{C}(\sigma) \quad . \tag{4.53}$$

Definition 4.5 Ideal (or linear) elastic medium. *An elastic medium is ideally (or linearly) elastic if the mapping between the stress space $\mathfrak{S}$ and the configuration space $\mathfrak{C}$ is geodesic.*[16]

Using the results derived in chapters 1 and 2, we easily obtain the

Property 4.1 Hooke's law (of asymmetric elasticity). *For an ideally elastic (or linearly elastic) medium, there is a positive definite*[17] *tensor $\mathbf{c} = \{c_{ijk\ell}\}$ with the symmetry*

$$c_{ijk\ell} = c_{k\ell ij} \tag{4.54}$$

such that the relation between the stress $\boldsymbol{\sigma}$ and the configuration $\mathbf{C}$ is

$$\sigma_{ij} = c_{ijk\ell}\, \varepsilon^{k\ell} \quad ; \quad \boldsymbol{\sigma} = \mathbf{c}\, \boldsymbol{\varepsilon} \quad , \tag{4.55}$$

where $\varepsilon^i{}_j = (\log \mathbf{C})^i{}_j$, i.e.,

$$\boldsymbol{\varepsilon} = \log \mathbf{C} \quad . \tag{4.56}$$

This immediately suggests the

Definition 4.6 Strain. *The geotensor $\boldsymbol{\varepsilon} = \log \mathbf{C}$ associated to the configuration $\mathbf{C} = \{C^i{}_j\}$ is called the* strain. *As this geotensor connects the configuration $\mathbf{I}$ to the configuration $\mathbf{C}$, we say that "$\boldsymbol{\varepsilon} = \log \mathbf{C}$ is the strain experienced by the body when transforming from the configuration $\mathbf{I}$ to the configuration $\mathbf{C}$".*

As we have seen above, it is the decomposition of the strain $\boldsymbol{\varepsilon}$ into a symmetric part $\mathbf{e}$ and an antisymmetric part $\mathbf{s}$ that allows the interpretation of the transformation from $\mathbf{I}$ to $\mathbf{C}$ in terms of a deformation $\mathbf{E} = \exp \mathbf{e}$ (in the sense of the theory of symmetric elasticity) and a micro-rotation $\mathbf{S} = \exp \mathbf{s}$.

Example 4.2 *If a 3D ideally elastic medium is* isotropic, *there are three positive constants (Jeffreys quantities) $\{c_\kappa, c_\mu, c_\theta\}$ such that the stiffness tensor takes the form (see appendix A.22)*

$$c_{ijk\ell} = \frac{c_\kappa}{3}\, g_{ij}\, g_{k\ell} + c_\mu \left(\tfrac{1}{2}\, (g_{ik}\, g_{j\ell} + g_{i\ell}\, g_{jk}) - \tfrac{1}{3}\, g_{ij}\, g_{k\ell} \right) + \frac{c_\theta}{2}\, (g_{ik}\, g_{j\ell} - g_{i\ell}\, g_{jk}) \tag{4.57}$$

[16] According to the metric structure induced on the configuration space by the Lie group manifold GL(3).

[17] The positive definiteness of $\mathbf{c}$ results from the expression of the elastic energy density (see section 4.3.5).

where g_{ij} are the components of the metric tensor. The three eigenvalues (eigenstiffnesses) of the tensor are c_κ (mutiplicity 1), c_μ (multiplicity 5), and c_θ (multiplicity 3). See appendix A.22 for details. The stress-strain relation then becomes

$$\bar{\sigma} = c_\kappa \, \bar{\varepsilon} \quad ; \quad \hat{\sigma} = c_\mu \, \hat{\varepsilon} \quad ; \quad \check{\sigma} = c_\theta \, \check{\varepsilon} \quad , \tag{4.58}$$

where a bar, a hat, and a check respectively denote the isotropic part, the symmetric traceless part and the antisymmetric part of a tensor (see equations (A.414)–(A.416)*). When the 'rotational eigenstiffness' c_θ is zero, the antisymmetric part of the stress vanishes: the stress is symmetric. The only configurations that are then accessible from the reference configuration are those suggested in figure 4.5. The quantity $\kappa = c_\kappa/3$ is usually called the* incompressibility modulus *(or "bulk" modulus), while the quantity $\mu = c_\mu/2$ is usually called the* shear modulus.

While the tensor $\mathbf{c}$ is called the *stiffness tensor*, its inverse

$$\mathbf{d} = \mathbf{c}^{-1} \tag{4.59}$$

is called the *compliance tensor*.

Consider two configurations, $\mathbf{C}_1$ and $\mathbf{C}_2$. We know that the stress corresponding to some configuration $\mathbf{C}_1$ is $\sigma_1 = \mathbf{c} \log \mathbf{C}_1$ while that corresponding to some other configuration $\mathbf{C}_2$ is $\sigma_2 = \mathbf{c} \log \mathbf{C}_2$. Any path (in the stress space) for changing from σ_1 to σ_2 will define a path in the configuration space for changing from $\mathbf{C}_1$ to $\mathbf{C}_2$. A linear change of stress $\sigma(\lambda) = \lambda\, \sigma_2 + (1-\lambda)\, \sigma_1$, i.e.,

$$\sigma(\lambda) = \mathbf{c}\,(\, \lambda \log \mathbf{C}_2 + (1-\lambda) \log \mathbf{C}_1 \,) \quad ; \quad (0 \le \lambda \le 1) \quad , \tag{4.60}$$

would produce in the configuration space the path $\mathbf{C}(\lambda) = \exp(\, \lambda \log \mathbf{C}_2 + (1-\lambda) \log \mathbf{C}_1 \,)$, that is not a geodesic path (remember equation (1.156), page 61). A linear change of stress would produce a geodesic path in the configuration space only if the initial stress σ_1 is zero.

The following question, then, makes sense: what is the value of the stress when the configuration of the body is changing from $\mathbf{C}_1$ to $\mathbf{C}_2$ following a geodesic path in the configuration space? I leave as an (easy) exercise[18] to the reader to demonstrate that the answer is

$$\sigma(\lambda) = \mathbf{c} \log(\, (\mathbf{C}_2\, \mathbf{C}_1^{-1})^\lambda \; \mathbf{C}_1 \,) \quad ; \quad (0 \le \lambda \le 1) \quad , \tag{4.61}$$

or, more explicitly, $\sigma(\lambda) = \mathbf{c} \log(\, \exp[\, \lambda \log(\mathbf{C}_2\, \mathbf{C}_1^{-1})\,]\; \mathbf{C}_1 \,)$.

[18]One way of demonstrating this requires rewriting equation (4.61) as $\varepsilon(\lambda) \equiv \mathbf{d}\, \sigma(\lambda) = \lambda \log(\mathbf{C}_2\, \mathbf{C}_1^{-1}) \oplus \log \mathbf{C}_1$.

4.3.5 Elastic Energy

The work that is necessary to deform an elastic medium is evaluated in appendix A.26. When the configuration is changed, following an arbitrary path[19] $\mathbf{C}(t)$ in the configuration space, from $\mathbf{C}(t_0) = \mathbf{C}_0$ to $\mathbf{C}(t_1) = \mathbf{C}_1$, the work that the external forces must perform is (equation A.470)

$$W(\mathbf{C}_1;\mathbf{C}_0)_\Gamma = V_0 \oint_{t_0}^{t_1} dt \, \det \mathbf{C}(t) \, \mathrm{tr}\Big(\hat{\sigma}(t) \, \boldsymbol{\nu}(t)^t + \check{\sigma}(t) \, \boldsymbol{\omega}(t)^t \Big) \quad . \tag{4.62}$$

Here, V_0 is the volume of the body in the undeformed configuration, $\hat{\sigma}$ and $\check{\sigma}$ are respectively the symmetric and antisymmetric part of the stress,

$$\nu \equiv \dot{\mathbf{E}} \, \mathbf{E}^{-1} \tag{4.63}$$

is the *deformation rate* (declinative of $\mathbf{E}$), and

$$\omega \equiv \dot{\mathbf{S}} \, \mathbf{S}^{-1} = \dot{\mathbf{S}} \, \mathbf{S}^{*} \tag{4.64}$$

is the *micro-rotation velocity* (declinative of $\mathbf{S}$). The deformation $\mathbf{E}$ and the micro-rotation $\mathbf{S}$ associated to a configuration $\mathbf{C}$ have been introduced in equations (4.39)–(4.42).

For isochoric transformations (i.e., transformations conserving volume), one obtains the result (demonstration in appendix A.26) that, in this theory, *the elastic forces are conservative*. This means that to every configuration $\mathbf{C} \in \mathfrak{C}$ we can associate an *elastic energy density*, say $U(\mathbf{C})$. Changes in configuration produce changes in the energy density that correspond to the work (positive or negative) produced by the forces inducing the configuration change.

The expression found for the energy density associated to a configuration $\mathbf{C}$ is (equation A.472)

$$\boxed{U(\mathbf{C}) = \tfrac{1}{2} \, \mathrm{tr} \, \sigma \, \varepsilon^t = \tfrac{1}{2} \, \sigma_{ij} \, \varepsilon^{ij} = \tfrac{1}{2} \, c_{ijk\ell} \, \varepsilon^{ij} \, \varepsilon^{k\ell}} \quad , \tag{4.65}$$

where $\varepsilon = \log \mathbf{C}$ is the strain associated with the configuration $\mathbf{C}$. The expression we have obtained for the elastic energy density is identical to that obtained in the infinitesimal theory (of small deformations). We also see that the expression is valid even when there may be micro-rotations. The simplicity of this result is a potent indication that the elastic theory developed here makes sense.

But this holds only for isochoric transformations. For transformations changing volume, we can either keep the theory as it is, and accept that the elastic forces changing the volume of the body are not conservative, or we can introduce a simple modification of the theory, replacing the Hooke's law $\sigma = \mathbf{c} \, \varepsilon$ by the law

[19]Here t is an arbitrary parameter. It may, for instance, be Newtonian time.

$$\sigma = \frac{1}{\exp \operatorname{tr} \varepsilon} \, \mathrm{c} \, \varepsilon \quad . \tag{4.66}$$

As $\exp \operatorname{tr} \varepsilon = \det \mathbf{C}$, this modification cancels the term $\det \mathbf{C}$ in equation 4.62. Then, the elastic forces are unconditionally conservative, and the energy density is unconditionally given by expression (4.65).

4.3.6 Examples

Let us now analyze here a few simple 3D transformations of an isotropic elastic body, all represented (in 2D) in figure 4.8. We assume an Euclidean space with Cartesian coordinates (so covariant and contravariant components of tensors are identical).

4.3.6.1 Homothecy

The body transforms from the configuration $\mathbf{I}$ to configuration

$$\mathbf{C} = \begin{pmatrix} \exp k & 0 & 0 \\ 0 & \exp k & 0 \\ 0 & 0 & \exp k \end{pmatrix} \quad . \tag{4.67}$$

The strain is

$$\varepsilon = \log \mathbf{C} = \begin{pmatrix} k & 0 & 0 \\ 0 & k & 0 \\ 0 & 0 & k \end{pmatrix} \quad , \tag{4.68}$$

and, as the strain is purely isotropic, the stress is (equation 4.58) $\sigma = c_\kappa \, \varepsilon = c_\kappa \, k \, \mathbf{I}$. Alternatively, using the stress function in equation (4.66), $\sigma = (c_\kappa \, k)/(\exp 3k) \, \mathbf{I}$.

4.3.6.2 Pure Shear

The body transforms from the configuration $\mathbf{I}$ to configuration

$$\mathbf{C} = \begin{pmatrix} 1/\exp k & 0 & 0 \\ 0 & \exp k & 0 \\ 0 & 0 & 1 \end{pmatrix} \quad , \tag{4.69}$$

and one has $\det \mathbf{C} = 1$. The strain is

$$\varepsilon = \log \mathbf{C} = \begin{pmatrix} -k & 0 & 0 \\ 0 & k & 0 \\ 0 & 0 & 0 \end{pmatrix} \quad , \tag{4.70}$$

and one has $\operatorname{tr} \varepsilon = 0$. As the strain is symmetric and traceless, the stress is (equation 4.58) $\sigma = c_\mu \, \varepsilon$.

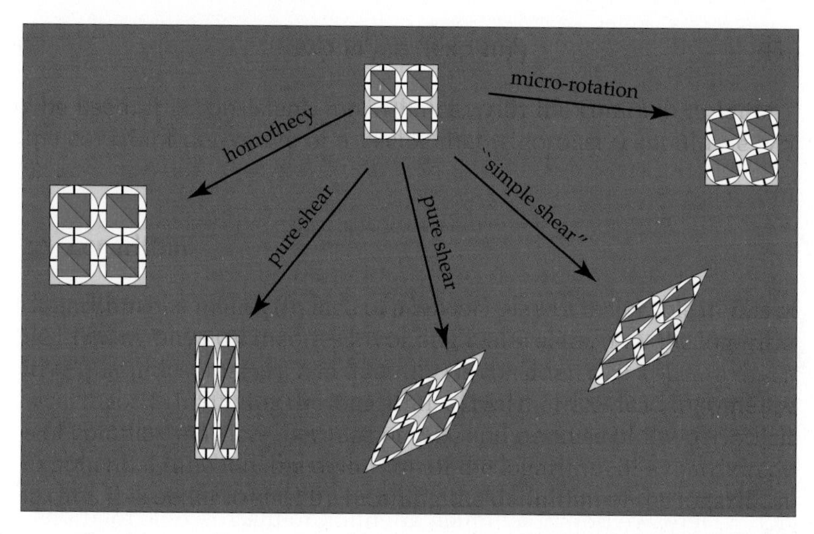

Fig. 4.8. *The five transformations explicitly analyzed in the text: homothecy, pure shear, "simple shear" (here meaning pure shear plus micro-rotation), and pure micro-rotation.*

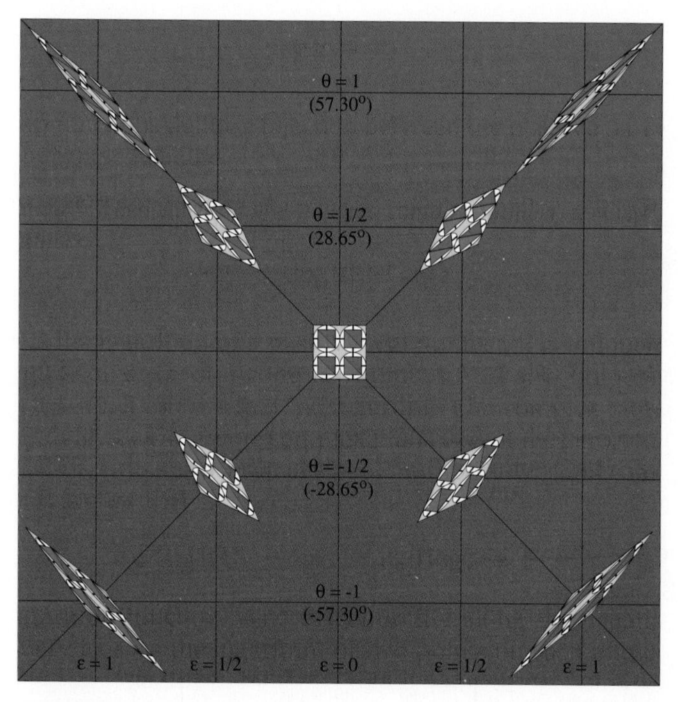

Fig. 4.9. *The configurations of the form expressed in equation (4.73) ("simple shears") belong to one of the light-cones of SL(2) (the angle θ is indicated). The configurations here represented can be interpreted as two-dimensional sections of three-dimensional configurations.*

Equivalently, in a pure shear the body transforms from the configuration **I** to configuration

$$\mathbf{C} = \begin{pmatrix} \cosh k & \sinh k & 0 \\ \sinh k & \cosh k & 0 \\ 0 & 0 & 1 \end{pmatrix} \quad . \tag{4.71}$$

The strain is

$$\varepsilon = \log \mathbf{C} = \begin{pmatrix} 0 & k & 0 \\ k & 0 & 0 \\ 0 & 0 & 0 \end{pmatrix} \quad , \tag{4.72}$$

and the stress is $\sigma = c_\mu \, \varepsilon$.

4.3.6.3 "Simple Shear"

In the standard theory, it is said that "a simple shear is a pure shear plus a rotation." Here, we don't pay much attention to macroscopic rotations, but we are interested in micro-rotations. We may then here modify the notion, and define a *simple shear* as *a pure shear plus a micro-rotation.*

Let a 3D body transform from the configuration **I** to configuration[20]

$$\mathbf{C} = \begin{pmatrix} 1 & 2\theta & 0 \\ 0 & 1 & 0 \\ 0 & 0 & 1 \end{pmatrix} \quad , \tag{4.73}$$

with $\det \mathbf{C} = 1$. The strain is

$$\varepsilon = \log \mathbf{C} = \begin{pmatrix} 0 & 2\theta & 0 \\ 0 & 0 & 0 \\ 0 & 0 & 0 \end{pmatrix} \quad , \tag{4.74}$$

and one has $\operatorname{tr} \varepsilon = 0$. The decomposition of the strain in its symmetric and antisymmetric parts gives

$$\varepsilon = \mathbf{e} + \mathbf{s} = \begin{pmatrix} 0 & \theta & 0 \\ \theta & 0 & 0 \\ 0 & 0 & 0 \end{pmatrix} + \begin{pmatrix} 0 & \theta & 0 \\ -\theta & 0 & 0 \\ 0 & 0 & 0 \end{pmatrix} \quad . \tag{4.75}$$

The value of $\mathbf{s}$ shows that the micro-rotation is of angle θ .

Using equation (4.58) we find the stress $\sigma = c_\mu \, \mathbf{e} + c_\theta \, \mathbf{s}$, i.e.,

$$\sigma = \begin{pmatrix} 0 & (c_\mu + c_\theta)\,\theta & 0 \\ (c_\mu - c_\theta)\,\theta & 0 & 0 \\ 0 & 0 & 0 \end{pmatrix} \quad . \tag{4.76}$$

[20]We take here a 3D version of expressions (1.200) and (1.201), with $\varphi = 0$ and $\varepsilon = \theta$.

To obtain such a transformation, a moment-force density $\chi_{ij} = \sigma_{ij} - \sigma_{ji}$, must act on the body. It has the value

$$\chi = \begin{pmatrix} 0 & 2\,c_\theta\,\theta & 0 \\ -2\,c_\theta\,\theta & 0 & 0 \\ 0 & 0 & 0 \end{pmatrix} \quad . \tag{4.77}$$

While the "simple shear" transformation is represented in figure 4.8, figure 4.9 represents the (2D) simple shear configurations as points of the configuration space SL(2) (the points are along the light-cone $\varepsilon = \theta$).

4.3.6.4 Pure Micro-rotation

The body transforms from the configuration $\mathbf{I}$ to configuration

$$\mathbf{C} = \begin{pmatrix} \cos\theta & \sin\theta & 0 \\ -\sin\theta & \cos\theta & 0 \\ 0 & 0 & 1 \end{pmatrix} \quad . \tag{4.78}$$

and one has $\det \mathbf{C} = 1$. The strain is

$$\varepsilon = \log \mathbf{C} = \begin{pmatrix} 0 & \theta & 0 \\ -\theta & 0 & 0 \\ 0 & 0 & 0 \end{pmatrix} \quad , \tag{4.79}$$

and one has $\operatorname{tr} \varepsilon = 0$. As the strain is antisymmetric, the stress is (equation 4.58) $\sigma = c_\theta\, \varepsilon$.

4.3.7 Material Coordinates and Heterogeneous Transformations

Let us now briefly return to heterogeneous transformations, and let us change from the laboratory system of coordinates $\{x^i\}$ used above, to a *material system of coordinates*, i.e., to a system of coordinates $\{X^\alpha\}$ that is attached to the body (and deforms with it).

The two relations

$$X^\alpha = X^\alpha(x^1, x^2, x^3) \quad ; \quad x^i = x^i(X^1, X^2, X^3) \tag{4.80}$$

expressing the change of coordinates are the same relations written in equation 4.30, although there they had a different interpretation. To avoid possible misunderstandings, let us use Latin indices for the laboratory coordinates (and the components of tensors) and Greek indices for the material coordinates.

Introducing the coefficients

$$S^\alpha{}_i = \frac{\partial X^\alpha}{\partial x^i} \quad ; \quad T^i{}_\alpha = \frac{\partial x^i}{\partial X^\alpha} \tag{4.81}$$

we can relate the components $A^{ij\dots}{}_{k\ell\dots}$ of a tensor $\mathbf{A}$ in the laboratory coordinates to the components $A^{\alpha\beta\dots}{}_{\mu\nu\dots}$ in the material coordinates:

$$A^{\alpha\beta\dots}{}_{\mu\nu\dots} = S^{\alpha}{}_{i}\, S^{\beta}{}_{j} \dots A^{ij\dots}{}_{k\ell\dots}\, T^{k}{}_{\mu}\, T^{\ell}{}_{\nu} \dots \quad . \tag{4.82}$$

In particular, the covariant components of the metric in the material coordinates can be expressed as

$$g_{\alpha\beta} = T^{i}{}_{\alpha}\, g_{ij}\, T^{j}{}_{\beta} \quad . \tag{4.83}$$

One typically chooses for the material coordinates the "imprint" of the laboratory coordinates at some time t_0 on the material body. Then one has the time-varying metric components $g_{\alpha\beta}(t)$ (space variables omitted), the components $g_{\alpha\beta}(t_0)$ being identical to the components of the metric in the laboratory coordinates (one should realize that it is not the metric that is changing, it is the coordinate system that is evolving). With this in mind, one can rewrite equation 4.83 as

$$g_{\alpha\beta}(t) = T^{\mu}{}_{\alpha}\, g_{\mu\nu}(t_0)\, T^{\nu}{}_{\beta} \quad . \tag{4.84}$$

Disregarding rotations (micro or macro), it is clear that the deformation of the body can be represented by the functions $g_{\alpha\beta}(X^1, X^2, X^3, t)$. A question arises: can any field $g_{\alpha\beta}(X^1, X^2, X^3, t)$ be interpreted as the components of the metric in the material coordinates of a deforming body? The answer is obviously negative, as too many degrees of freedom are involved: a (symmetric) field $g_{\alpha\beta}$ consists of six independent functions, while to define a deformation the three displacement functions at the left in equation (4.80) suffice.

The restriction to be imposed on a metric field $g_{\alpha\beta}(X^1, X^2, X^3, t)$ is that the Riemann tensor $R_{\alpha\beta\gamma\delta}(X^1, X^2, X^3, t)$ computed from these components has to be time-invariant (as the metric of the space is not changing). In particular, when working with bodies deforming inside an Euclidean space, *the components of the Riemann tensor evaluated from the components $g_{\alpha\beta}(t)$ must vanish.*

As demonstrated in appendix A.27, this condition is equivalent to the condition that the metric components $g_{\alpha\beta}(t)$ must satisfy

$$\boxed{\nabla_i \nabla_j\, g_{k\ell} + \nabla_k \nabla_\ell\, g_{ij} - \nabla_i \nabla_\ell\, g_{kj} - \nabla_k \nabla_j\, g_{i\ell} = \tfrac{1}{2}\, g^{pq} \left(G_{i\ell p}\, G_{kjq} - G_{k\ell p}\, G_{ijq} \right) ,} \tag{4.85}$$

where

$$G_{ijk} = \nabla_i g_{jk} + \nabla_j g_{ik} - \nabla_k g_{ij} \tag{4.86}$$

and where the ad-hoc operator ∇ is a covariant derivative, but defined using the metric components $g_{\alpha\beta}(t_0)$ (instead of the actual metric components $g_{\alpha\beta}(t)$).

In the absence of micro-rotations, the strain was defined above (equation 4.41) as $\boldsymbol{\varepsilon} = \log \mathbf{E} = \log \sqrt{\mathbf{g}^{-1}\,\mathbf{T}^t\,\mathbf{g}\,\mathbf{T}}$. It follows from equation 4.84 that, in terms of the changing metric components, one has[21] $\varepsilon^\alpha{}_\beta = \log \sqrt{g^{\alpha\sigma}(t_0)\, g_{\sigma\beta}(t)}$ or, for short,

$$\boldsymbol{\varepsilon} = \log \sqrt{\mathbf{g}^{-1}(t_0)\,\mathbf{g}(t)} \quad . \tag{4.87}$$

If the strain is small, one may keep only the first-order terms in the compatibility condition (4.85). This gives (see appendix A.27)

$$\nabla_i \nabla_j \varepsilon_{k\ell} + \nabla_k \nabla_\ell \varepsilon_{ij} - \nabla_i \nabla_\ell \varepsilon_{kj} - \nabla_k \nabla_j \varepsilon_{i\ell} = 0 \quad . \tag{4.88}$$

This is the well-known *Saint-Venant condition*: a tensor field $\boldsymbol{\varepsilon}(\mathbf{x})$ can be interpreted as a (small) strain field only if it satisfies this equation. We see that the Saint-Venant condition is just a linearized version of the actual condition, equation (4.85).

4.3.8 Comments on the Different Measures of Strain

Suggestions to use a logarithmic measure of strain can be traced back to the beginning of the century[22] and its 1D version is used, today, by material scientists contemplating large deformations.[23] In theoretical expositions of the theory of finite deformations, the logarithmic measure of strain is often proposed, and subsequently dismissed, with unconvincing arguments that always come from misunderstandings of the mathematics of tensor exponentiation.

For instance, Truesdell and Toupin's treatise on Classical Field Theories (1960) that has strongly influenced two generations of scholars, says that *"while logarithmic measures of strain are a favorite in one-dimensional or semi-qualitative treatment, they have never been successfully applied in general. Such simplicity for certain problems as may result from a particular strain measure is bought at the cost of complexity for other problems. In a Euclidean space, distances are measured by a quadratic form, and attempt to elude this fact is unlikely to succeed"*. It seems that "having never been successfully applied in general" means "a complete, consistent mathematical

---

[21]Using the notation $f(M^\alpha{}_\beta) \equiv f(\mathbf{M})^\alpha{}_\beta$.

[22]In the Truesdell and Toupin treatise (1960) there are references, among others, to the works of Ludwik (1909) and Hencky (1928, 1929), for infinitesimal strains, and to Murnaghan (1941) and Richter (1948, 1949) for finite strain. Nadai (1937) used the term *natural strain*.

[23]See, for instance, Means (1976), Malvern (1969) and Poirier (1985). Here is how the argument goes. A body of length ℓ is in a state of strain ε. When the body increases its length by $\Delta\ell$, the ratio $\Delta\varepsilon = \Delta\ell/\ell$ is interpreted as the strain increment, so the strain becomes $\varepsilon + \Delta\varepsilon$. The total strain when the body passes from length ℓ_0 to length ℓ is then obtained by integration, $\varepsilon = \int_{\ell_0}^{\ell} d\varepsilon = \int_{\ell_0}^{\ell} d\ell/\ell$, this giving a true finite measure of strain $\varepsilon = \log(\ell/\ell_0)$.

theory having never been proposed". I hope that the step proposed here goes in the right direction. That "in a Euclidean space, distances are measured by a quadratic form, and attempt to elude this fact is unlikely to succeed" seems to mean that a deformation theory will probably use the metric tensor as a fundamental element. It is true that the strain must be a simple function of the metric, but this simple function is[24] $\varepsilon = \log \sqrt{\mathbf{g}} = \frac{1}{2} \log \mathbf{g}$, not $\varepsilon = \frac{1}{2}(\mathbf{g} - \mathbf{I})$, an expression that is only a first order approximation to the actual (logarithmic) strain.

A more recent point of view on the problem is that of Rougée (1997). The book has a mathematical nature, and is quite complete in recounting all the traditional measures of strain. The author clearly shows his preference for the logarithmic measure. But, quite honestly, he declares his perplexity. While *"among all possible measures of strain, [the logarithmic measure] is the least bad, [. . .] what prevents the [general] use [of the logarithmic measure of deformation] is that its computation, and the computation of the associated stress [. . .] is not simple"*. I disagree with this. The computation of the logarithm of a tensor is a very simple matter, if the mathematics are well understood. And the computation of stresses is as simple as the computation of strains.

[24]This is equation (4.87), written in the case where the coordinates at time t_0 are Cartesian, and formally writing $\mathbf{g}(t_0) = \mathbf{I}$.

A Appendices

A.1 Adjoint and Transpose of a Linear Operator

A.1.1 Transpose

Let $\mathbb{E}$ denote a finite-dimensional linear space, with vectors $\mathbf{a} = a^\alpha \, \mathbf{e}_\alpha$, $\mathbf{b} = b^\alpha \, \mathbf{e}_\alpha , \dots$, and let $\mathbb{F}$ denote another finite-dimensional linear space, with vectors $\mathbf{v} = v^i \, \mathbf{e}_i$, $\mathbf{w} = w^i \, \mathbf{e}_i , \dots$. The duals of the two spaces are denoted $\mathbb{E}^*$ and $\mathbb{F}^*$ respectively, and their vectors (forms) are respectively denoted $\widehat{\mathbf{a}} = \widehat{a}_\alpha \, \mathbf{e}^\alpha$, $\widehat{\mathbf{b}} = \widehat{b}_\alpha \, \mathbf{e}^\alpha , \dots$ and $\widehat{\mathbf{v}} = \widehat{v}_i \, \mathbf{e}^i$, $\widehat{\mathbf{w}} = \widehat{w}_i \, \mathbf{e}^i , \dots$. The duality product in each space is respectively denoted

$$\langle \, \widehat{\mathbf{a}} \, , \, \mathbf{b} \, \rangle_{\mathbb{E}} = \widehat{a}_\alpha \, b^\alpha \qquad ; \qquad \langle \, \widehat{\mathbf{v}} \, , \, \mathbf{w} \, \rangle_{\mathbb{F}} = \widehat{v}_i \, w^i \quad . \tag{A.1}$$

Let $\mathbf{K}$ be a linear mapping that maps $\mathbb{E}$ into $\mathbb{F}$:

$$\mathbf{K} \; : \; \mathbb{E} \; \mapsto \; \mathbb{F} \qquad ; \qquad \mathbf{v} = \mathbf{K} \, \mathbf{a} \qquad ; \qquad v^i = K^i{}_\alpha \, a^\alpha \quad . \tag{A.2}$$

The *transpose* of $\mathbf{K}$, denoted $\mathbf{K}^t$, is (Taylor and Lay, 1980) the linear mapping that maps $\mathbb{F}^*$ into $\mathbb{E}^*$,

$$\mathbf{K}^t \; : \; \mathbb{F}^* \; \mapsto \; \mathbb{E}^* \qquad ; \qquad \widehat{\mathbf{a}} = \mathbf{K}^t \, \widehat{\mathbf{v}} \qquad ; \qquad \widehat{a}_\alpha = (K^t)_\alpha{}^i \, \widehat{v}_i \quad , \tag{A.3}$$

such that for any $\mathbf{a} \in \mathbb{E}$ and any $\widehat{\mathbf{v}} \in \mathbb{F}^*$,

$$\boxed{\langle \, \widehat{\mathbf{v}} \, , \, \mathbf{K} \, \mathbf{a} \, \rangle_{\mathbb{F}} = \langle \, \mathbf{K}^t \, \widehat{\mathbf{v}} \, , \, \mathbf{a} \, \rangle_{\mathbb{E}} \quad .} \tag{A.4}$$

Using the notation in equation (A.1) and those on the right in equations (A.2) and (A.3) one obtains

$$\boxed{(K^t)_\alpha{}^i = K^i{}_\alpha \quad ,} \tag{A.5}$$

this meaning that the two operators $\mathbf{K}$ and $\mathbf{K}^t$ have the same components. In matrix terminology, the matrices representing $\mathbf{K}$ and $\mathbf{K}^t$ are the transpose (in the ordinary sense) of each other.

Note that the transpose of an operator is always defined, irrespectively of the fact that the linear spaces under consideration have or not a scalar product defined.

A.1.2 Metrics

Let $\mathbf{g}_{\mathbb{E}}$ and $\mathbf{g}_{\mathbb{F}}$ be two *metric tensors*, i.e., two symmetric,[1] invertible operators mapping the spaces $\mathbb{E}$ and $\mathbb{F}$ into their respective duals:

$$\begin{aligned} \mathbf{g}_{\mathbb{E}} &: \mathbb{E} \mapsto \mathbb{E}^* \quad ; \quad \widehat{\mathbf{a}} = \mathbf{g}_{\mathbb{E}}\,\mathbf{a} \quad ; \quad a_\alpha = (g_{\mathbb{E}})_{\alpha\beta}\, a^\beta \\ \mathbf{g}_{\mathbb{F}} &: \mathbb{F} \mapsto \mathbb{F}^* \quad ; \quad \widehat{\mathbf{v}} = \mathbf{g}_{\mathbb{F}}\,\mathbf{v} \quad ; \quad v_i = (g_{\mathbb{F}})_{ij}\, v^j \quad . \end{aligned} \tag{A.6}$$

In the two equations on the right, one should have written $\widehat{a}_\alpha$ and $\widehat{v}_i$ instead of a_α and v_i but it is usual to drop the hats, as the position of the indices indicates if one has an element of the 'primal' spaces $\mathbb{E}$ and $\mathbb{F}$ or an element of the dual spaces $\mathbb{E}^*$ and $\mathbb{F}^*$. Reciprocally, one writes

$$\begin{aligned} \mathbf{g}_{\mathbb{E}}^{-1} &: \mathbb{E}^* \mapsto \mathbb{E} \quad ; \quad \mathbf{a} = \mathbf{g}_{\mathbb{E}}^{-1}\,\widehat{\mathbf{a}} \quad ; \quad a^\alpha = (g_{\mathbb{E}})^{\alpha\beta}\, a_\beta \\ \mathbf{g}_{\mathbb{F}}^{-1} &: \mathbb{F}^* \mapsto \mathbb{F} \quad ; \quad \mathbf{v} = \mathbf{g}_{\mathbb{F}}^{-1}\,\widehat{\mathbf{v}} \quad ; \quad v^i = (g_{\mathbb{F}})^{ij}\, v_j \quad , \end{aligned} \tag{A.7}$$

with $(g_{\mathbb{E}})_{\alpha\beta}\,(g_{\mathbb{E}})^{\beta\gamma} = \delta_\alpha{}^\gamma$ and $(g_{\mathbb{F}})_{ij}\,(g_{\mathbb{F}})^{jk} = \delta_i{}^k$.

A.1.3 Scalar Products

Given $\mathbf{g}_{\mathbb{E}}$ and $\mathbf{g}_{\mathbb{F}}$ we can define, in addition to the duality products (equation A.1) the *scalar products*

$$(\,\mathbf{a}\,,\,\mathbf{b}\,)_{\mathbb{E}} = \langle\,\widehat{\mathbf{a}}\,,\,\mathbf{b}\,\rangle_{\mathbb{E}} \quad ; \quad (\,\mathbf{v}\,,\,\mathbf{w}\,)_{\mathbb{F}} = \langle\,\widehat{\mathbf{v}}\,,\,\mathbf{w}\,\rangle_{\mathbb{F}} \quad , \tag{A.8}$$

i.e.,

$$\boxed{(\,\mathbf{a}\,,\,\mathbf{b}\,)_{\mathbb{E}} = \langle\,\mathbf{g}_{\mathbb{E}}\,\mathbf{a}\,,\,\mathbf{b}\,\rangle_{\mathbb{E}} \quad ; \quad (\,\mathbf{v}\,,\,\mathbf{w}\,)_{\mathbb{F}} = \langle\,\mathbf{g}_{\mathbb{F}}\,\mathbf{v}\,,\,\mathbf{w}\,\rangle_{\mathbb{F}} \quad .} \tag{A.9}$$

Using indices, the definition of scalar product gives

$$\boxed{(\,\mathbf{a}\,,\,\mathbf{b}\,)_{\mathbb{E}} = (g_{\mathbb{E}})_{\alpha\beta}\, a^\alpha\, b^\beta \quad ; \quad (\,\mathbf{v}\,,\,\mathbf{w}\,)_{\mathbb{F}} = (g_{\mathbb{F}})_{ij}\, v^i\, v^j \quad .} \tag{A.10}$$

A.1.4 Adjoint

If a scalar product has been defined over the linear spaces $\mathbb{E}$ and $\mathbb{F}$, one can introduce, in addition to the transpose of an operator, its adjoint. Letting $\mathbf{K}$ the linear mapping introduced above (equation A.2), its *adjoint*, denoted $\mathbf{K}^*$, is (Taylor and Lay, 1980) the linear mapping that maps $\mathbb{F}$ into $\mathbb{E}$,

$$\mathbf{K}^* : \mathbb{F} \mapsto \mathbb{E} \quad ; \quad \mathbf{a} = \mathbf{K}^*\,\mathbf{v} \quad ; \quad a^\alpha = (K^*)^\alpha{}_i\, v^i \quad , \tag{A.11}$$

[1] A metric tensor $\mathbf{g}$ maps a linear space into its dual. So does its transpose $\mathbf{g}^t$. The condition that $\mathbf{g}$ is symmetric corresponds to $\mathbf{g} = \mathbf{g}^t$. This simply amounts to say that, using any basis, $g_{\alpha\beta} = g_{\beta\alpha}$.

such that for any $\mathbf{a} \in \mathbb{E}$ and any $\mathbf{v} \in \mathbb{F}$,

$$(\mathbf{v} , \mathbf{K} \mathbf{a})_{\mathbb{F}} = (\mathbf{K}^* \mathbf{v} , \mathbf{a})_{\mathbb{E}} \quad . \tag{A.12}$$

Using the notation in equation (A.10) and those on the right in equations (A.11) and (A.12) one obtains $(K^*)^\alpha{}_i = (g_{\mathbb{F}})_{ij} \, K^j{}_\beta \, (g_{\mathbb{E}})^{\beta\alpha}$, where, as usual, $g^{\alpha\beta}$ is defined by the condition $g_{\alpha\beta} \, g^{\beta\gamma} = \delta_\alpha{}^\gamma$. Equivalently, using equation (A.5), $(K^*)^\alpha{}_i = (g_{\mathbb{E}})^{\alpha\beta} \, (K^t)_\beta{}^j \, (g_{\mathbb{F}})_{ji}$ an expression that can be written

$$\mathbf{K}^* = \mathbf{g}_{\mathbb{E}}^{-1} \, \mathbf{K}^t \, \mathbf{g}_{\mathbb{F}} \quad , \tag{A.13}$$

this showing the formal relation linking the adjoint and the transpose of a linear operator.

A.1.5 Transjoint Operator

The operator

$$\widetilde{\mathbf{K}} = \mathbf{g}_{\mathbb{F}} \, \mathbf{K} \, \mathbf{g}_{\mathbb{E}}^{-1} \tag{A.14}$$

called the *transjoint* of $\mathbf{K}$, clearly maps $\mathbb{E}^*$ into $\mathbb{F}^*$. Using the index notation, $\widetilde{K}_i{}^\alpha = (g_{\mathbb{F}})_{ij} \, K^j{}_\beta \, (g_{\mathbb{E}})^{\beta\alpha}$. We have now a complete set of operators associated to an operator $\mathbf{K}$:

$$\begin{aligned} \mathbf{K} &: \mathbb{E} \mapsto \mathbb{F} \quad ; \quad \mathbf{K}^* : \mathbb{F} \mapsto \mathbb{E} \\ \mathbf{K}^t &: \mathbb{F}^* \mapsto \mathbb{E}^* \quad ; \quad \widetilde{\mathbf{K}} : \mathbb{E}^* \mapsto \mathbb{F}^* \quad . \end{aligned} \tag{A.15}$$

A.1.6 Associated Endomorphisms

Note that using the pair $\{\mathbf{K}, \mathbf{K}^*\}$ one can define two different endomorphisms $\mathbf{K}^* \mathbf{K} : \mathbb{E} \mapsto \mathbb{F}$ and $\mathbf{K} \mathbf{K}^* : \mathbb{F} \mapsto \mathbb{E}$. It is easy to see that the components of the two endomorphisms are

$$\begin{aligned} (\mathbf{K}^* \mathbf{K})^\alpha{}_\beta &= (g_{\mathbb{E}})^{\alpha\gamma} \, K^i{}_\gamma \, (g_{\mathbb{F}})_{ij} \, K^j{}_\beta \\ (\mathbf{K} \mathbf{K}^*)^i{}_j &= K^i{}_\alpha \, (g_{\mathbb{E}})^{\alpha\beta} \, K^k{}_\beta \, (g_{\mathbb{F}})_{kj} \quad . \end{aligned} \tag{A.16}$$

One has, in particular, $(\mathbf{K} \mathbf{K}^*)^i{}_i = (\mathbf{K}^* \mathbf{K})^\alpha{}_\alpha = (g_{\mathbb{E}})^{\beta\gamma} \, (g_{\mathbb{F}})_{jk} \, K^j{}_\beta \, K^k{}_\gamma$, this demonstrating the property

$$\operatorname{tr} (\mathbf{K} \mathbf{K}^*) = \operatorname{tr} (\mathbf{K}^* \mathbf{K}) \quad . \tag{A.17}$$

The Frobenius norm of the operator $\mathbf{K}$ is defined as

$$\| \mathbf{K} \| = \sqrt{\operatorname{tr} (\mathbf{K} \mathbf{K}^*)} = \sqrt{\operatorname{tr} (\mathbf{K}^* \mathbf{K})} \quad . \tag{A.18}$$

A.1.7 Formal Identifications

Let us collect here equations (A.13) and (A.14):

$$(K^*)^\alpha{}_i = (g_{\mathbb{E}})^{\alpha\beta} \, (K^t)_\beta{}^j \, (g_{\mathbb{F}})_{ji} \quad ; \quad \widetilde{K}_i{}^\alpha = (g_{\mathbb{F}})_{ij} \, K^j{}_\beta \, (g_{\mathbb{E}})^{\beta\alpha} \quad . \tag{A.19}$$

As it is customary to use the same letter for a vector and for the form associated to it by the metric, we could extend the rule to operators. Then, these two equations show that $\mathbf{K}^*$ is obtained from $\mathbf{K}^t$ (and, respectively, $\widetilde{\mathbf{K}}$ is obtained from $\mathbf{K}$) by "raising and lowering indices", so one could use an unique symbol for K^* and K^t (and, respectively, for $\widetilde{K}$ and K). As there is sometimes confusion between between the notion of adjoint and of transpose, it is better to refrain from using such notation.

A.1.8 Orthogonal Operators (for Endomorphisms)

Consider an operator $\mathbf{K}$ mapping a linear space $\mathbb{E}$ into itself, and let $\mathbf{K}^{-1}$ be the inverse operator (defined as usual). The condition

$$\mathbf{K}^* = \mathbf{K}^{-1} \tag{A.20}$$

makes sense. An operator satisfying this condition is called *orthogonal*. Then, $K^i{}_j \, (K^*)^j{}_k = \delta^i{}_k$. Adapting equation (A.13) to this particular situation, and denoting as g_{ij} the components of the metric (remember that there is a single space here), gives $K^i{}_j \, g^{jk} \, (K^t)_k{}^\ell \, g_{\ell m} = \delta^i{}_m$. Using (A.5) this gives the expression

$$K^i{}_j \, g^{jk} \, K^\ell{}_k \, g_{\ell m} = \delta^i{}_m \quad , \tag{A.21}$$

which one could take directly as the condition defining an orthogonal operator. Raising and lowering indices this can also be written

$$K^{ik} \, K_{mk} = \delta^i{}_m \quad . \tag{A.22}$$

A.1.9 Self-adjoint Operators (for Endomorphisms)

Consider an operator $\mathbf{K}$ mapping a linear space $\mathbb{E}$ into itself. The condition

$$\mathbf{K}^* = \mathbf{K} \tag{A.23}$$

makes sense. An operator satisfying this condition is called *self-adjoint*. Adapting equation (A.13) to this particular situation, and denoting $\mathbf{g}$ the metric (remember that there is a single space here), gives $\mathbf{K}^t \, \mathbf{g} = \mathbf{g} \, \mathbf{K}$, i.e.,

$$g_{ij} \, K^j{}_k = g_{kj} \, K^j{}_i \quad , \tag{A.24}$$

expression that one could directly take as the condition defining a self-adjoint operator. Lowering indices this can also be written

$$K_{ij} = K_{ji} \quad . \tag{A.25}$$

Such an operator (i.e., such a tensor) is usually called 'symmetric', rather than self-adjoint. This is not correct, as a symmetric operator should be defined by the condition $\mathbf{K} = \mathbf{K}^t$, an expression that would make sense only when the operator $\mathbf{K}$ maps a space into its dual (see footnote 1).

A.2 Elementary Properties of Groups (in Additive Notation)

Setting $\mathbf{w} = \mathbf{v}$ in the group property (1.49) and using the third of the properties (1.41), one sees that for any $\mathbf{u}$ and $\mathbf{v}$ in a group, the *oppositivity property*

$$\mathbf{v} \ominus \mathbf{u} = -(\mathbf{u} \ominus \mathbf{v}) \tag{A.26}$$

holds (see figure 1.7 for a discussion on this property.) From the group property (1.49) and the oppositivity property (A.26), follows that for any $\mathbf{u}$, $\mathbf{v}$ and $\mathbf{w}$ in a group, $(\mathbf{v} \ominus \mathbf{w}) \ominus (\mathbf{u} \ominus \mathbf{w}) = -(\mathbf{u} \ominus \mathbf{v})$. Using the equivalence (1.36) between the operation $\oplus$ and the operation $\ominus$, this gives $\mathbf{v} \ominus \mathbf{w} = (-(\mathbf{u} \ominus \mathbf{v})) \oplus (\mathbf{u} \ominus \mathbf{w})$. When setting $\mathbf{u} = \mathbf{0}$, this gives $\mathbf{v} \ominus \mathbf{w} = (-(\mathbf{0} \ominus \mathbf{v})) \oplus (\mathbf{0} \ominus \mathbf{w})$, or, when using the third of equations (1.41), $\mathbf{v} \ominus \mathbf{w} = (-(-\mathbf{v})) \oplus (-\mathbf{w})$. Finally, using the property that the opposite of an anti-element is the element itself ((1.39)), one arrives to the conclusion that for any $\mathbf{v}$ and $\mathbf{w}$ of a group,

$$\mathbf{v} \ominus \mathbf{w} = \mathbf{v} \oplus (-\mathbf{w}) \quad . \tag{A.27}$$

Setting $\mathbf{w} = -\mathbf{u}$ in this equation gives $\mathbf{v} \ominus (-\mathbf{u}) = \mathbf{v} \oplus (-(-\mathbf{u}))$, i.e., for any $\mathbf{u}$ and $\mathbf{v}$ in a group,

$$\mathbf{v} \ominus (-\mathbf{u}) = \mathbf{v} \oplus \mathbf{u} \quad . \tag{A.28}$$

Let us see that in a group, the equation $\mathbf{w} = \mathbf{v} \oplus \mathbf{u}$ cannot only be solved for $\mathbf{v}$, as postulated for a troupe, but it can also be solved for $\mathbf{u}$. Solving first $\mathbf{w} = \mathbf{v} \oplus \mathbf{u}$ for $\mathbf{v}$ gives (postulate (1.36)) $\mathbf{v} = \mathbf{w} \ominus \mathbf{u}$, i.e., using the oppositivity property (A.26) $\mathbf{v} = -(\mathbf{u} \ominus \mathbf{w})$, equation that, because of the property (1.39) is equivalent to $-\mathbf{v} = \mathbf{u} \ominus \mathbf{w}$. Using again the postulate (1.36) then gives $\mathbf{u} = (-\mathbf{v}) \oplus \mathbf{w}$. We have thus demonstrated that in a group one has the equivalence

$$\mathbf{w} = \mathbf{v} \oplus \mathbf{u} \quad \Longleftrightarrow \quad \mathbf{u} = (-\mathbf{v}) \oplus \mathbf{w} \quad . \tag{A.29}$$

Using this and the property (A.27), we see that condition (1.36) can, in a group, be completed and made explicit as

$$\mathbf{w} = \mathbf{v} \oplus \mathbf{u} \quad \Longleftrightarrow \quad \mathbf{v} = \mathbf{w} \oplus (-\mathbf{u}) \quad \Longleftrightarrow \quad \mathbf{u} = (-\mathbf{v}) \oplus \mathbf{w} \quad . \tag{A.30}$$

Using the oppositivity property of a group (equation A.26), as well as the property (A.27), one can write, for any $\mathbf{v}$ and $\mathbf{w}$ of a group, $\mathbf{v} \ominus \mathbf{w} = -(\mathbf{w} \oplus (-\mathbf{v}))$, or, setting $\mathbf{w} = -\mathbf{u}$, $\mathbf{v} \ominus (-\mathbf{u}) = -((-\mathbf{u}) \oplus (-\mathbf{v}))$. From the property (A.28) it then follows that for any $\mathbf{u}$ and $\mathbf{v}$ of a group,

$$\mathbf{v} \oplus \mathbf{u} = -((-\mathbf{u}) \oplus (-\mathbf{v})) \quad . \tag{A.31}$$

With the properties so far demonstrated it is easy to give to the homogeneity property (1.49) some equivalent expressions. Among them,

$$(\mathbf{v} \ominus \mathbf{w}) \oplus (\mathbf{w} \ominus \mathbf{u}) = (\mathbf{v} \oplus \mathbf{w}) \ominus (\mathbf{u} \oplus \mathbf{w}) = \mathbf{v} \ominus \mathbf{u} \quad . \tag{A.32}$$

Writing the homogeneity property (1.49) with $\mathbf{u} = -\mathbf{x}$, $\mathbf{v} = \mathbf{z} \oplus \mathbf{y}$, and $\mathbf{w} = \mathbf{y}$, one obtains (for any $\mathbf{x}$, $\mathbf{y}$ and $\mathbf{z}$) $((\mathbf{z} \oplus \mathbf{y}) \ominus \mathbf{y}) \ominus ((-\mathbf{x}) \ominus \mathbf{y}) = (\mathbf{z} \oplus \mathbf{y}) \ominus (-\mathbf{x})$, or, using the property (A.28) $\mathbf{z} \oplus (-((-\mathbf{x}) \ominus \mathbf{y})) = (\mathbf{z} \oplus \mathbf{y}) \oplus \mathbf{x}$. Using now the oppositivity property (A.26), $\mathbf{z} \oplus (\mathbf{y} \ominus (-\mathbf{x})) = (\mathbf{z} \oplus \mathbf{y}) \oplus \mathbf{x}$, i.e., using again (A.28), $\mathbf{z} \oplus (\mathbf{y} \oplus \mathbf{x}) = (\mathbf{z} \oplus \mathbf{y}) \oplus \mathbf{x}$. We thus arrive, relabeling $(\mathbf{x}, \mathbf{y}, \mathbf{z}) = (\mathbf{u}, \mathbf{v}, \mathbf{w})$, at the following property: in a group (i.e., in a troupe satisfying the property (1.49)) the *associativity* property holds, i.e., for any three elements $\mathbf{u}$, $\mathbf{v}$ and $\mathbf{w}$,

$$\mathbf{w} \oplus (\mathbf{v} \oplus \mathbf{u}) = (\mathbf{w} \oplus \mathbf{v}) \oplus \mathbf{u} \quad . \tag{A.33}$$

A.3 Troupe Series

The demonstrations in this section were kindly worked by Georges Jobert (pers. commun.).

A.3.1 Sum of Autovectors

We have seen in section 1.2.4 that the axioms for the o-sum imply the form (equation 1.69)

$$\boxed{\mathbf{w} \oplus \mathbf{v} = (\mathbf{w} + \mathbf{v}) + \mathbf{e}(\mathbf{w}, \mathbf{v}) + \mathbf{q}(\mathbf{w}, \mathbf{w}, \mathbf{v}) + \mathbf{r}(\mathbf{w}, \mathbf{v}, \mathbf{v}) + \dots \quad ,} \tag{A.34}$$

the tensors $\mathbf{e}$, $\mathbf{q}$ and $\mathbf{r}$ having the symmetries (equation 1.70)

$$\mathbf{q}(\mathbf{w}, \mathbf{v}, \mathbf{u}) = \mathbf{q}(\mathbf{v}, \mathbf{w}, \mathbf{u}) \quad ; \quad \mathbf{r}(\mathbf{w}, \mathbf{v}, \mathbf{u}) = \mathbf{r}(\mathbf{w}, \mathbf{u}, \mathbf{v}) \tag{A.35}$$

and (equations 1.71)

$$\begin{aligned} \mathbf{e}(\mathbf{v}, \mathbf{u}) + \mathbf{e}(\mathbf{u}, \mathbf{v}) &= 0 \\ \mathbf{q}(\mathbf{w}, \mathbf{v}, \mathbf{u}) + \mathbf{q}(\mathbf{v}, \mathbf{u}, \mathbf{w}) + \mathbf{q}(\mathbf{u}, \mathbf{w}, \mathbf{v}) &= 0 \\ \mathbf{r}(\mathbf{w}, \mathbf{v}, \mathbf{u}) + \mathbf{r}(\mathbf{v}, \mathbf{u}, \mathbf{w}) + \mathbf{r}(\mathbf{u}, \mathbf{w}, \mathbf{v}) &= 0 \quad . \end{aligned} \tag{A.36}$$

A.3.2 Difference of Autovectors

It is easy to see that the series for the o-difference necessarily has the form

$$(\mathbf{w} \ominus \mathbf{u}) \;=\; (\mathbf{w} - \mathbf{u}) + \mathbf{W}_2(\mathbf{w}, \mathbf{u}) + \mathbf{W}_3(\mathbf{w}, \mathbf{u}) + \cdots \quad , \tag{A.37}$$

where $\mathbf{W}_n$ indicates a term of order n. The two operations $\oplus$ and $\ominus$ are linked through $\mathbf{w} = \mathbf{v} \oplus \mathbf{u} \Leftrightarrow \mathbf{v} = \mathbf{w} \ominus \mathbf{u}$ (equation 1.36), so one must have $\mathbf{w} = (\mathbf{w} \ominus \mathbf{u}) \oplus \mathbf{u}$. Using the expression (A.34), this condition is written

$$\mathbf{w} \;=\; (\,(\mathbf{w} \ominus \mathbf{u}) + \mathbf{u}\,) + \mathbf{e}(\mathbf{w} \ominus \mathbf{u}, \mathbf{u}) + \mathbf{q}(\mathbf{w} \ominus \mathbf{u}, \mathbf{w} \ominus \mathbf{u}, \mathbf{u}) + \mathbf{r}(\mathbf{w} \ominus \mathbf{u}, \mathbf{u}, \mathbf{u}) + \dots \quad , \tag{A.38}$$

and inserting here expression (A.37) we obtain, making explicit only the terms up to third order,

$$\begin{aligned} \mathbf{w} \;=& (\,(\mathbf{w} - \mathbf{u}) + \mathbf{W}_2(\mathbf{w}, \mathbf{u}) + \mathbf{W}_3(\mathbf{w}, \mathbf{u}) + \mathbf{u}\,) + \mathbf{e}(\,(\mathbf{w} - \mathbf{u}) \\ & + \mathbf{W}_2(\mathbf{w}, \mathbf{u})\,,\; \mathbf{u}\,) + \mathbf{q}(\mathbf{w} - \mathbf{u}, \mathbf{w} - \mathbf{u}, \mathbf{u}) + \mathbf{r}(\mathbf{w} - \mathbf{u}, \mathbf{u}, \mathbf{u}) + \dots \quad , \end{aligned} \tag{A.39}$$

i.e., developing and using properties (A.36)–(A.35)

$$\begin{aligned} \mathbf{0} \;=& \;\mathbf{W}_2(\mathbf{w}, \mathbf{u}) + \mathbf{W}_3(\mathbf{w}, \mathbf{u}) + \mathbf{e}(\mathbf{w}, \mathbf{u}) + \mathbf{e}(\mathbf{W}_2(\mathbf{w}, \mathbf{u}), \mathbf{u}) \\ & + \mathbf{q}(\mathbf{w}, \mathbf{w}, \mathbf{u}) + (\mathbf{r} - 2\,\mathbf{q})(\mathbf{w}, \mathbf{u}, \mathbf{u}) + \dots \quad . \end{aligned} \tag{A.40}$$

As the series has to vanish for every $\mathbf{u}$ and $\mathbf{w}$, each term has to vanish. For the second-order terms this gives

$$\mathbf{W}_2(\mathbf{w}, \mathbf{u}) \;=\; \text{-}\mathbf{e}(\mathbf{w}, \mathbf{u}) \quad , \tag{A.41}$$

and the condition (A.39) then simplifies to $\mathbf{0} = \mathbf{W}_3(\mathbf{w}, \mathbf{u}) - \mathbf{e}(\mathbf{e}(\mathbf{w}, \mathbf{u}), \mathbf{u}) + \mathbf{q}(\mathbf{w}, \mathbf{w}, \mathbf{u}) + (\mathbf{r} - 2\,\mathbf{q})(\mathbf{w}, \mathbf{u}, \mathbf{u}) + \dots$. The condition that the third-order term must vanish then gives

$$\mathbf{W}_3(\mathbf{w}, \mathbf{u}) \;=\; \mathbf{e}(\mathbf{e}(\mathbf{w}, \mathbf{u}), \mathbf{u}) - \mathbf{q}(\mathbf{w}, \mathbf{w}, \mathbf{u}) + (2\,\mathbf{q} - \mathbf{r})(\mathbf{w}, \mathbf{u}, \mathbf{u}) \quad . \tag{A.42}$$

Introducing this and equation (A.41) into (A.37) gives

$$\boxed{\begin{aligned} (\mathbf{w} \ominus \mathbf{u}) \;=& \;(\mathbf{w} - \mathbf{u}) - \mathbf{e}(\mathbf{w}, \mathbf{u}) + \mathbf{e}(\mathbf{e}(\mathbf{w}, \mathbf{u}), \mathbf{u}) \\ & - \mathbf{q}(\mathbf{w}, \mathbf{w}, \mathbf{u}) + (2\,\mathbf{q} - \mathbf{r})(\mathbf{w}, \mathbf{u}, \mathbf{u}) + \dots \quad , \end{aligned}} \tag{A.43}$$

so we have now an expression for the o-difference in terms of the same tensors appearing in the o-sum.

A.3.3 Commutator

Using the two series (A.34) and (A.43) gives, when retaining only the terms up to second order $(\mathbf{v} \oplus \mathbf{u}) \ominus (\mathbf{u} \oplus \mathbf{v}) = \mathbf{e}(\mathbf{v}, \mathbf{u}) - \mathbf{e}(\mathbf{u}, \mathbf{v}) + \dots$, i.e., using the antisymmetry of $\mathbf{e}$ (first of conditions (A.36)), $(\mathbf{v} \oplus \mathbf{u}) \ominus (\mathbf{u} \oplus \mathbf{v}) = 2\,\mathbf{e}(\mathbf{v}, \mathbf{u}) + \dots$. Comparing this with the definition of the infinitesimal commutator (equation 1.77) gives

$$\boxed{[\mathbf{v}, \mathbf{u}] \;=\; 2\,\mathbf{e}(\mathbf{v}, \mathbf{u}) \quad .} \tag{A.44}$$

A.3.4 Associator

Using the series (A.34) for the o-sum and the series (A.43) for the o-difference, using the properties (A.36) and (A.35), and making explicit only the terms up to third order gives, after a long but easy computation, $(\mathbf{w}\oplus(\mathbf{v}\oplus\mathbf{u}))\ominus((\mathbf{w}\oplus\mathbf{v})\oplus\mathbf{u}) = \mathbf{e}(\mathbf{w},\mathbf{e}(\mathbf{v},\mathbf{u})) - \mathbf{e}(\mathbf{e}(\mathbf{w},\mathbf{v}),\mathbf{u}) - 2\,\mathbf{q}(\mathbf{w},\mathbf{v},\mathbf{u}) + 2\,\mathbf{r}(\mathbf{w},\mathbf{v},\mathbf{u}) + \cdots$. Comparing this with the definition of the infinitesimal associator (equation 1.78) gives

$$\boxed{[\mathbf{w},\mathbf{v},\mathbf{u}] = \mathbf{e}(\mathbf{w},\mathbf{e}(\mathbf{v},\mathbf{u})) - \mathbf{e}(\mathbf{e}(\mathbf{w},\mathbf{v}),\mathbf{u}) - 2\,\mathbf{q}(\mathbf{w},\mathbf{v},\mathbf{u}) + 2\,\mathbf{r}(\mathbf{w},\mathbf{v},\mathbf{u}) \quad .} \tag{A.45}$$

A.3.5 Relation Between Commutator and Associator

As the circular sums of $\mathbf{e}$, $\mathbf{q}$ and $\mathbf{r}$ vanish (relations A.36) we immediately obtain $[\mathbf{w},\mathbf{v},\mathbf{u}] + [\mathbf{v},\mathbf{u},\mathbf{w}] + [\mathbf{u},\mathbf{w},\mathbf{v}] = 2\Big(\mathbf{e}(\mathbf{w},\mathbf{e}(\mathbf{v},\mathbf{u})) + \mathbf{e}(\mathbf{v},\mathbf{e}(\mathbf{u},\mathbf{w})) + \mathbf{e}(\mathbf{u},\mathbf{e}(\mathbf{w},\mathbf{v}))\Big)$, i.e., using (A.45),

$$\boxed{\begin{aligned} &[\,\mathbf{w},[\mathbf{v},\mathbf{u}]\,] + [\,\mathbf{v},[\mathbf{u},\mathbf{w}]\,] + [\,\mathbf{u},[\mathbf{w},\mathbf{v}]\,] \\ &\qquad = 2(\,[\mathbf{w},\mathbf{v},\mathbf{u}] + [\mathbf{v},\mathbf{u},\mathbf{w}] + [\mathbf{u},\mathbf{w},\mathbf{v}]\,) \quad . \end{aligned}} \tag{A.46}$$

This demonstrates the property 1.6 of the main text.

A.3.6 Inverse Relations

We have obtained the expression of the infinitesimal commutator and of the infinitesimal associator in terms of $\mathbf{e}$, $\mathbf{q}$ and $\mathbf{r}$. Let us obtain the inverse relations. Equation (A.45) directly gives

$$\boxed{\mathbf{e}(\mathbf{v},\mathbf{u}) = \tfrac{1}{2}\,[\mathbf{v},\mathbf{u}] \quad .} \tag{A.47}$$

Because of the different symmetries satisfied by $\mathbf{q}$ and $\mathbf{r}$, the single equation (A.45) can be solved to give both $\mathbf{q}$ and $\mathbf{r}$. This is done by writing equation (A.45) exchanging the "slots" of $\mathbf{u}$, $\mathbf{v}$ and $\mathbf{w}$ and reiterately using the properties (A.36) and (A.35) and the property (A.46). This gives (the reader may just verify that inserting these values into (A.45) gives an identity)

$$\boxed{\begin{aligned} \mathbf{q}(\mathbf{w},\mathbf{v},\mathbf{u}) &= \frac{1}{24}\Big(\,[\,\mathbf{w}\,,\,[\mathbf{v},\mathbf{u}]\,] - [\,\mathbf{v}\,,\,[\mathbf{u},\mathbf{w}]\,]\,\Big) \\ &\qquad + \frac{1}{6}\Big(\,[\mathbf{w},\mathbf{u},\mathbf{v}] + [\mathbf{v},\mathbf{u},\mathbf{w}] - [\mathbf{w},\mathbf{v},\mathbf{u}] - [\mathbf{v},\mathbf{w},\mathbf{u}]\,\Big) \\ \mathbf{r}(\mathbf{w},\mathbf{v},\mathbf{u}) &= \frac{1}{24}\Big(\,[\,\mathbf{v}\,,\,[\mathbf{u},\mathbf{w}]\,] - [\,\mathbf{u}\,,\,[\mathbf{w},\mathbf{v}]\,]\,\Big) \\ &\qquad + \frac{1}{6}\Big(\,[\mathbf{w},\mathbf{v},\mathbf{u}] + [\mathbf{w},\mathbf{u},\mathbf{v}] - [\mathbf{u},\mathbf{w},\mathbf{v}] - [\mathbf{v},\mathbf{w},\mathbf{u}]\,\Big) \quad . \end{aligned}} \tag{A.48}$$

Using this and equation (A.45) allows one to write the series (A.34) and (A.43) respectively as equations (1.84) and (1.85) in the main text.

A.3.7 Torsion and Anassociativity

The torsion tensor and the anassociativity tensor have been defined respectively through (equations 1.86–(1.87))

$$\begin{aligned} [\mathbf{v},\mathbf{u}]^k &= T^k{}_{ij}\, v^i\, u^j \\ [\mathbf{w},\mathbf{v},\mathbf{u}]^\ell &= \tfrac{1}{2}\, A^\ell{}_{ijk}\, w^i\, v^j\, u^k \quad . \end{aligned} \tag{A.49}$$

Obtaining explicit expressions is just a matter of writing the index equivalent of the two equations (A.44)–(A.45). This gives

$$\boxed{\begin{aligned} T^k{}_{ij} &= 2\, e^k{}_{ij} \\ A^\ell{}_{ijk} &= 2\,(e^\ell{}_{ir}\, e^r{}_{jk} + e^\ell{}_{kr}\, e^r{}_{ij}) - 4\, q^\ell{}_{ijk} + 4\, r^\ell{}_{ijk} \quad . \end{aligned}} \tag{A.50}$$

A.4 Cayley-Hamilton Theorem

It is important to realize that, thanks to the Cayley-Hamilton theorem, any infinite series concerning an $n \times n$ matrix can always be rewritten as a polynomial of, at most, degree $n-1$. This, in particular, is true for the series expressing the exponential and the logarithm of the function.

The *characteristic polynomial* of a square matrix $\mathbf{M}$ is the polynomial in the scalar variable x defined as

$$\varphi(x) = \det(x\,\mathbf{I} - \mathbf{M}) \, . \tag{A.51}$$

Any eigenvalue λ of a matrix $\mathbf{M}$ satisfies the property $\varphi(\lambda) = 0$. The Cayley-Hamilton theorem states that the matrix $\mathbf{M}$ satisfies the matrix equivalent of this equation:

$$\varphi(\mathbf{M}) = \mathbf{0} \quad . \tag{A.52}$$

Explicitly, given an $n \times n$ matrix $\mathbf{M}$, and writing

$$\det(x\,\mathbf{I} - \mathbf{M}) = x^n + \alpha_{n-1}\, x^{n-1} + \cdots + \alpha_1\, x + \alpha_0 \quad , \tag{A.53}$$

then, the eigenvalues of $\mathbf{M}$ satisfy

$$\lambda^n + \alpha_{n-1}\, \lambda^{n-1} + \cdots + \alpha_1\, \lambda + \alpha_0 = 0 \quad , \tag{A.54}$$

while the matrix $\mathbf{M}$ itself satisfies

$$\mathbf{M}^n + \alpha_{n-1}\, \mathbf{M}^{n-1} + \cdots + \alpha_1\, \mathbf{M} + \alpha_0\, \mathbf{I} = \mathbf{0} \quad . \tag{A.55}$$

In particular, this implies that one can always express the nth power of an $n \times n$ matrix as a function of all the lower order powers:

$$\mathbf{M}^n = -(\alpha_{n-1}\,\mathbf{M}^{n-1} + \cdots + \alpha_1\,\mathbf{M} + \alpha_0\,\mathbf{I}) \quad . \tag{A.56}$$

Example A.1 *For* 2×2 *matrices,* $\mathbf{M}^2 = (\operatorname{tr}\mathbf{M})\,\mathbf{M} - (\det\mathbf{M})\,\mathbf{I} = (\operatorname{tr}\mathbf{M})\,\mathbf{M} - \frac{1}{2}\,(\operatorname{tr}^2\mathbf{M} - \operatorname{tr}\mathbf{M}^2)\,\mathbf{I}$. *For* 3×3 *matrices,* $\mathbf{M}^3 = (\operatorname{tr}\mathbf{M})\,\mathbf{M}^2 - \frac{1}{2}\,(\operatorname{tr}^2\mathbf{M} - \operatorname{tr}\mathbf{M}^2)\,\mathbf{M} + (\det\mathbf{M})\,\mathbf{I} = (\operatorname{tr}\mathbf{M})\,\mathbf{M}^2 - (\det\mathbf{M}\operatorname{tr}\mathbf{M}^{-1})\,\mathbf{M} + (\det\mathbf{M})\,\mathbf{I}$. *For* 4×4 *matrices,* $\mathbf{M}^4 = (\operatorname{tr}\mathbf{M})\,\mathbf{M}^3 - \frac{1}{2}\,(\operatorname{tr}^2\mathbf{M} - \operatorname{tr}\mathbf{M}^2)\,\mathbf{M}^2 + (\det\mathbf{M}\operatorname{tr}\mathbf{M}^{-1})\,\mathbf{M} - (\det\mathbf{M})\,\mathbf{I}$.

Therefore,

Property A.1 *A series expansion of any analytic function* $f(\mathbf{M})$ *of a matrix* $\mathbf{M}$, *i.e., a series* $f(\mathbf{M}) = \sum_{p=0}^{\infty} a_p\,\mathbf{M}^p$, *only contains, in fact, terms of order less or equal to* n:

$$f(\mathbf{M}) = \alpha_{n-1}\,\mathbf{M}^{n-1} + \alpha_{n-2}\,\mathbf{M}^{n-2} + \cdots + \alpha_2\,\mathbf{M}^2 + \alpha_1\,\mathbf{M} + \alpha_0\,\mathbf{I}\,, \tag{A.57}$$

where $\alpha_{n-1}\,,\,\alpha_{n-2}\,\ldots\,\alpha_1\,,\,\alpha_0$ *are complex numbers.*

Example A.2 *Let* $\mathbf{r}$ *be an antisymmetric* 3×3 *matrix. Thanks to the Cayley-Hamilton theorem, the exponential series collapses into the second-degree polynomial (see equation A.266)*

$$\exp\mathbf{r} = \mathbf{I} + \frac{\sinh r}{r}\,\mathbf{r} + \frac{\cosh r - 1}{r^2}\,\mathbf{r}^2 \quad , \tag{A.58}$$

where $r = \sqrt{(\operatorname{tr}\mathbf{r}^2)/2}$.

A.5 Function of a Matrix

A.5.1 Function of a Jordan Block Matrix

A *Jordan block matrix* is an $n\times n$ matrix with the special form (λ being a complex number)

$$\mathbf{J} = \begin{pmatrix} \lambda & 1 & 0 & \cdots & 0 \\ 0 & \lambda & 1 & \ddots & \vdots \\ 0 & 0 & \lambda & \ddots & 0 \\ \vdots & \ddots & \ddots & \ddots & 1 \\ 0 & 0 & 0 & 0 & \lambda \end{pmatrix} \quad . \tag{A.59}$$

Let $f(z)$ be a polynomial (of certain finite order k) of the complex variable z, $f(z) = \alpha_0 + \alpha_1\,z + \alpha_2\,z^2 + \cdots + \alpha_k\,z^k$, and $f(\mathbf{M})$ its direct generalization into a polynomial of a square complex matrix $\mathbf{M}$: $f(\mathbf{M}) = \alpha_0 \mathbf{I} + \alpha_1 \mathbf{M} + \alpha_2 \mathbf{M}^2 + \cdots + \alpha_k \mathbf{M}^k$. It is easy to verify that for a Jordan block matrix one has (for any order k of the polynomial)

$$f(\mathbf{J}) = \begin{pmatrix} f(\lambda) & f'(\lambda) & \frac{1}{2} f''(\lambda) & \cdots & \frac{1}{(n-1)!} f^{(n-1)}(\lambda) \\ 0 & f(\lambda) & f'(\lambda) & \ddots & \vdots \\ 0 & 0 & f(\lambda) & \ddots & \frac{1}{2} f''(\lambda) \\ \vdots & \ddots & \ddots & \ddots & f'(\lambda) \\ 0 & 0 & 0 & 0 & f(\lambda) \end{pmatrix} , \tag{A.60}$$

where $f', f'' \dots$ are the successive derivatives of the function f. This property suggests to introduce the following

Definition A.1 *Let $f(z)$ be an analytic function of the complex variable z and $\mathbf{J}$ a Jordan block matrix. The function $f(\mathbf{J})$ is, by definition, the matrix in equation* (A.60).

Example A.3 *For instance, for a 5×5 Jordan block matrix, when $\lambda \neq 0$,*

$$\log \begin{pmatrix} \lambda & 1 & 0 & 0 & 0 \\ 0 & \lambda & 1 & 0 & 0 \\ 0 & 0 & \lambda & 1 & 0 \\ 0 & 0 & 0 & \lambda & 1 \\ 0 & 0 & 0 & 0 & \lambda \end{pmatrix} = \begin{pmatrix} \log\lambda & 1/\lambda & \text{-}1/(2\,\lambda^2) & 1/(3\,\lambda^3) & \text{-}1/(4\,\lambda^4) \\ 0 & \log\lambda & 1/\lambda & \text{-}1/(2\,\lambda^2) & 1/(3\,\lambda^3) \\ 0 & 0 & \log\lambda & 1/\lambda & \text{-}1/(2\,\lambda^2) \\ 0 & 0 & 0 & \log\lambda & 1/\lambda \\ 0 & 0 & 0 & 0 & \log\lambda \end{pmatrix} . \tag{A.61}$$

A.5.2 Function of an Arbitrary Matrix

Any invertible square matrix $\mathbf{M}$ accepts the *Jordan decomposition*

$$\mathbf{M} = \mathbf{U}\,\mathbf{J}\,\mathbf{U}^{-1} , \tag{A.62}$$

where $\mathbf{U}$ is a matrix of $GL(n, \mathbb{C})$ (even when $\mathbf{M}$ is a real matrix) and where the *Jordan matrix* $\mathbf{J}$ is a matrix made by Jordan blocks (note the "diagonal" made with ones)

$$\mathbf{J} = \begin{pmatrix} \mathbf{J}_1 & & & & \\ & \mathbf{J}_2 & 0 & & \\ & & \mathbf{J}_3 & & \\ & 0 & & \mathbf{J}_4 & \\ & & & & \ddots \end{pmatrix} , \quad \mathbf{J}_i = \begin{pmatrix} \lambda_i & 1 & 0 & 0 & \cdots \\ 0 & \lambda_i & 1 & 0 & \cdots \\ 0 & 0 & \lambda_i & 1 & \cdots \\ 0 & 0 & 0 & \lambda_i & \cdots \\ \vdots & \vdots & \vdots & \vdots & \ddots \end{pmatrix} , \tag{A.63}$$

$\{\lambda_1, \lambda_2 \dots\}$ being the eigenvalues of $\mathbf{M}$ (arbitrarily ordered). In the special case where all the eigenvalues are distinct, all the matrices $\mathbf{J}_i$ are 1×1 matrices, so $\mathbf{J}$ is diagonal.

Definition A.2 *Let $f(z)$ be an analytic function of the complex variable z and $\mathbf{M}$ an arbitrary $n \times n$ real or complex matrix, with the Jordan decomposition as in equations* (A.62) *and* (A.63)*. When it makes sense,*[2] *the function $\mathbf{M} \mapsto f(\mathbf{M})$ is defined as*

$$f(\mathbf{M}) \;=\; \mathbf{U}\, f(\mathbf{J})\, \mathbf{U}^{-1} \quad , \tag{A.64}$$

where, by definition,

$$f(\mathbf{J}) \;=\; \begin{pmatrix} f(\mathbf{J}_1) & & & & \\ & f(\mathbf{J}_2) & & 0 & \\ & & f(\mathbf{J}_3) & & \\ & 0 & & f(\mathbf{J}_4) & \\ & & & & \ddots \end{pmatrix} \quad , \tag{A.65}$$

the function f of a Jordan block having been introduced in definition A.1.

It is easy to see that the function of $\mathbf{M}$ so calculated is independent of the particular ordering of the eigenvalues used to define $\mathbf{U}$ and $\mathbf{J}$.

For the logarithm function, $f(z) = \log z$ the above definition makes sense for all invertible matrices (e.,g., Horn and Johnson, 1999), so we can use the following

Definition A.3 *The logarithm of an invertible matrix with Jordan decomposition $\mathbf{M} = \mathbf{U}\,\mathbf{J}\,\mathbf{U}^{-1}$ is defined as*

$$\log \mathbf{M} \;=\; \mathbf{U}\,(\log \mathbf{J})\,\mathbf{U}^{-1} \quad . \tag{A.66}$$

One has the property $\exp(\log \mathbf{M}) = \mathbf{M}$, this showing that one has actually defined the logarithm of $\mathbf{M}$.

Example A.4 *The matrix*

$$\mathbf{M} \;=\; \begin{pmatrix} 2 & 4 & -6 & 0 \\ 4 & 6 & -3 & -4 \\ 0 & 0 & 4 & 0 \\ 0 & 4 & -6 & 2 \end{pmatrix} \tag{A.67}$$

has the four eigenvalues $\{2,2,4,6\}$. Having a repeated eigenvalue, its Jordan decomposition

$$\mathbf{M} \;=\; \begin{pmatrix} 1 & -1/4 & 0 & 1 \\ 0 & 1/4 & 3 & 1 \\ 0 & 0 & 2 & 0 \\ 1 & 0 & 0 & 1 \end{pmatrix} \cdot \begin{pmatrix} 2 & 1 & 0 & 0 \\ 0 & 2 & 0 & 0 \\ 0 & 0 & 4 & 0 \\ 0 & 0 & 0 & 6 \end{pmatrix} \cdot \begin{pmatrix} 1 & -1/4 & 0 & 1 \\ 0 & 1/4 & 3 & 1 \\ 0 & 0 & 2 & 0 \\ 1 & 0 & 0 & 1 \end{pmatrix}^{-1} \tag{A.68}$$

contains a Jordan matrix (in the middle) that is not diagonal. The logarithm of $\mathbf{M}$ is, then,

[2] For instance, for the exponential series of a tensor to make sense, the tensor must be adimensional (must not have physical dimensions).

$$\log \mathbf{M} = \begin{pmatrix} 1 & -1/4 & 0 & 1 \\ 0 & 1/4 & 3 & 1 \\ 0 & 0 & 2 & 0 \\ 1 & 0 & 0 & 1 \end{pmatrix} \cdot \begin{pmatrix} \log 2 & 1/2 & 0 & 0 \\ 0 & \log 2 & 0 & 0 \\ 0 & 0 & \log 4 & 0 \\ 0 & 0 & 0 & \log 6 \end{pmatrix} \cdot \begin{pmatrix} 1 & -1/4 & 0 & 1 \\ 0 & 1/4 & 3 & 1 \\ 0 & 0 & 2 & 0 \\ 1 & 0 & 0 & 1 \end{pmatrix}^{-1} \quad . \tag{A.69}$$

It is easy to see that if $\mathbf{P}(\mathbf{M})$ and $\mathbf{Q}(\mathbf{M})$ are two polynomials of the matrix $\mathbf{M}$, then $\mathbf{P}(\mathbf{M})\,\mathbf{Q}(\mathbf{M}) = \mathbf{Q}(\mathbf{M})\,\mathbf{P}(\mathbf{M})$. It follows that the functions of a same matrix commute. For instance,

$$\frac{d}{d\lambda}(\exp \lambda\mathbf{M}) = \mathbf{M}\,(\exp \lambda\mathbf{M}) = (\exp \lambda\mathbf{M})\,\mathbf{M} \quad . \tag{A.70}$$

A.5.3 Alternative Definitions of exp and log

$$\exp \mathbf{t} = \lim_{n\to\infty} \left(\mathbf{I} + \frac{1}{n}\,\mathbf{t} \right)^n \quad ; \quad \log \mathbf{T} = \lim_{x\to 0} \frac{1}{x}\,(\,\mathbf{T}^x - \mathbf{I}\,) \quad . \tag{A.71}$$

A.5.4 A Series for the Logarithm

The Taylor series (1.124) is not the best series for computing the logarithm. Based on the well-known scalar formula $\log s = \sum_{n=0}^{\infty} \frac{2}{2n+1} (\frac{s-1}{s+1})^{2n+1}$, one may use (Lastman and Sinha, 1991) the series

$$\log \mathbf{T} = \sum_{n=0}^{\infty} \frac{2}{2n+1} \left((\mathbf{T}-\mathbf{I})\,(\mathbf{T}+\mathbf{I})^{-1} \right)^{2n+1} \quad . \tag{A.72}$$

As $(\mathbf{T}-\mathbf{I})\,(\mathbf{T}+\mathbf{I})^{-1} = (\mathbf{T}+\mathbf{I})^{-1}\,(\mathbf{T}-\mathbf{I}) = (\mathbf{I}+\mathbf{T}^{-1})^{-1} - (\mathbf{I}+\mathbf{T})^{-1}$ different expressions can be given to this series. The series has a wider domain of convergence than the Taylor series, and converges more rapidly than it. Letting $\ell_K(\mathbf{T})$ be the partial sum $\sum_{n=0}^{K}$, one also has the property

$$\mathbf{T} \text{ orthogonal} \;\Rightarrow\; \ell_K(\mathbf{T}) \text{ skew-symmetric} \tag{A.73}$$

(for a demonstration, see Dieci, 1996). It is easy to verify that one has the property

$$\ell_K(\mathbf{T}^{-1}) = -\ell_K(\mathbf{T}) \quad . \tag{A.74}$$

There is another well-known series for the logarithm, $\log s = \sum_{n=1}^{\infty} \frac{1}{n} (\frac{s-1}{s})^n$ (Gradshteyn and Ryzhik, 1980). It also generalizes to matrices, but it does not seem to have any particular advantage with respect to the two series already considered.

A.5.5 Cayley-Hamilton Polynomial for the Function of a Matrix

A.5.5.1 Case with All Eigenvalues Distinct

If all the eigenvalues λ_i of an $n \times n$ matrix $\mathbf{M}$ are distinct, the Cayley-Hamilton polynomial (see appendix A.4) of degree $(n-1)$ expressing an analytical function $f(\mathbf{M})$ of the matrix is given by Sylvester's formula (Sylvester, 1883; Hildebrand, 1952; Moler and Van Loan, 1978):

$$f(\mathbf{M}) = \sum_{i=1}^{n} \left[\frac{f(\lambda_i)}{\prod_{j\neq i} (\lambda_i - \lambda_j)} \prod_{j\neq i} (\mathbf{M} - \lambda_j \mathbf{I}) \right] . \tag{A.75}$$

We can write an alternative version of this formula, that uses the notion of *adjoint* of a matrix (definition recalled in footnote[3]). It is possible to demonstrate the relation $\prod_{j\neq i}(\mathbf{M} - \lambda_j \mathbf{I}) = (-1)^{n-1} \operatorname{ad}(\mathbf{M} - \lambda_i \mathbf{I})$. Using it, Sylvester formula (A.75) accepts the equivalent expression (the change of the signs of the terms in the denominator absorbing the factor $(-1)^{n-1}$)

$$f(\mathbf{M}) = \sum_{i=1}^{n} \frac{f(\lambda_i)}{\prod_{j\neq i}(\lambda_j - \lambda_i)} \operatorname{ad}(\mathbf{M} - \lambda_i \mathbf{I}) . \tag{A.76}$$

Example A.5 *For a* 3×3 *matrix with distinct eigenvalues, Sylvester's formula gives*

$$\begin{aligned} f(\mathbf{M}) &= \frac{f(\lambda_1)}{(\lambda_1 - \lambda_2)(\lambda_1 - \lambda_3)} (\mathbf{M} - \lambda_2 \mathbf{I}) (\mathbf{M} - \lambda_3 \mathbf{I}) \\ &+ \frac{f(\lambda_2)}{(\lambda_2 - \lambda_3)(\lambda_2 - \lambda_1)} (\mathbf{M} - \lambda_3 \mathbf{I}) (\mathbf{M} - \lambda_1 \mathbf{I}) \\ &+ \frac{f(\lambda_3)}{(\lambda_3 - \lambda_1)(\lambda_3 - \lambda_2)} (\mathbf{M} - \lambda_1 \mathbf{I}) (\mathbf{M} - \lambda_2 \mathbf{I}) \quad , \end{aligned} \tag{A.77}$$

while formula (A.76) *gives*

$$\begin{aligned} f(\mathbf{M}) = & \frac{f(\lambda_1)}{(\lambda_2 - \lambda_1)(\lambda_3 - \lambda_1)} \operatorname{ad}(\mathbf{M} - \lambda_1 \mathbf{I}) + \frac{f(\lambda_2)}{(\lambda_3 - \lambda_2)(\lambda_1 - \lambda_2)} \operatorname{ad}(\mathbf{M} - \lambda_2 \mathbf{I}) \\ & + \frac{f(\lambda_3)}{(\lambda_1 - \lambda_3)(\lambda_2 - \lambda_3)} \operatorname{ad}(\mathbf{M} - \lambda_3 \mathbf{I}) \end{aligned} \tag{A.78}$$

[3]The *minor* of an element A_{ij} of a square matrix $\mathbf{A}$, denoted $\operatorname{minor}(A_{ij})$, is the determinant of the matrix obtained by deleting the ith row and jth column of $\mathbf{A}$. The *cofactor* of the element A_{ij} is the number $\operatorname{cof}(A_{ij}) = (-1)^{i+j} \operatorname{minor}(A_{ij})$. The *adjoint* of a matrix $\mathbf{A}$, denoted $\operatorname{ad}\mathbf{A}$, is the transpose of the matrix formed by replacing each entry of the matrix by its cofactor. This can be written $(\operatorname{ad}\mathbf{A})^{i_1}{}_{j_1} = \frac{1}{(n-1)!} \epsilon^{i_1 i_2 i_3 \dots i_n} \epsilon_{j_1 j_2 j_3 \dots j_n} A^{j_2}{}_{i_2} A^{j_3}{}_{i_3} \dots A^{j_n}{}_{i_n}$, where $\epsilon^{ij\dots}$ and $\epsilon_{ij\dots}$ are the totally antisymmetric symbols of order n. When a matrix $\mathbf{A}$ is invertible, then $\mathbf{A}^{-1} = \frac{1}{(\det \mathbf{A})} \operatorname{ad}(\mathbf{A})$.

A.5.5.2 Case with Repeated Eigenvalues

The two formulas above can be generalized to the case where there are repeated eigenvalues. In what follows, let us denote m_i the multiplicity of the eigenvalue λ_i. Concerning the generalization of Sylvester formula, Buchheim (1886) uses obscure notation and Rinehart (1955) gives his version of Buchheim result but, unfortunately, with two mistakes. When these are corrected (Loring Tu, pers. commun.), one finds the expression

$$\boxed{f(\mathbf{M}) = \sum_i \left[\left(\sum_{k=0}^{m_i-1} b_k(\lambda_i)\,(\mathbf{M}-\lambda_i\,\mathbf{I})^k \right) \prod_{j\neq i} (\mathbf{M}-\lambda_j\,\mathbf{I})^{m_j} \right] \quad ,} \tag{A.79}$$

where the sum is performed over all distinct eigenvalues λ_i, and where the $b_k(\lambda_i)$ are the scalars

$$b_k(\lambda_i) = \frac{1}{k!}\,\frac{d^k}{d\lambda^k}\,\frac{f(\lambda)}{\prod_{j\neq i}(\lambda-\lambda_j)^{m_j}}\Bigg|_{\lambda=\lambda_i} \quad . \tag{A.80}$$

When all the eigenvalues are distinct, equation (A.79) reduces to the Sylvester formula (A.75). Formula (A.76) can also be generalized to the case with repeated eigenvalues: if the eigenvalue λ_i has multiplicity m_i, then (White, pers. commun.[4])

$$\boxed{f(\mathbf{M}) = (-1)^{n-1} \sum_i \frac{1}{(m_i-1)!} \left[\frac{d^{m_i-1}}{d\lambda^{m_i-1}} \left(\frac{f(\lambda)}{\prod_{j\neq i}(\lambda-\lambda_j)^{m_j}}\;\mathrm{ad}(\mathbf{M}-\lambda\,\mathbf{I}) \right) \right]_{\lambda=\lambda_i} \quad ,} \tag{A.81}$$

the sum being again performed for all distinct eigenvalues. If all eigenvalues are distinct, all the m_i take the value 1, and this formula reduces to (A.76).

Example A.6 *Applying formula* (A.79) *to the diagonal matrix*

$$\mathbf{M} = \mathrm{diag}(\alpha,\beta,\beta,\gamma,\gamma,\gamma) \quad , \tag{A.82}$$

gives $f(\mathbf{M}) = b_0(\alpha)\,(\mathbf{M}-\beta\,\mathbf{I})^2\,(\mathbf{M}-\gamma\,\mathbf{I})^3 + (\,b_0(\beta)\,\mathbf{I}+b_1(\beta)\,(\mathbf{M}-\beta\,\mathbf{I})\,)\,(\mathbf{M}-\alpha\,\mathbf{I})\,(\mathbf{M}-\gamma\,\mathbf{I})^3 + (\,b_0(\gamma)\,\mathbf{I}+b_1(\gamma)\,(\mathbf{M}-\gamma\,\mathbf{I})+b_2(\gamma)\,(\mathbf{M}-\gamma\,\mathbf{I})^2\,)\,(\mathbf{M}-\alpha\,\mathbf{I})\,(\mathbf{M}-\beta\,\mathbf{I})^2$ *and, when using the value of the scalars (as defined in equation* (A.80)*), one obtains*

$$f(\mathbf{M}) = \mathrm{diag}(f(\alpha), f(\beta), f(\beta), f(\gamma), f(\gamma), f(\gamma)) \quad , \tag{A.83}$$

as it should for a diagonal matrix. Alternatively, when applying formula (A.81) *to the same matrix* $\mathbf{M}$ *one has* $\mathrm{ad}(\mathbf{M}-\lambda\,\mathbf{I}) = \mathrm{diag}((\beta-\lambda)^2(\gamma-\lambda)^3, (\alpha-\lambda)(\beta-$

[4]In the web site http://chemical.caeds.eng.uml.edu/onlinec/onlinec.htm (see also http://profjrwhite.com/courses.htm), John R. White proposes this formula in his course on System Dynamics, but the original source of the formula is unknown.

$\lambda)(\gamma-\lambda)^3, (\alpha-\lambda)(\beta-\lambda)(\gamma-\lambda)^3, (\alpha-\lambda)(\beta-\lambda)^2(\gamma-\lambda)^2, (\alpha-\lambda)(\beta-\lambda)^2(\gamma-\lambda)^2, (\alpha-\lambda)(\beta-\lambda)^2(\gamma-\lambda)^2)$, *and one obtains*

$$\begin{aligned} f(\mathbf{M}) &= -\frac{1}{0!}\left[\frac{f(\lambda)}{(\lambda-\beta)^2(\lambda-\gamma)^3}\,\mathrm{ad}(\mathbf{M}-\lambda\,\mathbf{I})\right]_{\lambda=\alpha} \\ &\quad -\frac{1}{1!}\left[\frac{d}{d\lambda}\left(\frac{f(\lambda)}{(\lambda-\alpha)(\lambda-\gamma)^3}\,\mathrm{ad}(\mathbf{M}-\lambda\,\mathbf{I})\right)\right]_{\lambda=\beta} \\ &\quad -\frac{1}{2!}\left[\frac{d^2}{d\lambda^2}\left(\frac{f(\lambda)}{(\lambda-\alpha)(\lambda-\beta)^2}\,\mathrm{ad}(\mathbf{M}-\lambda\,\mathbf{I})\right)\right]_{\lambda=\gamma} \quad . \end{aligned} \tag{A.84}$$

When the computations are done, this leads to the result already expressed in equation (A.83).

Example A.7 *When applying formula* (A.81) *to the matrix* **M** *in example A.4, one obtains*

$$\begin{aligned} f(\mathbf{M}) &= -\frac{1}{1!}\left[\frac{d}{d\lambda}\left(\frac{f(\lambda)}{(\lambda-4)\,(\lambda-6)}\,\mathrm{ad}(\mathbf{M}-\lambda\,\mathbf{I})\right)\right]_{\lambda=2} \\ &\quad -\frac{1}{0!}\left[\frac{f(\lambda)}{(\lambda-2)^2\,(\lambda-6)}\,\mathrm{ad}(\mathbf{M}-\lambda\,\mathbf{I})\right]_{\lambda=4} \\ &\quad -\frac{1}{0!}\left[\frac{f(\lambda)}{(\lambda-2)^2\,(\lambda-4)}\,\mathrm{ad}(\mathbf{M}-\lambda\,\mathbf{I})\right]_{\lambda=6} \quad . \end{aligned} \tag{A.85}$$

This result is, of course, identical to that obtained using a Jordan decomposition. In particular, when $f(\mathbf{M}) = \log\mathbf{M}$, *one obtains the result expressed in equation* (A.69).

A.5.5.3 Formula not Requiring Eigenvalues

Cardoso (2004) has developed a formula for the logarithm of a matrix that is directly based on the coefficients of the characteristic polynomial. Let **A** be an $n\times n$ matrix, and let

$$p(\lambda) = \lambda^n + c_1\,\lambda^{n-1} + \cdots + c_{n-1}\,\lambda + c_n \tag{A.86}$$

be the characteristic polynomial of the matrix $\mathbf{I}-\mathbf{A}$. Defining $q(s) = s^n\,p(1/s)$ gives

$$q(s) = 1 + c_1\,s + \cdots + c_{n-1}\,s^{n-1} + c_n\,s^n \quad , \tag{A.87}$$

and one has (Cardoso, 2004)

$$\log\mathbf{A} = f_1\,\mathbf{I} + f_2\,(\mathbf{I}-\mathbf{A}) + \cdots + f_n\,(\mathbf{I}-\mathbf{A})^n \quad , \tag{A.88}$$

where

$$f_1 = c_n \int_0^1 ds \, \frac{s^{n-1}}{q(s)} \quad ; \quad f_n = - \int_0^1 ds \, \frac{s^{n-2}}{q(s)} \quad ;$$
$$f_i = - \int_0^1 ds \, \frac{s^{i-2} + c_1 \, s^{i-1} + \cdots + c_{n-i} \, s^{n-i}}{q(s)} \qquad (i = 2, \ldots, n-1) \quad . \tag{A.89}$$

In fact, Cardoso's formula is more general in two aspects, it is valid for any polynomial $p(\lambda)$ such that $p(\mathbf{I} - \mathbf{A}) = \mathbf{0}$ (special matrices may satisfy $p(\mathbf{I} - \mathbf{A}) = \mathbf{0}$ for polynomials of degree lower than the degree of the characteristic polynomial), and it gives the logarithm of a matrix of the form $\mathbf{I} - t\,\mathbf{B}$.

Example A.8 *For a* 2×2 *matrix* $\mathbf{A}$ *one gets*

$$q(s) = 1 + c_1 \, s + c_2 \, s^2 \quad , \tag{A.90}$$

with $c_1 = \mathrm{tr}(\mathbf{A}) - 2$ *and* $c_2 = \det(\mathbf{A}) - \mathrm{tr}(\mathbf{A}) + 1$, *and one obtains*

$$\log \mathbf{A} = f_1 \, \mathbf{I} + f_2 \, (\mathbf{I} - \mathbf{A}) \quad , \tag{A.91}$$

where $f_1 = c_2 \int_0^1 ds \; s/q(s)$ *and* $f_2 = - \int_0^1 ds \; 1/q(s)$.

Example A.9 *For a* 3×3 *matrix* $\mathbf{A}$ *one gets*

$$q(s) = 1 + c_1 \, s + c_2 \, s^2 + c_3 \, s^3 \quad , \tag{A.92}$$

with $c_1 = \mathrm{tr}(\mathbf{A}) - 3$, $c_2 = -(2\,\mathrm{tr}(\mathbf{A}) + \tau(\mathbf{A}) - 3)$, *and* $c_3 = \det(\mathbf{A}) + \tau(\mathbf{A}) + \mathrm{tr}(\mathbf{A}) - 1$, *where* $\tau(\mathbf{A}) \equiv (1/2)\,(\mathrm{tr}(\mathbf{A}^2) - \mathrm{tr}(\mathbf{A})^2]$, *and one obtains*

$$\log \mathbf{A} = f_1 \, \mathbf{I} + f_2 \, (\mathbf{I} - \mathbf{A}) + f_3 \, (\mathbf{I} - \mathbf{A})^2 \quad , \tag{A.93}$$

where $f_1 = c_3 \int_0^1 ds \; s^2/q(s)$, $f_2 = - \int_0^1 ds \, (1 + c_1 \, s)/q(s)$ *and, finally,* $f_3 = - \int_0^1 ds \; s/q(s)$.

A.6 Logarithmic Image of SL(2)

Let us examine with some detail the bijection between SL(2) and $\ell i\,$SL(2). The following matrix spaces appear (see figure A.1):

- SL(2) is the space of all 2×2 real matrices with unit determinant. It is the exponential of $\ell i\,$SL(2). It is made by the union of the subsets $\mathrm{SL}(2)_{\mathrm{I}}$ and $\mathrm{SL}(2)_{-}$ defined as follows.
  - $\mathrm{SL}(2)_{\mathrm{I}}$ is the subset of SL(2) consisting of the matrices with the two eigenvalues real and positive or the two eigenvalues complex mutually conjugate. It is the exponential of $\mathfrak{sl}(2)_0$. When these matrices are interpreted as points of the SL(2) Lie group manifold, they corresponds to all the points geodesically connected to the origin.

- $\mathrm{SL}(2)_+$ is the subset of $\mathrm{SL}(2)_\mathrm{I}$ consisting of matrices with the two eigenvalues real and positive. It is the exponential of $\mathfrak{sl}(2)_+$.
- $\mathrm{SL}(2)_{<\pi}$ is the subset of $\mathrm{SL}(2)_\mathrm{I}$ consisting of matrices with the two eigenvalues complex mutually conjugate. It is the exponential of $\mathfrak{sl}(2)_{<\pi}$.

– $\mathrm{SL}(2)_-$ is the subset of $\mathrm{SL}(2)$ consisting of matrices with the two eigenvalues real and negative. It is the exponential of $\ell i\,\mathrm{SL}(2)_-$.

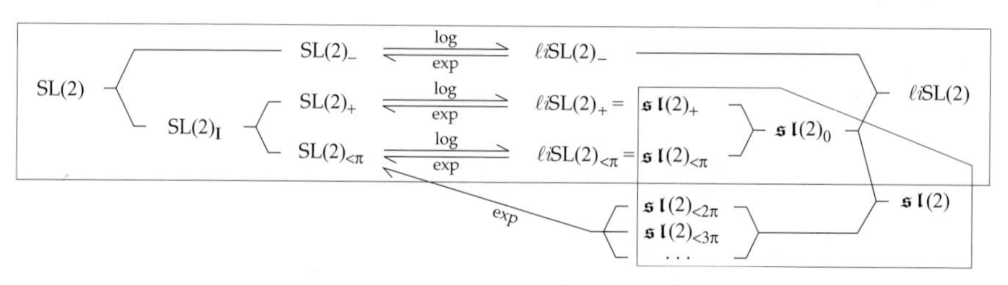

Fig. A.1. *The different matrix subspaces appearing when "taking the logarithm" of* $\mathrm{SL}(2)$.

– $\ell i\,\mathrm{SL}(2)$ is the space of all 2×2 matrices (real and complex) that are the logarithm of the matrices in $\mathrm{SL}(2)$. It is made by the union of the subsets $\mathfrak{sl}(2)_0$ and $\ell i\,\mathrm{SL}(2)_-$ defined as follows.
  – $\mathfrak{sl}(2)_0$ is the set of all 2×2 real traceless matrices with real norm (positive or zero) or with imaginary norm smaller than $i\pi$. It is the logarithm of $\mathrm{SL}(2)_\mathrm{I}$. When these matrices are interpreted as oriented geodesic segments in the Lie group manifold $\mathrm{SL}(2)$, they correspond to segments leaving the origin.
    - $\mathfrak{sl}(2)_+$ is the subset of $\mathfrak{sl}(2)_0$ consisting of matrices with real norm (positive or zero). It is the exponential of $\mathrm{SL}(2)_+$.
    - $\mathfrak{sl}(2)_{<\pi}$ is the subset of $\mathfrak{sl}(2)_0$ consisting of matrices with imaginary norm smaller than $i\pi$. It is the logarithm of $\mathrm{SL}(2)_{<\pi}$.
  – $\ell i\,\mathrm{SL}(2)_-$ is the subset of $\ell i\,\mathrm{SL}(2)$ consisting of matrices with imaginary norm larger or equal than $i\pi$. It is the logarithm of $\mathrm{SL}(2)_-$.

These are the sets naturally appearing in the bijection between $\mathrm{SL}(2)$ and $\ell i\,\mathrm{SL}(2)$. Note that the set $\ell i\,\mathrm{SL}(2)$ is different from $\mathfrak{sl}(2)$, the set of all $2{\times}2$ real traceless matrices: $\mathfrak{sl}(2)$ is the union of $\mathfrak{sl}(2)_0$ and the sets $\mathfrak{sl}(2)_{<n\pi}$ consisting of $2{\times}2$ real traceless matrices $\mathbf{s}$ with imaginary norm $(n-1)\,i\,\pi \leq \|\,\mathbf{s}\,\| < n\,i\,\pi$, for $n>1$. There is no simple relation between $\mathfrak{sl}(2)$ and $\mathrm{SL}(2)$: because of the periodic character of the exponential function, every space $\mathfrak{sl}(2)_{<n\pi}$ is mapped into $\mathrm{SL}(2)_{<\pi}$, which is already the image of $\mathfrak{sl}(2)_{<\pi}$ through the exponential function.

A.7 Logarithmic Image of SO(3)

Although the goal here is to characterize $\ell i\,\mathrm{SO}(3)$ let us start with the (much simpler) characterization of $\ell i\,\mathrm{SO}(2)$. SO(2), is the set of matrices

$$\mathbf{R}(\alpha) = \begin{pmatrix} \cos\alpha & \sin\alpha \\ -\sin\alpha & \cos\alpha \end{pmatrix} \quad ; \quad (-\pi < \alpha \leq \pi) \ . \tag{A.94}$$

One easily obtains (using, for instance, a Jordan decomposition)

$$\log \mathbf{R}(\alpha) = \frac{\log e^{i\alpha}}{2} \begin{pmatrix} 1 & -i \\ i & 1 \end{pmatrix} + \frac{\log e^{-i\alpha}}{2} \begin{pmatrix} 1 & i \\ -i & 1 \end{pmatrix} \quad ; \quad (-\pi < \alpha \leq \pi) \ . \tag{A.95}$$

A complex number can be written as $z = |z|\, e^{i \arg z}$, where, by convention, $-\pi < \arg z \leq \pi$. The logarithm of a complex number has been defined as $\log z = \log|z| + i \arg z$. Therefore, while $\alpha < \pi$, $\log e^{\pm i\alpha} = \pm i\alpha$, but when α reaches the value π, $\log e^{\pm i\pi} = +i\pi$ (see figure A.2). Therefore, for $-\pi < \alpha < \pi$, $\log \mathbf{R}(\alpha) = \begin{pmatrix} 0 & \alpha \\ -\alpha & 0 \end{pmatrix}$, and, for $\alpha = \pi$, $\log \mathbf{R}(\pi) = \begin{pmatrix} i\pi & 0 \\ 0 & i\pi \end{pmatrix}$. Note that $\log \mathbf{R}(\pi)$ is different from the matrix obtained from $\log \mathbf{R}(\alpha)$ by continuity when $\alpha \to \pi$.

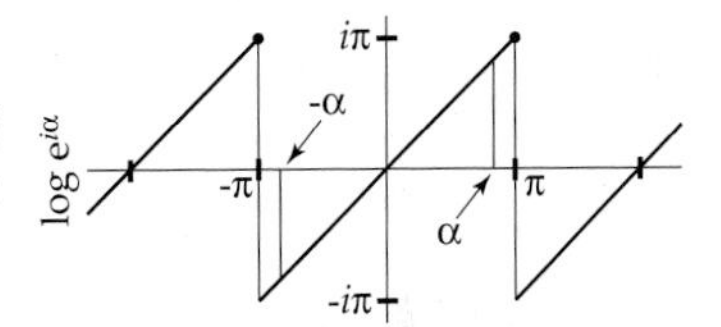

Fig. A.2. *Evaluation of the function* $\log e^{i\alpha}$, *for real* α. *While* $\alpha < \pi$, $\log e^{\pm i\alpha} = \pm i\alpha$, *but when* α *reaches the value* π, $\log e^{i\pi} = \log e^{-i\pi} = +i\pi$.

Therefore, the set $\ell i\,\mathrm{SO}(2)$, image of SO(2) by the logarithm function consists two subsets: $\ell i\,\mathrm{SO}(2) = \mathfrak{so}(2)_0 \cup \ell i\,\mathrm{SO}(2)_\pi$.

- The set $\mathfrak{so}(2)_0$ consists of all 2×2 real antisymmetric matrices $\mathbf{r}$ with $\|\,\mathbf{r}\,\| = \sqrt{(\mathrm{tr}\,\mathbf{r}^2)/2} < i\pi$.
- The set $\ell i\,\mathrm{SO}(2)_\pi$ contains the single matrix $\mathbf{r}(\pi) = \begin{pmatrix} i\pi & 0 \\ 0 & i\pi \end{pmatrix}$.

We can now turn to the the problem of characterizing $\ell i\,\mathrm{SO}(3)$. Any matrix $\mathbf{R}$ of SO(3) can be written in the special form

$$\mathbf{R} = \mathbf{S}\,\Lambda(\alpha)\,\mathbf{S}^{-1} \quad ; \quad (0 \leq \alpha \leq \pi) \ , \tag{A.96}$$

where $\Lambda(\alpha) = \begin{pmatrix} 1 & 0 & 0 \\ 0 & \cos\alpha & \sin\alpha \\ 0 & -\sin\alpha & \cos\alpha \end{pmatrix}$ and where $\mathbf{S}$ is an orthogonal matrix representing a rotation "whose axis is on $\{yz\}$ plane". For any rotation vector can be brought to the x axis by such a rotation. One has

$$\log \mathbf{R} \; = \; \mathbf{S} \, (\log \Lambda(\alpha)) \, \mathbf{S}^{-1} \quad . \tag{A.97}$$

From the 2D results just obtained it immediately follows that for $0 \leq \alpha < \pi$, $\log \Lambda(\alpha) = \begin{pmatrix} 0 & 0 & 0 \\ 0 & 0 & \alpha \\ 0 & -\alpha & 0 \end{pmatrix}$, while for $\alpha = \pi$, $\log \Lambda(\pi) = \begin{pmatrix} 0 & 0 & 0 \\ 0 & i\pi & 0 \\ 0 & 0 & i\pi \end{pmatrix}$. So, in $\ell i\,\mathrm{SO}(3)$ we have the set $\mathfrak{so}(3)_0$, consisting of the real matrices

$$\mathbf{r} \; = \; \alpha \, \mathbf{S} \begin{pmatrix} 0 & 0 & 0 \\ 0 & 0 & 1 \\ 0 & -1 & 0 \end{pmatrix} \mathbf{S}^{-1} \quad ; \qquad (0 \leq \alpha < \pi) \quad , \tag{A.98}$$

and the set $\ell i\,\mathrm{SO}(3)_\pi$, consisting of the imaginary matrices

$$\mathbf{r} \; = \; i\pi \, \mathbf{S} \begin{pmatrix} 0 & 0 & 0 \\ 0 & 1 & 0 \\ 0 & 0 & 1 \end{pmatrix} \mathbf{S}^{-1} \quad , \tag{A.99}$$

where $\mathbf{S}$ is an arbitrary rotation in the $\{yz\}$ plane. Equation (A.98) displays a pseudo-vector aligned along the x axis combined with a rotation on the $\{y,z\}$ plane: this gives a pseudo-vector arbitrarily oriented, i.e., an arbitrarily oriented real antisymmetric matrix $\mathbf{r}$. The norm of $\mathbf{r}$ is here $\| \mathbf{r} \| = \sqrt{(\mathrm{tr}\,\mathbf{r}^2)/2} = i\alpha$. As $\alpha < \pi$, $\sqrt{(\mathrm{tr}\,\mathbf{r}^2)/2} < i\alpha$. Equation (A.99) corresponds to an arbitrary diagonalizable matrix with eigenvalues $\{0, i\pi, i\pi\}$. The norm is here $\| \mathbf{r} \| = \sqrt{(\mathrm{tr}\,\mathbf{r}^2)/2} = i\pi$.

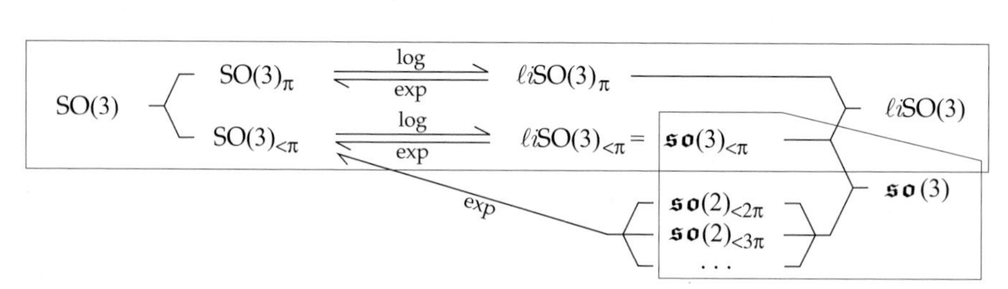

Fig. A.3. *The different matrix subspaces appearing when "taking the logarithm" of* SO(3). *Also sketched is the set* $\mathfrak{so}(3)$ *of all real antisymmetric matrices.*

Therefore, the set $\ell i\,\mathrm{SO}(3)$, image of SO(3) by the logarithm function consists of two subsets (see figure A.3): $\ell i\,\mathrm{SO}(3) = \ell i\,\mathrm{SO}(3)_{<\pi} \cup \ell i\,\mathrm{SO}(3)_\pi$.

- The set $\ell i\,\mathrm{SO}(3)_{<\pi}$ consists of all 3×3 real antisymmetric matrices $\mathbf{r}$ with $\sqrt{(\mathrm{tr}\,\mathbf{r}^2)/2} < i\pi$.
- The set $\ell i\,\mathrm{SO}(3)_\pi$ consists of all imaginary diagonalizable matrices with eigenvalues $\{0, i\pi, i\pi\}$. For all the matrices of this set, $\sqrt{(\mathrm{tr}\,\mathbf{r}^2)/2} = i\pi$.

A matrix of SO(3) has eigenvalues $\{1, e^{\pm\alpha}\}$, where α is the rotation angle ($0 \leq \alpha \leq \pi$). The image of $\ell i\,\mathrm{SO}(3)_{<\pi}$ by the exponential function is clearly the subset $\mathrm{SO}(3)_{<\pi}$ of SO(3) with rotation angle $\alpha < \pi$. The image of $\ell i\,\mathrm{SO}(3)_{\pi}$ by the exponential function is the subset $\mathrm{SO}(3)_{\pi}$ of SO(3) with rotation angle $\alpha = \pi$, i.e., with eigenvalues $\{1, e^{\pm\pi}\}$. Figure A.3 also sketches the set $\mathfrak{so}(3)$ of all real antisymmetric matrices. It is divided in subsets where the norm of the matrices verifies $(n-1)\,\pi \leq \sqrt{(\mathrm{tr}\,\mathbf{r}^2)/2} < n\,\pi$. All these subsets are mapped into $\mathrm{SO}(3)_{<\pi}$ by the exponential function.

A.8 Central Matrix Subsets as Autovector Spaces

In this appendix it is demonstrated that the central matrix subsets introduced in definitions 1.40 and 1.41 (page 57) actually are autovector spaces. As the two spaces are isomorphic, let us just make a single demonstration, using the o-additive representation.

The axioms that must satisfy a set of elements to be an autovector spaces are in definitions 1.19 (page 23, global autovector space) and 1.21 (page 25, local autovector space).

Let $\mathbb{M}$ by a multiplicative group of matrices, $\ell i\,\mathbb{M}$ its logarithmic image, and $\mathfrak{m}_0$ the subset introduced in definition 1.41, a subset that we must verify is a local autovector space. By definition, $\mathfrak{m}_0$ is the subset of matrices of $\ell i\,\mathbb{M}$ such that for any real $\lambda \in [-1, 1]$ and for any matrix $\mathbf{a}$ of the subset, $\lambda\,\mathbf{a}$ belongs to $\ell i\,\mathbb{M}$. Over the algebra $\mathfrak{m}$, the sum of two matrices $\mathbf{b} + \mathbf{a}$ and the product $\lambda\,\mathbf{a}$ are defined, so they are also defined over $\mathfrak{m}_0$, that is a subset of $\mathfrak{m}$ (these may not be internal operations in $\mathfrak{m}_0$). The group operation being $\mathbf{b} \oplus \mathbf{a} = \log(\exp\mathbf{b}\,\exp\mathbf{a})$ it is clear that the zero matrix $\mathbf{0}$ is the neutral element for both, the operation $+$ and the operation $\oplus$. The first condition in definition 1.19 is, therefore, satisfied.

For colinear matrices near the origin, one has[5] $\mathbf{b} \oplus \mathbf{a} = \mathbf{b} + \mathbf{a}$, so the second condition in the definition 1.19 is also (locally) satisfied.

Finally, the operation $\oplus$ is analytic in terms of the operation $+$ inside a finite neighborhood of the origin (BCH series), so the third condition is also satisfied.

It remains to be checked if the precise version of the locality conditions (definition 1.21) is also satisfied.

The first condition is that for any matrix $\mathbf{a}$ of $\mathfrak{m}_0$ there is a finite interval of the real line around the origin such that for any λ in the interval, the element $\lambda\,\mathbf{a}$ also belongs to $\mathfrak{m}_0$. This is obviously implied by the very definition of $\mathfrak{m}_0$.

[5] Two matrices $\mathbf{a}$ and $\mathbf{b}$ are colinear if $\mathbf{b} = \lambda\,\mathbf{a}$. Then, $\mathbf{b} \oplus \mathbf{a} = \lambda\,\mathbf{a} \oplus \mathbf{a} = \log(\exp(\lambda\,\mathbf{a})\,\exp\mathbf{a}) = \log(\exp((\lambda+1)\,\mathbf{a})) = (\lambda+1)\,\mathbf{a} = \lambda\,\mathbf{a} + \mathbf{a} = \mathbf{b} + \mathbf{a}$.

The second condition is that for any two matrices $\mathbf{a}$ and $\mathbf{b}$ of $\mathfrak{m}_0$ there is a finite interval of the real line around the origin such that for any λ and μ in the interval, the matrix $\mu\,\mathbf{b} \oplus \lambda\,\mathbf{a}$ also belongs to $\mathfrak{m}_0$.

As $\mu\,\mathbf{b} \oplus \lambda\,\mathbf{a} = \log(\exp(\mu\,\mathbf{b})\,\exp(\lambda\,\mathbf{a}))$, what we have to verify is that for any two matrices $\mathbf{a}$ and $\mathbf{b}$ of $\mathfrak{m}_0$, there is a finite interval of the real line around the origin such that for any λ and μ in the interval, the matrix $\log(\exp(\mu\,\mathbf{b})\,\exp(\lambda\,\mathbf{a}))$ also belongs to $\mathfrak{m}_0$. Let be $\mathbf{A} = \exp \mathbf{a}$ and $\mathbf{B} = \exp \mathbf{b}$. For small enough (but finite) λ and μ, $\exp(\lambda\,\mathbf{a}) = \mathbf{A}^\lambda$, $\exp(\mu\,\mathbf{b}) = \mathbf{B}^\mu$, $\mathbf{C} = \mathbf{B}^\mu\,\mathbf{A}^\lambda$ exists, and its logarithm belongs to $\mathfrak{m}_0$.

A.9 Geometric Sum on a Manifold

A.9.1 Connection

The notion of connection has been introduced in section 1.3.1 in the main text. With the connection available, one may then introduce the notion of covariant derivative of a vector field,[6] to obtain

$$\nabla_i w^j = \partial_i w^j + \Gamma^j{}_{is}\, w^s \quad . \tag{A.100}$$

This is far from being an acceptable introduction to the covariant derivative, but this equation unambiguously fixes the notation. It follows from this expression, using the definition of dual basis, $\langle\, \mathbf{e}^i \,,\, \mathbf{e}_j \,\rangle = \delta^i{}_j$, that the covariant derivative of a form is given by the expression

$$\nabla_i f_j = \partial_i f_j - \Gamma^s{}_{ij}\, f_s \quad . \tag{A.101}$$

More generally, it is well-known that the covariant derivative of a tensor is

$$\begin{aligned} \nabla_m T^{ij\dots}{}_{k\ell\dots} = \partial_m T^{ij\dots}{}_{k\ell\dots} &+ \Gamma^i{}_{ms}\, T^{sj\dots}{}_{k\ell\dots} + \Gamma^j{}_{ms}\, T^{is\dots}{}_{k\ell\dots} + \dots \\ &- \Gamma^s{}_{mk}\, T^{ij\dots}{}_{s\ell\dots} - \Gamma^s{}_{m\ell}\, T^{ij\dots}{}_{ks\dots} - \dots \end{aligned} \tag{A.102}$$

A.9.2 Autoparallels

Consider a curve $x^i = x^i(\lambda)$, parameterized with an arbitrary parameter λ, at any point along the curve define the *tangent vector* (associated to the particular parameter λ) as the vector whose components (in the local natural basis at the given point) are

$$v^i(\lambda) \equiv \frac{dx^i}{d\lambda}(\lambda) \quad . \tag{A.103}$$

[6]Using poor notation, equation (1.97) can be written $\partial_i\,\mathbf{e}_j = \Gamma^k{}_{ij}\,\mathbf{e}_k$. When considering a vector field $\mathbf{w(x)}$, then, formally, $\partial_i \mathbf{w} = \partial_i\,(w^j\,\mathbf{e}_j) = (\partial_i\,w^j)\,\mathbf{e}_j + w^j\,(\partial_i\,\mathbf{e}_j) = (\partial_i\,w^j)\,\mathbf{e}_j + w^j\,\Gamma^k{}_{ij}\,\mathbf{e}_k$, i.e., $\partial_i \mathbf{w} = (\nabla_i\,w^k)\,\mathbf{e}_k$ where $\nabla_i\,w^k = \partial_i\,w^k + \Gamma^k{}_{ij}\,w^j$.

The covariant derivative $\nabla_j v^i$ is not defined, as v^i is only defined along the curve, but it is easy to give sense (see below) to the expression $v^j \nabla_j v^i$ as the *covariant derivative along the curve*.

Definition A.4 *The curve* $x^i = x^i(\lambda)$ *is called* autoparallel *(with respect to the connection* $\Gamma^k{}_{ij}$*), if the covariant derivative along the curve of the tangent vector* $v^i = dx^i/d\lambda$ *is zero at every point.*

Therefore, the curve is autoparallel iff

$$v^j \nabla_j v^i = 0 \quad . \tag{A.104}$$

As $\frac{d}{d\lambda} = \frac{dx^i}{d\lambda}\frac{\partial}{\partial x^i} = v^i \frac{\partial}{\partial x^i}$, one has the property

$$\frac{d}{d\lambda} = v^i \frac{\partial}{\partial x^i} \quad , \tag{A.105}$$

useful for subsequent developments. Equation (A.104) is written, more explicitly, $v^j (\partial_j v^i + \Gamma^i{}_{jk} v^k) = 0$, i.e., $v^j \partial_j v^i + \Gamma^i{}_{jk} v^j v^k = 0$. The use of (A.105) allows one then to write the condition for autoparallelism as $\frac{dv^i}{d\lambda} + \Gamma^i{}_{jk} v^j v^k = 0$, or, more symmetrically,

$$\frac{dv^i}{d\lambda} + \gamma^i{}_{jk} v^j v^k = 0 \quad , \tag{A.106}$$

where $\gamma^i{}_{jk}$ is the symmetric part of the connection,

$$\gamma^i{}_{jk} = \tfrac{1}{2}(\Gamma^i{}_{jk} + \Gamma^i{}_{kj}) \quad . \tag{A.107}$$

The equation defining the coordinates of an autoparallel curve are obtained by using again $v^i = dx^i/d\lambda$ in equation (A.106):

$$\boxed{\frac{d^2x^i}{d\lambda^2} + \gamma^i{}_{jk} \frac{dx^j}{d\lambda}\frac{dx^k}{d\lambda} = 0 \quad .} \tag{A.108}$$

Clearly, the autoparallels are defined by the symmetric part of the connection only. If there exists a parameter λ with respect to which a curve is autoparallel, then any other parameter $\mu = \alpha\,\lambda + \beta$ (where α and β are two constants) also satisfies the condition (A.108). Any such parameter defining an autoparallel curve is called an *affine parameter*.

Taking the derivative of (A.106) gives

$$\frac{d^3x^i}{d\lambda^3} + A^i{}_{jk\ell}\, v^j v^k v^\ell = 0 \quad , \tag{A.109}$$

where the following circular sum has been introduced:

$$A^i{}_{jk\ell} = \tfrac{1}{3} \textstyle\sum_{(jk\ell)} (\partial_j \gamma^i{}_{k\ell} - 2\,\gamma^i{}_{js}\,\gamma^s{}_{k\ell}) \quad . \tag{A.110}$$

To be more explicit, let us, from now on, denote as $x^i(\lambda\|\lambda_0)$ the coordinates of the point reached when describing an autoparallel started at point λ_0 . From the Taylor expansion

$$\begin{aligned} x^i(\lambda\|\lambda_0) &= x^i(\lambda_0) + \frac{dx^i}{d\lambda}(\lambda_0)\,(\lambda-\lambda_0) + \frac{1}{2}\,\frac{d^2x^i}{d\lambda^2}(\lambda_0)\,(\lambda-\lambda_0)^2 \\ &+ \frac{1}{3!}\,\frac{d^3x^i}{d\lambda^3}(\lambda_0)\,(\lambda-\lambda_0)^3 + \dots \quad , \end{aligned} \tag{A.111}$$

one gets, using the results above (setting $\lambda = 0$ and writing x^i , v^i , $\gamma^i{}_{jk}$ and $A^i{}_{jk\ell}$ instead of $x^i(0)$, $v^i(0)$, $\gamma^i{}_{jk}(0)$ and $A^i{}_{jk\ell}(0)$),

$$\boxed{x^i(\lambda\|0) = x^i + \lambda\, v^i - \frac{\lambda^2}{2}\,\gamma^i{}_{jk}\, v^j\, v^k - \frac{\lambda^3}{3!}\, A^i{}_{jk\ell}\, v^j\, v^k\, v^\ell + \dots \quad .} \tag{A.112}$$

A.9.3 Parallel Transport of a Vector

Let us now transport a vector along this autoparallel curve $x^i = x^i(\lambda)$ with affine parameter λ and with tangent $v^i = dx^i/d\lambda$. So, given a vector w^i at every point along the curve, we wish to characterize the fact that all these vectors are deduced one from the other by parallel transport along the curve. We shall use the notation $w^i(\lambda\|\lambda_0)$ to denote the components (in the local basis at point λ) of the vector obtained at point λ by parallel transport of some initial vector $w^i(\lambda_0)$ given at point λ_0 .

Definition A.5 *The vectors* $w^i(\lambda\|\lambda_0)$ *are* parallel-transported *along the curve* $x^i = x^i(\lambda)$ *with affine parameter* λ *and with tangent* $v^i = dx^i/d\lambda$ *iff the covariant derivative along the curve of* $w^i(\lambda)$ *is zero at every point.*

Explicitly, this condition is written (equation similar to A.104),

$$v^j\, \nabla_j\, w^i = 0 \quad . \tag{A.113}$$

The same developments that transformed equation (A.104) into equation (A.106) now transform this equation into

$$\boxed{\frac{dw^i}{d\lambda} + \Gamma^i{}_{jk}\, v^j\, w^k = 0 \quad .} \tag{A.114}$$

Given a vector $\mathbf{w}(\lambda_0)$ at a given point λ_0 of an autoparallel curve, whose components are $w^i(\lambda_0)$ on the local basis at the given point, then, the components $w^i(\lambda\|\lambda_0)$ of the vector transported at another point λ along the curve are (in the local basis at that point) those obtained from (A.114) by integration from λ_0 to λ .

Taking the derivative of expression (A.114), using equations (A.106), (A.105) and (A.114) again one easily obtains

$$\frac{d^2 w^i}{d\lambda^2} + H_-{}^i{}_{\ell jk}\, v^j\, v^k\, w^\ell = 0 \quad , \tag{A.115}$$

where the following circular sum has been introduced:

$$H_\pm{}^i{}_{\ell jk} = \tfrac{1}{2} \textstyle\sum_{(jk)} (\,\partial_j\, \Gamma^i{}_{k\ell} \pm \Gamma^i{}_{s\ell}\, \Gamma^s{}_{jk} \pm \Gamma^i{}_{js}\, \Gamma^s{}_{k\ell}\,) \tag{A.116}$$

(the coefficients $H_+{}^i{}_{jk\ell}$ are used below). From the Taylor expansion

$$w^i(\lambda\|\lambda_0) = w^i(\lambda_0) + \frac{dw^i}{d\lambda}(\lambda_0)\,(\lambda-\lambda_0) + \tfrac{1}{2}\,\frac{d^2w^i}{d\lambda^2}(\lambda_0)\,(\lambda-\lambda_0)^2 + \dots \quad , \tag{A.117}$$

one gets, using the results above (setting $\lambda_0 = 0$ and writing v^i, w^i and $\Gamma^i{}_{jk}$ instead of $v^i(0)$, $w^i(0)$ and $\Gamma^i{}_{jk}(0)$),

$$\boxed{w^i(\lambda\|0) = w^i - \lambda\, \Gamma^i{}_{jk}\, v^j\, w^k - \frac{\lambda^2}{2}\, H_-{}^i{}_{\ell jk}\, v^j\, v^k\, w^\ell + \dots \quad .} \tag{A.118}$$

Should one have transported a form instead of a vector, one would have obtained, instead,

$$f_j(\lambda\|0) = f_j + \lambda\, \Gamma^k{}_{ij}\, v^i\, f_k + \frac{\lambda^2}{2}\, H_+{}^\ell{}_{jki}\, v^i\, v^k\, f_\ell + \dots \quad , \tag{A.119}$$

an equation that essentially is a higher order version of the expression (1.97) used above to introduce the connection coefficients.

A.9.4 Autoparallel Coordinates

Geometrical computations are simplified when using coordinates adapted to the problem in hand. It is well-known that many computations in differential geometry are better done in 'geodesic coordinates'. We don't have here such coordinates, as we are not assuming that we deal with a metric manifold. But thanks to the identification we have just defined between vectors and autoparallel lines, we can introduce a system of 'autoparallel coordinates'.

Definition A.6 *Consider an n-dimensional manifold, an arbitrary origin* $\mathcal{O}$ *in the manifold and the linear space tangent to the manifold at* $\mathcal{O}$. *Given an arbitrary basis* $\{\mathbf{e}_1, \dots, \mathbf{e}_n\}$ *in the linear space, any vector can be decomposed as* $\mathbf{v} = v^1\, \mathbf{e}_1 + \dots + v^n\, \mathbf{e}_n$. *Inside the finite region around the origin where the association between vectors and autoparallel segments is invertible, to any point* $\mathcal{P}$ *of the manifold we attribute the coordinates* $\{v^1, \dots, v^n\}$, *and call this an* autoparallel coordinate system.

We may remember here equation (A.112)

$$x^i(\lambda\|0) = x^i + \lambda\, v^i - \frac{\lambda^2}{2}\, \Gamma^i{}_{jk}\, v^j\, v^k - \frac{\lambda^3}{3!}\, A^i{}_{jk\ell}\, v^j\, v^k\, v^\ell + \dots \quad , \tag{A.120}$$

giving the coordinates of an autoparallel line, where (equations (A.107) and (A.110))

$$\gamma^i{}_{jk} = \tfrac{1}{2}\,(\Gamma^i{}_{jk} + \Gamma^i{}_{kj}) \qquad ; \qquad A^i{}_{jk\ell} = \tfrac{1}{3}\textstyle\sum_{(jk\ell)}(\partial_j\gamma^i{}_{k\ell} - 2\,\gamma^i{}_{js}\,\gamma^s{}_{k\ell}) \quad . \tag{A.121}$$

But if the coordinates are autoparallel, then, by definition,

$$x^i(\lambda\|0) = \lambda\,v^i \quad , \tag{A.122}$$

so we have the

Property A.2 *At the origin of an autoparallel system of coordinates, the symmetric part of the connection,* $\gamma^k{}_{ij}$*, vanishes.*

More generally, we have

Property A.3 *At the origin of an autoparallel system of coordinates, the coefficients* $A^i{}_{jk\ell}$ *vanish, as do all the similar coefficients appearing in the series* (A.120).

A.9.5 Geometric Sum

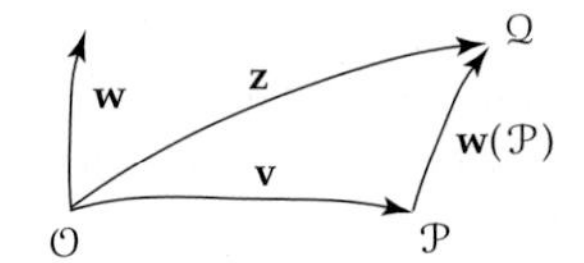

Fig. A.4. *Geometrical setting for the evaluation of the geometric sum* $\mathbf{z} = \mathbf{w} \oplus \mathbf{v}$.

We wish to evaluate the geometric sum

$$\mathbf{z} = \mathbf{w} \oplus \mathbf{v} \tag{A.123}$$

to third order in the terms containing $\mathbf{v}$ and $\mathbf{w}$.

To evaluate this sum, we choose a system of autoparallel coordinates. In such a system, the coordinates of the point $\mathcal{P}$ can be obtained as (equation A.122)

$$x^i(\mathcal{P}) = v^i \quad , \tag{A.124}$$

while the (unknown) coordinates of the point $\mathcal{Q}$ are

$$x^i(\mathcal{Q}) = z^i \quad . \tag{A.125}$$

The coordinates of the point $\mathcal{Q}$ can also be written using the autoparallel that starts at point $\mathcal{P}$. As this point is not at the origin of the autoparallel coordinates, we must use the general expression (A.112),

$$x^i(\mathcal{Q}) = x^i(\mathcal{P}) + {}_{\mathcal{P}}w^i - \tfrac{1}{2}\,{}_{\mathcal{P}}\gamma^i{}_{jk}\,{}_{\mathcal{P}}w^j\,{}_{\mathcal{P}}w^k - \tfrac{1}{6}\,{}_{\mathcal{P}}A^i{}_{jk\ell}\,{}_{\mathcal{P}}w^j\,{}_{\mathcal{P}}w^k\,{}_{\mathcal{P}}w^\ell + O(4) \quad , \tag{A.126}$$

where ${}_{\mathcal{P}}w^i$ are the components (on the local basis at $\mathcal{P}$) of the vector obtained at $\mathcal{P}$ by parallel transport of the vector w^i at $\mathcal{O}$. These components can be obtained, using equation (A.118), as

$$ {}_{\mathcal{P}}w^i = w^i - \Gamma^i{}_{jk}\, v^j\, w^k - \tfrac{1}{2}\, H_-{}^i{}_{\ell jk}\, v^j\, v^k\, w^\ell + O(4) \quad , \tag{A.127} $$

where $\Gamma^i{}_{jk}$ is the connection and $B_-{}^i{}_{\ell jk}$ is the circular sum defined in equation (A.116). The symmetric part of the connection at point $\mathcal{P}$ is easily obtained as ${}_{\mathcal{P}}\gamma^i{}_{jk} = \gamma^i{}_{jk} + v^\ell\, \partial_\ell \gamma^i{}_{jk} + O(2)$, but, as the symmetric part of the connection vanishes at the origin of an autoparallel system of coordinates (property A.2), we are left with

$$ {}_{\mathcal{P}}\gamma^i{}_{jk} = v^\ell\, \partial_\ell \gamma^i{}_{jk} + O(2) \quad , \tag{A.128} $$

while ${}_{\mathcal{P}}A^i{}_{jk\ell} = A^i{}_{jk\ell} + O(1)$. The coefficients $A^i{}_{jk\ell}$ also vanish at the origin (property A.2), and we are left with ${}_{\mathcal{P}}A^i{}_{jk\ell} = O(1)$, this showing that the last (explicit) term in the series (A.126) is, in fact (in autoparallel coordinates) fourth-order, and can be dropped. Inserting then (A.124) and (A.125) into (A.126) gives

$$ z^i = v^i + {}_{\mathcal{P}}w^i - \tfrac{1}{2}\, {}_{\mathcal{P}}\gamma^i{}_{jk}\, {}_{\mathcal{P}}w^j\, {}_{\mathcal{P}}w^k + O(4) \quad . \tag{A.129} $$

It only remains to insert here (A.127) and (A.128), this giving (dropping high order terms) $z^i = w^i + v^i - \Gamma^i{}_{jk}\, v^j\, w^k - \tfrac{1}{2}\, B_-{}^i{}_{jk\ell}\, v^j\, v^k\, w^\ell - \tfrac{1}{2}\, \partial_\ell \gamma^i{}_{jk}\, v^\ell\, w^j\, w^k + O(4)$. As we have defined $\mathbf{z} = \mathbf{w} \oplus \mathbf{v}$, we can write, instead,

$$ \boxed{(\mathbf{w} \oplus \mathbf{v})^i = w^i + v^i - \Gamma^i{}_{jk}\, v^j\, w^k - \tfrac{1}{2}\, H_-{}^i{}_{\ell jk}\, v^j\, v^k\, w^\ell - \tfrac{1}{2}\, \partial_\ell \gamma^i{}_{jk}\, v^\ell\, w^j\, w^k + O(4) \,.} \tag{A.130} $$

To compare this result with expression (A.34),

$$ (\mathbf{w} \oplus \mathbf{v})^i = w^i + v^i + e^i{}_{jk}\, w^j\, v^k + q^i{}_{jk\ell}\, w^j\, w^k\, v^\ell + r^i{}_{jk\ell}\, w^j\, v^k\, v^\ell + \dots \quad , \tag{A.131} $$

that was used to introduce the coefficients $e^i{}_{jk}$, $q^i{}_{jk\ell}$ and $r^i{}_{jk\ell}$, we can change indices and use the antisymmetry of $\Gamma^i{}_{jk}$ at the origin of autoparallel coordinates, to write

$$ (\mathbf{w} \oplus \mathbf{v})^i = w^i + v^i + \Gamma^i{}_{jk}\, w^j\, v^k - \tfrac{1}{2}\, \partial_\ell \gamma^i{}_{jk}\, w^j\, w^k\, v^\ell - \tfrac{1}{2}\, H_-{}^i{}_{jk\ell}\, w^j\, v^k\, v^\ell + \dots \quad , \tag{A.132} $$

this giving

$$ e^i{}_{jk} = \tfrac{1}{2}\, (\Gamma^i{}_{jk} - \Gamma^i{}_{kj}) \quad ; \quad q^i{}_{jk\ell} = -\tfrac{1}{2}\, \partial_\ell \gamma^i{}_{jk} \quad ; \quad r^i{}_{jk\ell} = -\tfrac{1}{2}\, H_-{}^i{}_{jk\ell} \,, \tag{A.133} $$

where the $H_-{}^i{}_{jk\ell}$ have been defined in (equation A.116). In autoparallel coordinates the term containing the symmetric part of the connection vanishes, and we are left with

$$ H_-{}^i{}_{\ell jk} = \tfrac{1}{2} \textstyle\sum_{(jk)} (\, \partial_j \Gamma^i{}_{k\ell} - \Gamma^i{}_{js}\, \Gamma^s{}_{k\ell} \,) \quad . \tag{A.134} $$

The torsion tensor and the anassociativity tensor are (equations A.50)

$$\begin{aligned} T^k{}_{ij} &= 2\, e^k{}_{ij} \\ A^\ell{}_{ijk} &= 2\,(e^\ell{}_{ir}\, e^r{}_{jk} + e^\ell{}_{kr}\, e^r{}_{ij}) - 4\, q^\ell{}_{ijk} + 4\, r^\ell{}_{ijk} \quad . \end{aligned} \tag{A.135}$$

For the torsion this gives (remembering that the connection is antisymmetric at the origin of autoparallel coordinates) $T^k{}_{ij} = -2\, e^k{}_{ij} = -2\,\Gamma^k{}_{ji} = 2\,\Gamma^k{}_{ij} = \Gamma^k{}_{ij} - \Gamma^k{}_{ji}$, i.e.,

$$\boxed{T^k{}_{ij} = \Gamma^k{}_{ij} - \Gamma^k{}_{ji} \quad .} \tag{A.136}$$

This is the usual relation between torsion and connection, this demonstrating that our definition or torsion (as the first order of the finite commutator) matches the the usual one. For the anassociativity tensor this gives

$$\boxed{A^\ell{}_{ijk} = R^\ell{}_{ijk} + \nabla_k T^\ell{}_{ij} \quad ,} \tag{A.137}$$

where

$$R^\ell{}_{ijk} = \partial_k \Gamma^\ell{}_{ji} - \partial_j \Gamma^\ell{}_{ki} + \Gamma^\ell{}_{ks}\, \Gamma^s{}_{ji} - \Gamma^\ell{}_{js}\, \Gamma^s{}_{ki} \quad , \tag{A.138}$$

and

$$\nabla_k T^\ell{}_{ij} = \partial_k T^\ell{}_{ij} + \Gamma^\ell{}_{ks}\, T^s{}_{ij} - \Gamma^s{}_{ki}\, T^\ell{}_{sj} - \Gamma^s{}_{kj}\, T^\ell{}_{is} \quad . \tag{A.139}$$

It is clear that expression (A.138) corresponds to the usual Riemann tensor while expression (A.139) corresponds to the covariant derivative of the torsion.

As the expression (A.137) only involves tensors, it is the same as would be obtained by performing the computation in an arbitrary system of coordinates (not necessarily autoparallel).

A.10 Bianchi Identities

A.10.1 Connection, Riemann, Torsion

We have found the torsion tensor and the Riemann tensor in equations (A.136) and (A.138):

$$\begin{aligned} T^k{}_{ij} &= \Gamma^k{}_{ij} - \Gamma^k{}_{ji} \\ R^\ell{}_{ijk} &= \partial_k \Gamma^\ell{}_{ji} - \partial_j \Gamma^\ell{}_{ki} + \Gamma^\ell{}_{ks}\, \Gamma^s{}_{ji} - \Gamma^\ell{}_{js}\, \Gamma^s{}_{ki} \quad . \end{aligned} \tag{A.140}$$

For an arbitrary vector field, one easily obtains

$$(\nabla_i \nabla_j - \nabla_j \nabla_i)\, v^\ell = R^\ell{}_{kji}\, v^k + T^k{}_{ji}\, \nabla_k v^\ell \quad , \tag{A.141}$$

a well-known property relating Riemann, torsion, and covariant derivatives. With the conventions being used, the covariant derivatives of vectors and forms are written

$$\nabla_i v^j = \partial_i v^j + \Gamma^j{}_{is}\, v^s \qquad ; \qquad \nabla_i f_j = \partial_i f_j - f_s\, \Gamma^s{}_{ij} \quad . \tag{A.142}$$

A.10.2 Basic Symmetries

Expressions (A.140) show that torsion and Riemann have the symmetries

$$T^k{}_{ij} = -T^k{}_{ji} \quad ; \quad R^\ell{}_{kij} = -R^\ell{}_{kji} \quad . \tag{A.143}$$

(the Riemann has, in metric spaces, another symmetry[7]). The two symmetries above translate into the following two properties for the anassociativity (expressed in equation (A.137)):

$$\textstyle\sum_{(ij)} A^\ell{}_{ijk} = \sum_{(ij)} R^\ell{}_{ijk} \quad ; \quad \sum_{(jk)} A^\ell{}_{ijk} = \sum_{(jk)} \nabla_k T^\ell{}_{ij} \quad . \tag{A.144}$$

A.10.3 The Bianchi Identities (I)

A direct computation, using the relations (A.140) shows that one has the two identities

$$\boxed{\begin{aligned} \textstyle\sum_{(ijk)} (R^r{}_{ijk} + \nabla_i T^r{}_{jk}) &= \textstyle\sum_{(ijk)} T^r{}_{is}\, T^s{}_{jk} \\ \textstyle\sum_{(ijk)} \nabla_i R^r{}_{\ell jk} &= \textstyle\sum_{(ijk)} R^r{}_{\ell is}\, T^s{}_{jk} \quad , \end{aligned}} \tag{A.145}$$

where, here and below, the notation $\sum_{(ijk)}$ represents a sum with circular permutation of the three indices: $ijk + jki + kij$.

A.10.4 The Bianchi Identities (II)

The first Bianchi identity becomes simpler when written in terms of the anassociativity instead of the Riemann. For completeness, the two Bianchi identities can be written

$$\boxed{\begin{aligned} \textstyle\sum_{(ijk)} A^r{}_{ijk} &= \textstyle\sum_{(ijk)} T^r{}_{is}\, T^s{}_{jk} \\ \textstyle\sum_{(ijk)} \nabla_i R^r{}_{\ell jk} &= \textstyle\sum_{(ijk)} R^r{}_{\ell is}\, T^s{}_{jk} \quad , \end{aligned}} \tag{A.146}$$

where

$$R^\ell{}_{ijk} = A^\ell{}_{ijk} - \nabla_k T^\ell{}_{ij} \quad . \tag{A.147}$$

If the Jacobi tensor $J^\ell{}_{ijk} = \sum_{(ijk)} T^\ell{}_{is}\, T^s{}_{jk}$ is introduced, then the first Bianchi identity becomes

$$\textstyle\sum_{(ijk)} A^\ell{}_{ijk} = J^\ell{}_{ijk} \quad . \tag{A.148}$$

Of course, this is nothing but the index version of (A.46).

[7]Hehl (1974) demonstrates that $g_{\ell s}\, R^s{}_{kij} = -g_{ks}\, R^s{}_{\ell ij}$.

A.11 Total Riemann Versus Metric Curvature

A.11.1 Connection, Metric Connection and Torsion

The metric postulate (that the parallel transport conserves lengths) is[8]

$$\nabla_k g_{ij} = 0 \quad . \tag{A.149}$$

This gives $\partial_k g_{ij} - \Gamma^s{}_{ki}\, g_{sj} - \Gamma^s{}_{kj}\, g_{is} = 0$, i.e.,

$$\partial_k g_{ij} = \Gamma_{jki} + \Gamma_{ikj} \quad . \tag{A.150}$$

The *Levi-Civita connection*, or *metric connection* is defined as

$$\{^k{}_{ij}\} = \tfrac{1}{2}\, g^{ks}\, (\partial_i g_{js} + \partial_j g_{is} - \partial_s g_{ij}) \tag{A.151}$$

(the $\{^k{}_{ij}\}$ are also called the 'Christoffel symbols'). Using equation (A.150), one easily obtains $\{_{kij}\} = \Gamma_{kij} + \frac{1}{2}\,(T_{kji} + T_{jik} + T_{ijk})$, i.e.,

$$\Gamma_{kij} = \{_{kij}\} + \tfrac{1}{2}\, V_{kij} + \tfrac{1}{2}\, T_{kij} \quad , \tag{A.152}$$

where

$$\boxed{V_{kij} = T_{ikj} + T_{jki} \quad .} \tag{A.153}$$

The tensor $-\frac{1}{2}\,(T_{kij} + V_{kij})$ is named 'contortion' by Hehl (1973). Note that while T_{kij} is antisymmetric in its two last indices, V_{kij} is symmetric in them. Therefore, defining the symmetric part of the connection as

$$\gamma^k{}_{ij} \equiv \tfrac{1}{2}\,(\Gamma^k{}_{ij} + \Gamma^k{}_{ji}) \quad , \tag{A.154}$$

gives

$$\boxed{\gamma^k{}_{ij} = \{^k{}_{ij}\} + \tfrac{1}{2}\, V^k{}_{ij} \quad ,} \tag{A.155}$$

and the decomposition of $\Gamma^k{}_{ij}$ in symmetric and antisymmetric part is

$$\boxed{\Gamma^k{}_{ij} = \gamma^k{}_{ij} + \tfrac{1}{2}\, T^k{}_{ij} \quad .} \tag{A.156}$$

A.11.2 The Metric Curvature

The (total) Riemann $R^\ell{}_{ijk}$ is defined in terms of the (total) connection $\Gamma^k{}_{ij}$ by equation (A.138). The *metric curvature*, or *curvature*, here denoted $C^\ell{}_{ijk}$ has the same definition, but using the metric connection $\{^k{}_{ij}\}$ instead of the total connection:

$$\boxed{C^\ell{}_{ijk} = \partial_k\{^\ell{}_{ji}\} - \partial_j\{^\ell{}_{ki}\} + \{^\ell{}_{ks}\}\,\{^s{}_{ji}\} - \{^\ell{}_{js}\}\,\{^s{}_{ki}\}} \tag{A.157}$$

---

[8]For any transported vector one must have $\| \mathbf{v}(\mathbf{x} + \delta\mathbf{x}\|\mathbf{x}) \| = \| \mathbf{v}(\mathbf{x}) \|$, i.e., $g_{ij}(\mathbf{x} + \delta\mathbf{x})\, v^i(\mathbf{x}+\delta\mathbf{x}\|\mathbf{x})\, v^j(\mathbf{x}+\delta\mathbf{x}\|\mathbf{x}) = g_{ij}(\mathbf{x})\, v^i(\mathbf{x})\, v^j(\mathbf{x})$. Writing $g_{ij}(\mathbf{x}+\delta\mathbf{x}) = g_{ij}(\mathbf{x}) + (\partial_k g_{ij})(\mathbf{x})\, \delta x^k + \dots$ and (see equation 1.97) $v^i(\mathbf{x} + \delta\mathbf{x}\|\mathbf{x}) = v^i(\mathbf{x}) - \Gamma^i{}_{k\ell}\, \delta x^k\, v^\ell + \dots$, easily leads to $\partial_k g_{ij} - \Gamma^s{}_{kj}\, g_{is} - \Gamma^s{}_{ki}\, g_{sj} = 0$, that is (see equation A.102) the condition (A.149).

A.11.3 Totally Antisymmetric Torsion

In a manifold with coordinates $\{x^i\}$, with metric g_{ij}, and with (total) connection $\Gamma^k{}_{ij}$, consider a smooth curve parameterized by a metric coordinate s: $x^i = x^i(s)$, and, at any point along the curve, define

$$v^i \equiv \frac{dx^i}{ds} \quad . \tag{A.158}$$

The curve is called *autoparallel* (with respect to the connection $\Gamma^k{}_{ij}$) if $v^i \nabla_i v^k = 0$, i.e., if $v^i (\partial_i v^k + \Gamma^k{}_{ij} v^j) = 0$. This can be written $v^i \partial_i v^k + \Gamma^k{}_{ij} v^i v^j = 0$, or, equivalently, $dv^k/ds + \Gamma^k{}_{ij} v^i v^j = 0$. Using (A.158) then gives

$$\frac{d^2x^k}{ds^2} + \Gamma^k{}_{ij} \frac{dx^i}{ds} \frac{dx^j}{ds} = 0 \quad , \tag{A.159}$$

which is the equation defining an autoparallel curve.

Similarly, a line $x^i = x^i(s)$ is called geodesic[9] if it satisfies the condition

$$\frac{d^2x^k}{ds^2} + \{^k{}_{ij}\} \frac{dx^i}{ds} \frac{dx^j}{ds} = 0 \quad , \tag{A.160}$$

where $\{^k{}_{ij}\}$ is the metric connection (see equation (A.151)).

Expressing the connection $\Gamma^k{}_{ij}$ in terms of the metric connection and the torsion (equations (A.152)–(A.153)), the condition for autoparallels is $d^2x^k/ds^2 + (\{^k{}_{ij}\} + \frac{1}{2}(T^k{}_{ji} + T_{ji}{}^k + T_{ij}{}^k))(dx^i/ds)(dx^j/ds) = 0$. As $T^k{}_{ij}$ is antisymmetric in $\{i,j\}$ and $dx^i dx^j$ is symmetric, this simplifies to

$$\frac{d^2x^k}{ds^2} + \left(\{^k{}_{ij}\} + \frac{1}{2}(T_{ij}{}^k + T_{ji}{}^k)\right) \frac{dx^i}{ds} \frac{dx^j}{ds} = 0 \quad . \tag{A.161}$$

We see that a necessary and sufficient condition for the lines defined by this last equation (the autoparallels) to be identical to the lines defined by equation (A.160) (the geodesics) is $T_{ij}{}^k + T_{ji}{}^k = 0$. As the torsion is, by definition, antisymmetric in its two last indices, we see that, *when geodesics and autoparallels coincide, the torsion* $\mathbf{T}$ *is a totally antisymmetric tensor*:

$$\boxed{T_{ijk} = -T_{jik} = -T_{ikj} \quad .} \tag{A.162}$$

When the torsion is totally antisymmetric, it follows from the definition (A.153) that one has

[9]When a geodesic is defined this way one must prove that it has minimum length, i.e., that the integral $\int ds = \int \sqrt{g_{ij}\, dx^i\, dx^j}$ reaches its minimum along the line. This is easily demonstrated using standard variational techniques (see, for instance, Weinberg, 1972).

$$\boxed{V_{ijk} = 0 \quad .} \tag{A.163}$$

Then,

$$\boxed{\Gamma^k{}_{ij} = \{^k{}_{ij}\} + \tfrac{1}{2}\, T^k{}_{ij} \quad ,} \tag{A.164}$$

and

$$\boxed{\{^k{}_{ij}\} = \tfrac{1}{2}\left(\Gamma^k{}_{ij} + \Gamma^k{}_{ji}\right) = \gamma^k{}_{ij} \quad ,} \tag{A.165}$$

i.e., when autoparallels and geodesics coincide, the metric connection is the symmetric part of the total connection.

If the torsion is totally antisymmetric, one may introduce the tensor $\mathbf{J}$ as $J^\ell{}_{ijk} = T^\ell{}_{is}\, T^s{}_{jk} + T^\ell{}_{js}\, T^s{}_{ki} + T^\ell{}_{ks}\, T^s{}_{ij}$, i.e.,

$$\boxed{J^\ell{}_{ijk} = \textstyle\sum_{(ijk)} T^\ell{}_{is}\, T^s{}_{jk} \quad .} \tag{A.166}$$

It is easy to see that $\mathbf{J}$ is totally antisymmetric in its three lower indices,

$$J^\ell{}_{ijk} = -J^\ell{}_{jik} = -J^\ell{}_{ikj} \quad . \tag{A.167}$$

A.12 Basic Geometry of GL(n)

A.12.1 Bases for Linear Subspaces

To start, we need to make a distinction between the "entries" of a matrix and its components in a given matrix basis. When one works with matrices of the n^2-dimensional linear space $\mathfrak{gl}(n)$, one can always choose the canonical basis $\{\mathbf{e}_\alpha \otimes \mathbf{e}^\beta\}$. The entries $a^\alpha{}_\beta$ of a matrix $\mathbf{a}$ are then its components, as, by definition, $\mathbf{a} = a^\alpha{}_\beta\, \mathbf{e}_\alpha \otimes \mathbf{e}^\beta$. But when one works with matrices of a p-dimensional ($1 \leq p \leq n^2$) linear subspace $\mathfrak{gl}(n)_p$ of $\mathfrak{gl}(n)$, one often needs to consider a basis of the subspace, say $\{\mathbf{e}_1 \dots \mathbf{e}_p\}$, and decompose any matrix as $\mathbf{a} = a^i\, \mathbf{e}_i$. Then,

$$\mathbf{a} = a^\alpha{}_\beta\, \mathbf{e}_\alpha \otimes \mathbf{e}^\beta = a^i\, \mathbf{e}_i \quad , \tag{A.168}$$

this defining the *entries* $a^\alpha{}_\beta$ of the matrix $\mathbf{a}$ and its *components* a^i on the basis $\mathbf{e}_i$. Of course, if $p = n^2$, the subspace $\mathfrak{gl}(n)_p$ is the whole space $\mathfrak{gl}(n)$, the two bases $\mathbf{e}_i$ and $\mathbf{e}_\alpha \otimes \mathbf{e}^\beta$ are both bases of $\mathfrak{gl}(n)$ and the $a^\alpha{}_\beta$ and the a^i are both components.

Example A.10 *The group* $\mathfrak{sl}(2)$ *(real* 2×2 *traceless matrices) is three-dimensional. One possible basis for* $\mathfrak{sl}(2)$ *is*

$$\mathbf{e}_1 = \frac{1}{\sqrt{2}}\begin{pmatrix} 1 & 0 \\ 0 & -1 \end{pmatrix} \; ; \quad \mathbf{e}_2 = \frac{1}{\sqrt{2}}\begin{pmatrix} 0 & 1 \\ 1 & 0 \end{pmatrix} \; ; \quad \mathbf{e}_3 = \frac{1}{\sqrt{2}}\begin{pmatrix} 0 & 1 \\ -1 & 0 \end{pmatrix} \; . \tag{A.169}$$

The matrix $\mathbf{a} = a^\alpha{}_\beta\, \mathbf{e}_\alpha \otimes \mathbf{e}^\beta = a^i\, \mathbf{e}_i$ *is then* $\begin{pmatrix} a^1{}_1 & a^1{}_2 \\ a^2{}_1 & a^2{}_2 \end{pmatrix} = \begin{pmatrix} a^1 & a^2 + a^3 \\ a^2 - a^3 & \text{-}a^1 \end{pmatrix}$*, the four numbers* $a^\alpha{}_\beta$ *are the entries of the matrix* $\mathbf{a}$*, and the three numbers* a^i *are its components on the basis* (A.169)*. To obtain a basis for the whole* $\mathfrak{gl}(2)$*, one may add the fourth basis vector* $\mathbf{e}_0$*, identical to* $\mathbf{e}_1$ *excepted in that it has* $\{1,1\}$ *in the diagonal.*

Let us introduce the coefficients $\Lambda^\alpha{}_{\beta i}$ that define the basis of the subspace $\mathfrak{gl}(n)_p$:

$$\mathbf{e}_i = \Lambda^\alpha{}_{\beta i}\, \mathbf{e}_\alpha \otimes \mathbf{e}^\beta \quad . \tag{A.170}$$

Here, the Greek indices belong to the set $\{1,2,\dots,n\}$ and the Latin indices to the set $\{1,2,\dots,p\}$, with $p \leq n^2$. The reciprocal coefficients $\Lambda_\alpha{}^{\beta i}$ can be introduced by the condition

$$\Lambda_\alpha{}^{\beta i}\, \Lambda^\alpha{}_{\beta j} = \delta^i_j \quad , \tag{A.171}$$

and the condition that the object $P^\alpha{}_{\beta\mu}{}^\nu$ defined as

$$P^\alpha{}_{\beta\mu}{}^\nu = \Lambda^\alpha{}_{\beta i}\, \Lambda_\mu{}^{\nu i} \tag{A.172}$$

is a projector over the subspace $\mathfrak{gl}(n)_p$ (i.e., for any $a^\alpha{}_\beta$ of $\mathfrak{gl}(n)$, $P^\alpha{}_{\beta\mu}{}^\nu\, a^\mu{}_\nu$ belongs to $\mathfrak{gl}(n)_p$). It is then easy to see that the components on the basis $\mathbf{e}_\alpha \otimes \mathbf{e}^\beta$ of a vector $\mathbf{a} = a^i\, \mathbf{e}_i$ of $\mathfrak{gl}(n)_p$ are

$$a^\alpha{}_\beta = \Lambda^\alpha{}_{\beta i}\, a^i \quad . \tag{A.173}$$

Reciprocally,

$$a^i = \Lambda_\alpha{}^{\beta i}\, a^\alpha{}_\beta \tag{A.174}$$

gives the components on the basis $\mathbf{e}_i$ of the projection on the subspace $\mathfrak{gl}(n)_p$ of a vector $\mathbf{a} = a^\alpha{}_\beta\, \mathbf{e}_\alpha \otimes \mathbf{e}^\beta$ of $\mathfrak{gl}(n)$.

When $p = n^2$, i.e., when the subspace $\mathfrak{gl}(n)_p$ is $\mathfrak{gl}(n)$ itself, the equations above can be interpreted as a change from a double-index notation to a single-index notation. Then, the coefficients $\Lambda^\alpha{}_{\beta i}$ are such that the projector in equation (A.172) is the identity operator:

$$P^\alpha{}_{\beta\mu}{}^\nu = \Lambda^\alpha{}_{\beta i}\, \Lambda_\mu{}^{\nu i} = \delta^\alpha_\mu\, \delta^\nu_\beta \quad . \tag{A.175}$$

A.12.2 Torsion and Metric

While in equation (1.148) we have found the commutator

$$[\mathbf{b}, \mathbf{a}]^\alpha{}_\beta = b^\alpha{}_\sigma\, a^\sigma{}_\beta - a^\alpha{}_\sigma\, b^\sigma{}_\beta \quad , \tag{A.176}$$

the torsion $T^i{}_{jk}$ was defined by expressing the commutator of two elements as (equation 1.86)

$$[\mathbf{b},\mathbf{a}]^i \;=\; T^i{}_{jk}\, b^j\, a^k \quad . \tag{A.177}$$

Equation (A.176) can be transformed into equation (A.177) by writing it in terms of components in the basis $\mathbf{e}_i$ of the subspace $\mathfrak{gl}(n)_p$. One can use[10] equations (A.173) and (A.174), writing $[\mathbf{b},\mathbf{a}]^i = \Lambda_\alpha{}^{\beta i}\,[\mathbf{b},\mathbf{a}]^\alpha{}_\beta$, $a^\alpha{}_\beta = \Lambda^\alpha{}_{\beta i}\, a^i$ and $b^\alpha{}_\beta = \Lambda^\alpha{}_{\beta i}\, b^i$. This leads to expression (A.177) with

$$\boxed{T^i{}_{jk} \;=\; \Lambda_\alpha{}^{\beta i}\,(\Lambda^\alpha{}_{\sigma j}\,\Lambda^\sigma{}_{\beta k} - \Lambda^\alpha{}_{\sigma k}\,\Lambda^\sigma{}_{\beta j}) \quad .} \tag{A.178}$$

Property A.4 *The torsion (at the origin) of any p-dimensional subgroup $\mathfrak{gl}(n)_p$ of the n^2-dimensional group $\mathfrak{gl}(n)$ is, when using a vector basis $\mathbf{e}_i = \Lambda^\alpha{}_{\beta i}\,\mathbf{e}_\alpha \otimes \mathbf{e}^\beta$, that given in equation* (A.178), *where the reciprocal coefficients $\Lambda_\alpha{}^{\beta i}$ are defined by expressions* (A.171) *and* (A.172).

The necessary antisymmetry of the torsion in its two lower indices is evident in the expression.

The universal metric that was introduced in equation (1.31) is, when interpreted as a metric at the origin of $\mathfrak{gl}(n)$, the metric that shall leave to the right properties. We set the

Definition A.7 *The metric (at the origin) of $\mathfrak{gl}(n)$ is (χ and ψ being two arbitrary positive constants)*

$$g_\alpha{}^\beta{}_\mu{}^\nu \;=\; \chi\,\delta^\nu_\alpha\,\delta^\beta_\mu + \frac{\psi-\chi}{n}\,\delta^\beta_\alpha\,\delta^\nu_\mu \quad . \tag{A.179}$$

The restriction of this metric to the subspace $\mathfrak{gl}(n)_p$ is immediately obtained as $g_{ij} = \Lambda^\alpha{}_{\beta i}\,\Lambda^\mu{}_{\nu j}\, g_\alpha{}^\beta{}_\mu{}^\nu$, this leading to the following

Property A.5 *The metric (at the origin) of any p-dimensional subgroup $\mathfrak{gl}(n)_p$ of the n^2-dimensional group $\mathfrak{gl}(n)$ is, when using a vector basis $\mathbf{e}_i = \Lambda^\alpha{}_{\beta i}\,\mathbf{e}_\alpha \otimes \mathbf{e}^\beta$,*

$$\boxed{g_{ij} \;=\; \chi\,\Lambda^\alpha{}_{\beta i}\,\Lambda^\beta{}_{\alpha j} + \frac{\psi-\chi}{n}\,\Lambda^\alpha{}_{\alpha i}\,\Lambda^\beta{}_{\beta j} \quad ,} \tag{A.180}$$

where χ and ψ are two arbitrary positive constants.

With the universal metric at hand, one can define the all-covariant components of the torsion as $T_{ijk} = g_{is}\,T^s{}_{jk}$. An easy computation then leads to

Property A.6 *The all-covariant expression of the torsion (at the origin) of any p-dimensional subgroup $\mathfrak{gl}(n)_p$ of the n^2-dimensional group $\mathfrak{gl}(n)$ is, when using a vector basis $\mathbf{e}_i = \Lambda^\alpha{}_{\beta i}\,\mathbf{e}_\alpha \otimes \mathbf{e}^\beta$,*

$$T_{ijk} \;=\; \chi\,(\Lambda^\alpha{}_{\beta i}\,\Lambda^\beta{}_{\gamma j}\,\Lambda^\gamma{}_{\alpha k} - \Lambda^\gamma{}_{\alpha i}\,\Lambda^\beta{}_{\gamma j}\,\Lambda^\alpha{}_{\beta k}) \quad . \tag{A.181}$$

---

[10] Equation (A.174) can be used because all the considered matrices belong to the subspace $\mathfrak{gl}(n)_p$.

We see, in particular, that T_{ijk} is independent of the parameter ψ appearing in the metric.

We already know that the torsion $T^i{}_{jk}$ is antisymmetric in its two lower indices. Now, using equation (A.181), it is easy to see that we have the extra (anti) symmetry $T_{ijk} = -T_{jik}$. Therefore we have

Property A.7 *The torsion (at the origin) of any p-dimensional subgroup* $\mathfrak{gl}(n)_p$ *of the* n^2*-dimensional group* $\mathfrak{gl}(n)$ *is totally antisymmetric:*

$$\boxed{T_{ijk} = -T_{jik} = -T_{ikj} \quad .} \tag{A.182}$$

A.12.3 Coordinates over the Group Manifold

As suggested in the main text (see section 1.4.5), the best coordinates for the study of the geometry of the Lie group manifold GL(n) are what was there called the 'exponential coordinates'. As the 'points' of the Lie group manifold are the matrices of GL(n), the coordinates of a matrix $\mathbf{M}$ are, by definition, the quantities $M^\alpha{}_\beta$ themselves (see the main text for some details).

A.12.4 Connection

With the coordinate system introduced above over the group manifold, it is easy to define a parallel transport. We require the parallel transport for which the associated geometrical sum of oriented autoparallel segments (with common origin) is the Lie group operation, that in terms of matrices of GL(n) is written $\mathbf{C} = \mathbf{B}\,\mathbf{A}$.

We could proceed in two ways. We could seek an expression giving the finite transport of a vector between two points of the manifold, a transport that should lead to equation (A.200) below for the geometric sum of two vectors (one would then directly arrive at expression (A.194) below for the transport). Then, it would be necessary to verify that such a transport is a parallel transport[11] and find the connection that characterizes it.[12]

Alternatively, one can do this work in the background and, once the connection is obtained, postulate it, then derive the associated expression for the transport and, finally, verify that the geometric sum that it defines is (locally) identical to the Lie group operation. Let us follow this second approach.

[11] I.e., that it is defined by a connection.

[12] For instance, by developing the finite transport equation into a series, and recognizing the connection in the first-order term of the series (see equation (A.118) in appendix A.9).

Definition A.8 *The connection associated to the manifold* GL(n) *has, at the point whose exponential coordinates are* $X^\alpha{}_\beta$, *the components*

$$\boxed{\Gamma^\alpha{}_{\beta\mu}{}^\nu{}_\rho{}^\sigma = -\overline{X}^\sigma{}_\mu \, \delta^\alpha_\rho \, \delta^\nu_\beta \quad ,} \tag{A.183}$$

where we use a bar to denote the inverse of a matrix:

$$\overline{\mathbf{X}} \equiv \mathbf{X}^{-1} \qquad ; \qquad \overline{X}^\alpha{}_\beta \equiv (\mathbf{X}^{-1})^\alpha{}_\beta \quad . \tag{A.184}$$

A.12.5 Autoparallels

An autoparallel line $X^\alpha{}_\beta = X^\alpha{}_\beta(\lambda)$ is characterized by the condition (see equation (A.108) in the appendix) $d^2X^\alpha{}_\beta/d\lambda^2 + \Gamma^\alpha{}_{\beta\mu}{}^\nu{}_\rho{}^\sigma \, (dX^\mu{}_\nu/d\lambda)\,(dX^\rho{}_\sigma/d\lambda) = 0$. Using the connection in equation (A.183), this gives $d^2X^\alpha{}_\beta/d\lambda^2 = (dX^\alpha{}_\rho/d\lambda)\,\overline{X}^\rho{}_\sigma\,(dX^\sigma{}_\beta/d\lambda)$, i.e., for short,

$$\frac{d^2\mathbf{X}}{d\lambda^2} = \frac{d\mathbf{X}}{d\lambda}\,\mathbf{X}^{-1}\,\frac{d\mathbf{X}}{d\lambda} \quad . \tag{A.185}$$

The solution of this equation for a line that goes from a point $\mathbf{A} = \{A^\alpha{}_\beta\}$ to a point $\mathbf{B} = \{B^\alpha{}_\beta\}$ is

$$\mathbf{X}(\lambda) = \exp(\,\lambda\,\log(\mathbf{B}\,\mathbf{A}^{-1})\,)\,\mathbf{A} \qquad ; \qquad (0 \le \lambda \le 1) \quad . \tag{A.186}$$

It is clear that $\mathbf{X}(0) = \mathbf{A}$ and $\mathbf{X}(1) = \mathbf{B}$, so we need only to verify that the differential equation is satisfied. As for any matrix $\mathbf{M}$, one has[13] $\frac{d}{d\lambda}(\exp\lambda\,\mathbf{M}) = \mathbf{M}\,\exp(\lambda\,\mathbf{M})$ it first follows from equation (A.186)

$$\frac{d\mathbf{X}}{d\lambda} = \log(\mathbf{B}\,\mathbf{A}^{-1})\,\mathbf{X} \quad . \tag{A.187}$$

Taking the derivative of this expression one immediately sees that the condition (A.185) is satisfied. We have thus demonstrated the following

Property A.8 *On the manifold* GL(n), *endowed with the connection* (A.183), *the equation of the autoparallel line from a point* $\mathbf{A}$ *to a point* $\mathbf{B}$ *is that in equation* (A.186).

[13]One also has $\frac{d}{d\lambda}(\exp\lambda\,\mathbf{M}) = \exp(\lambda\,\mathbf{M})\,\mathbf{M}$, but this is not useful here.

A.12.6 Components of an Autovector (I)

In section 1.3.5 we have associated autoparallel lines leaving a point $\mathbf{A}$ with vectors of the linear tangent space at $\mathbf{A}$. Let us now express the vector $\mathbf{b}_\mathbf{A}$ (of the linear tangent space at $\mathbf{A}$) associated to the autoparallel line from a point $\mathbf{A}$ to a point[14] $\mathbf{B}$ of a Lie group manifold.

The autoparallel line from a point $\mathbf{A}$ to a point $\mathbf{B}$, is expressed in equation (A.186). The vector tangent to the trajectory, at an arbitrary point along the trajectory, is expressed in equation (A.187). In particular, then, the vector tangent to the trajectory at the starting point $\mathbf{A}$ is $\mathbf{b}_\mathbf{A} = \log(\mathbf{B}\,\mathbf{A}^{-1})\,\mathbf{A}$. This is not only *a* tangent vector to the trajectory: because the affine parameter has been chosen to vary between zero and one, this is *the* vector associated to the whole autoparallel segment, according to the protocol defined in section 1.3.5. We therefore have arrived at

Property A.9 *Consider, in the Lie group manifold* GL(n)*, the coordinates* $X^\alpha{}_\beta$ *that are the components of the matrices of* GL(n)*. The components (on the natural basis) at point* $\mathbf{A}$ *of the vector associated to the autoparallel line from point* $\mathbf{A} = \{A^\alpha{}_\beta\}$ *to point* $\mathbf{B} = \{B^\alpha{}_\beta\}$ *are the components of the matrix*

$$\mathbf{b}_\mathbf{A} = \log(\mathbf{B}\,\mathbf{A}^{-1})\,\mathbf{A} \quad . \tag{A.188}$$

We shall mainly be interested in the autoparallel segments from the origin $\mathbf{I}$ to all the other points of the manifold that are connected to the origin by an autoparallel line. As a special case of the property A.9 we have

Property A.10 *Consider, in the Lie group manifold* GL$^+$(n)*, the coordinates* $X^\alpha{}_\beta$ *that are the components of the matrices of* GL$^+$(n)*. The components (on the natural basis) at 'the origin' point* $\mathbf{I}$ *of the vector associated to the autoparallel line from the origin* $\mathbf{I}$ *to a point* $\mathbf{A} = \{A^\alpha{}_\beta\}$ *are the components* $a^\alpha{}_\beta$ *of the matrix*

$$\boxed{\mathbf{a} = \log \mathbf{A}} \quad . \tag{A.189}$$

Equations (A.186) and (A.189) allow one to write the coordinates of the autoparallel segment from $\mathbf{I}$ to $\mathbf{A}$ as (remember that $(0 \le \lambda \le 1)$) $\mathbf{A}(\lambda) = \exp(\lambda \log \mathbf{A})$, i.e.,

$$\mathbf{A}(\lambda) = \mathbf{A}^\lambda \quad . \tag{A.190}$$

Associated to each point of this line is the vector (at $\mathbf{I}$) $\mathbf{a}(\lambda) = \log \mathbf{A}(\lambda)$, i.e.,

$$\mathbf{a}(\lambda) = \lambda\,\mathbf{a} \quad . \tag{A.191}$$

[14]This point $\mathbf{B}$ must, of course, be connected to the point $\mathbf{A}$ by an autoparallel line. We shall see that arbitrary pairs of points on the Lie group manifold GL(n) are not necessarily connected in this way.

While we are using the 'exponential' coordinates $\mathbf{A} = \{A^\alpha{}_\beta\}$ over the manifold, it is clear from equation (A.191) that the coordinates $\mathbf{a} = \{a^\alpha{}_\beta\}$ would define, as mentioned above, an autoparallel system of coordinates (as defined in appendix A.9.4).

The components of vectors mentioned in properties A.9 and A.10 are those of vectors of the linear tangent space, so the title of this section, 'components of autovectors' is not yet justified. It will be, when the autovector space are built: the general definition of autovector space has contemplated that two operations + and ⊕ are defined over the same elements. The naïve vision of an element of the tangent space of a manifold as living outside the manifold is not always the best: it is better to imagine that the 'vectors' *are* the oriented autoparallel segments themselves.

A.12.7 Parallel Transport

Along the autoparallel line considered above, that goes from point $\mathbf{A}$ to point $\mathbf{B}$ (equation A.186), consider now a vector $\mathbf{t}(\lambda)$ whose components on the local basis at point (whose affine parameter is) λ are $t^\alpha{}_\beta(\lambda)$. The condition expressing that the vector is transported along the autoparallel line $X^\alpha{}_\beta(\lambda)$ by parallel transport is (see equation (A.114) in the appendix) $dt^\alpha{}_\beta/d\lambda + \Gamma^\alpha{}_{\beta\mu}{}^\nu{}_\rho{}^\sigma\,(dX^\mu{}_\nu/d\lambda)\,t^\rho{}_\sigma = 0$. Using the connection in equation (A.183), this gives $dt^\alpha{}_\beta/d\lambda = t^\alpha{}_\rho\,\overline{X}{}^\rho{}_\sigma\,(dX^\sigma{}_\beta/d\lambda)$, i.e., for short,

$$\frac{d\mathbf{t}}{d\lambda} = \mathbf{t}\,\overline{\mathbf{X}}\,\frac{d\mathbf{X}}{d\lambda} \quad . \tag{A.192}$$

Integration of this equation gives

$$\mathbf{t}(\lambda) = \mathbf{t}(0)\,\overline{\mathbf{X}}(0)\,\mathbf{X}(\lambda) \quad , \tag{A.193}$$

for one has $d\mathbf{t}/d\lambda = \mathbf{t}(0)\,\overline{\mathbf{X}}(0)\,d\mathbf{X}/d\lambda$, from which equation (A.192) follows (using again expression (A.193) to replace $\mathbf{t}(0)\,\overline{\mathbf{X}}(0)$ by $\mathbf{t}(\lambda)\,\overline{\mathbf{X}}(\lambda)$). Using $\lambda = 1$ in equation (A.193) leads to the following

Property A.11 *The transport of a vector* $\mathbf{t_A}$ *from a point* $\mathbf{A} = \{A^\alpha{}_\beta\}$ *to a point* $\mathbf{B} = \{B^\alpha{}_\beta\}$ *gives, at point* $\mathcal{B}$, *the vector* $\mathbf{t_B} = \mathbf{t_A}\,(\mathbf{A}^{-1}\,\mathbf{B})$, *i.e., explicitly,*

$$\boxed{(\mathbf{t_B})^\alpha{}_\beta = (\mathbf{t_A})^\alpha{}_\rho\,\overline{A}{}^\rho{}_\sigma\,B^\sigma{}_\beta \quad .} \tag{A.194}$$

A.12.8 Components of an Autovector (II)

Equation (A.188) gives the components (on the natural basis) at point $\mathbf{A}$ of the vector associated to the autoparallel line from point $\mathbf{A} = \{A^\alpha{}_\beta\}$ to point

$\mathbf{B} = \{B^\alpha{}_\beta\}$: $\mathbf{b_A} = \log(\mathbf{B}\,\mathbf{A}^{-1})\,\mathbf{A}$. Equation (A.194) allows the transport of a vector from one point to another. Transporting $\mathbf{b_A}$ from point $\mathbf{A}$ to the origin, point $\mathbf{I}$, gives the vector with components $\log(\mathbf{B}\,\mathbf{A}^{-1})\,\mathbf{A}\,\mathbf{A}^{-1}\,\mathbf{I} = \log(\mathbf{B}\,\mathbf{A}^{-1})$. Therefore, we have the following

Property A.12 *The components (on the natural basis) at the origin (point $\mathbf{I}$) of the vector obtained by parallel transport to the origin of the autoparallel line from point $\mathbf{A} = \{A^\alpha{}_\beta\}$ to point $\mathbf{B} = \{B^\alpha{}_\beta\}$ are the components of the matrix*

$$\boxed{\mathbf{b_I} = \log(\mathbf{B}\,\mathbf{A}^{-1})} \quad . \tag{A.195}$$

A.12.9 Geometric Sum

Let us now demonstrate that the geometric sum of two oriented autoparallel segments is the group operation $\mathbf{C} = \mathbf{B}\,\mathbf{A}$.

Consider (left of figure 1.10) the origin $\mathbf{I}$ and two points $\mathbf{A}$, and $\mathbf{B}$. The autovectors from point $\mathbf{I}$ to respectively the points $\mathbf{A}$ and $\mathbf{B}$ are (according to equation (A.189))

$$\mathbf{a} = \log \mathbf{A} \quad ; \quad \mathbf{b} = \log \mathbf{B} \quad . \tag{A.196}$$

We wish to obtain the geometric sum $\mathbf{c} = \mathbf{b} \oplus \mathbf{a}$, as defined in section 1.3.7 (the geometric construction is recalled at the right of figure 1.10). One must first transport the segment $\mathbf{b}$ to the tip of $\mathbf{a}$ to obtain the segment denoted $\mathbf{c_A}$. This transport is made using (a particular case of) equation (A.194) and gives $\mathbf{c_A} = \mathbf{b}\,\mathbf{A}$, i.e., as $\mathbf{b} = \log \mathbf{B}$,

$$\mathbf{c_A} = (\log \mathbf{B})\,\mathbf{A} \quad . \tag{A.197}$$

But equation (A.188) says that the autovector connecting the point $\mathbf{A}$ to the point $\mathbf{C}$ is

$$\mathbf{c_A} = \log(\mathbf{C}\,\mathbf{A}^{-1})\,\mathbf{A} \quad , \tag{A.198}$$

and comparison of these two equations gives

$$\mathbf{C} = \mathbf{B}\,\mathbf{A} \quad , \tag{A.199}$$

so we have demonstrated the following

Property A.13 *With the connection introduced in definition A.8, the sum of oriented autoparallel segments of the Lie group manifold* GL(n) *is, wherever it is defined, identical to the group operation $\mathbf{C} = \mathbf{B}\,\mathbf{A}$.*

As the autovector from point $\mathbf{I}$ to point $\mathbf{C}$ is $\mathbf{c} = \log \mathbf{C}$, the geometric sum $\mathbf{b} \oplus \mathbf{a}$ has given the autovector $\mathbf{c} = \log \mathbf{C} = \log(\mathbf{B}\,\mathbf{A}) = \log(\exp \mathbf{b}\ \exp \mathbf{a})$, so we have

Property A.14 *The geometric sum* $\mathbf{b} \oplus \mathbf{a}$ *of oriented autoparallel segments of the Lie group manifold* GL(n) *is, wherever it is defined, identical to the group operation, and is expressed as*

$$\mathbf{b} \oplus \mathbf{a} \;=\; \log(\exp\mathbf{b}\;\exp\mathbf{a}) \quad . \tag{A.200}$$

This, of course, is equation (1.146).

The reader may easily verify that if instead of the connection (A.183) we had chosen its 'transpose' $G^\alpha{}_{\beta\mu}{}^\nu{}_\rho{}^\sigma = \Gamma^\alpha{}_{\beta\rho}{}^\sigma{}_\mu{}^\nu$, instead of $\mathbf{c} = \log(\exp\mathbf{b}\;\exp\mathbf{a})$, we would have obtained the 'transposed' expression $\mathbf{c} = \log(\exp\mathbf{a}\;\exp\mathbf{b})$. This is not what we want.

A.12.10 Autovector Space

Given the origin $\mathbf{I}$ in the Lie group manifold, to every point $\mathbf{A}$ in the neighborhood of $\mathbf{I}$ we have associated the oriented geodesic segment $\mathbf{a} = \log\mathbf{A}$. The geometric sum of two such segments is given by the two equivalent expressions (A.199) and (A.200).

Equations (A.190) and (A.191) define the second basic operation of an autovector space: given the origin $\mathbf{I}$ on the manifold, to the real number λ and to the point $\mathbf{A}$ it is associated the point $\mathbf{A}^\lambda$. Equivalently, to the real number λ and to the segment $\mathbf{a}$ is associated the segment $\lambda\,\mathbf{a}$.

It is clear the we have a (local) autovector space, and we have a double representation of this autovector space, in terms of the matrices $\mathbf{A}\,,\,\mathbf{B}\,\ldots$ of the set GL(n) (that represent the points of the Lie group manifold) and in terms of the matrices $\mathbf{a} = \log\mathbf{A}\,,\;\mathbf{b} = \log\mathbf{B}\,\ldots$ representing the components of the autovectors at the origin (in the natural basis associated to the exponential coordinates $X^\alpha{}_\beta$).

A.12.11 Torsion

The torsion at the origin of the Lie group manifold has already been found (equation A.178). We could calculate the torsion at an arbitrary point by using again its definition in terms of the anticommutativity of the geometric sum, but as we know that the torsion can also be obtained as the antisymmetric part of the connection (equation 1.112), we can simply write $T^\alpha{}_{\beta\mu}{}^\nu{}_\rho{}^\sigma = \Gamma^\alpha{}_{\beta\mu}{}^\nu{}_\rho{}^\sigma - \Gamma^\alpha{}_{\beta\rho}{}^\sigma{}_\mu{}^\nu$, to obtain $T^\alpha{}_{\beta\mu}{}^\nu{}_\rho{}^\sigma = \overline{X}^\nu{}_\rho\,\delta^\sigma_\beta\,\delta^\alpha_\mu - \overline{X}^\sigma{}_\mu\,\delta^\alpha_\rho\,\delta^\nu_\beta$. We have thus arrived at the following

Property A.15 *The torsion in the Lie group manifold* GL(n) *is, at the point whose exponential coordinates are* $X^\alpha{}_\beta$,

$$\boxed{T^\alpha{}_{\beta\mu}{}^\nu{}_\rho{}^\sigma \;=\; \overline{X}^\nu{}_\rho\,\delta^\sigma_\beta\,\delta^\alpha_\mu - \overline{X}^\sigma{}_\mu\,\delta^\alpha_\rho\,\delta^\nu_\beta} \quad . \tag{A.201}$$

A.12.12 Jacobi

We found, using general arguments, that in a Lie group manifold, the Jacobi tensor identically vanishes (property 1.4.1.1)

$$\boxed{\mathbf{J} = \mathbf{0} \quad .} \tag{A.202}$$

The single-index version of the equation relating the Jacobi to the torsion was (equation 1.90) $J^i{}_{jk\ell} = T^i{}_{js}\, T^s{}_{k\ell} + T^i{}_{ks}\, T^s{}_{\ell j} + T^i{}_{\ell s}\, T^s{}_{jk}$. It is easy to translate this expression using the double-index notation, and to verify that the expression (A.201) for the torsion leads to the property (A.202), as it should.

A.12.13 Derivative of the Torsion

We have already found the covariant derivative of the torsion when analyzing manifolds (equation 1.111). The translation of this equation using double-index notation is $\nabla_\epsilon{}^\pi\, T^\alpha{}_{\beta\mu}{}^\nu{}_\rho{}^\sigma = \partial T^\alpha{}_{\beta\mu}{}^\nu{}_\rho{}^\sigma / \partial X^\epsilon{}_\pi + \Gamma^\alpha{}_{\beta\epsilon}{}^\pi{}_\varphi{}^\phi\, T^\varphi{}_{\phi\mu}{}^\nu{}_\rho{}^\sigma - \Gamma^\varphi{}_{\phi\epsilon}{}^\pi{}_\mu{}^\nu\, T^\alpha{}_{\beta\phi}{}^\varphi{}_\rho{}^\sigma - \Gamma^\varphi{}_{\phi\epsilon}{}^\pi{}_\rho{}^\sigma\, T^\alpha{}_{\beta\mu}{}^\nu{}_\varphi{}^\phi$. A direct evaluation, using the torsion in equation (A.201) and the connection in equation (A.183), shows that this expression identically vanishes,

$$\boxed{\nabla \mathbf{T} = \mathbf{0} \quad .} \tag{A.203}$$

Property A.16 *In the Lie group manifold* GL(*n*)*, the covariant derivative of the torsion is identically zero.*

A.12.14 Anassociativity and Riemann

The general relation between the anassociativity tensor and the Riemann tensor is (equation 1.113) $\mathbf{A} = \mathbf{R} + \nabla\mathbf{T}$. As a group is associative, $\mathbf{A} = \mathbf{0}$. Using the property (A.203) (vanishing of the derivative of the torsion), one then immediately obtains

$$\boxed{\mathbf{R} = \mathbf{0} \quad .} \tag{A.204}$$

Property A.17 *In the Lie group manifold* GL(*n*)*, the Riemann tensor (of the connection) identically vanishes.*

Of course, it is also possible to obtain this result by a direct use of the expression of the Riemann of a manifold (equation 1.110) that, when using the double-index notation, becomes $R^\alpha{}_{\beta\mu}{}^\nu{}_\rho{}^\sigma{}_\epsilon{}^\pi = \partial \Gamma^\alpha{}_{\beta\rho}{}^\sigma{}_\mu{}^\nu / \partial X^\epsilon{}_\pi - \partial \Gamma^\alpha{}_{\beta\epsilon}{}^\pi{}_\mu{}^\nu / \partial X^\rho{}_\sigma + \Gamma^\alpha{}_{\beta\epsilon}{}^\pi{}_\varphi{}^\phi\, \Gamma^\varphi{}_{\phi\rho}{}^\sigma{}_\mu{}^\nu - \Gamma^\alpha{}_{\beta\rho}{}^\sigma{}_\varphi{}^\phi\, \Gamma^\varphi{}_{\phi\epsilon}{}^\pi{}_\mu{}^\nu$. Using expression (A.183) for the connection, this gives $R^\alpha{}_{\beta\mu}{}^\nu{}_\rho{}^\sigma{}_\epsilon{}^\pi = 0$, as it should.

A.12.15 Parallel Transport of Forms

We have obtained above the expression for the parallel transport of a vector $t^\alpha{}_\beta$ (equation A.194). We shall in a moment need the equation describing the parallel transport of a form $f_\alpha{}^\beta$. As one must have $(\mathbf{f_B})_\alpha{}^\beta\,(\mathbf{t_B})^\alpha{}_\beta = (\mathbf{f_A})_\alpha{}^\beta\,(\mathbf{t_A})^\alpha{}_\beta$, one easily obtains $(\mathbf{f_B})_\alpha{}^\beta = \overline{B}^\beta{}_\rho\,A^\rho{}_\sigma\,(\mathbf{f_A})_\alpha{}^\sigma$. So, we can now complete the property A.11 with the following

Property A.18 *The transport of a form* $\mathbf{f_A}$ *from a point* $\mathcal{A}$ *with coordinates* $\mathbf{A} = \{A^\alpha{}_\beta\}$ *to a point* $\mathcal{B}$ *with coordinates* $\mathbf{B} = \{B^\alpha{}_\beta\}$ *gives, at point* $\mathcal{B}$, *the form*

$$(\mathbf{f_B})_\alpha{}^\beta \;=\; \overline{B}^\beta{}_\rho\,A^\rho{}_\sigma\,(\mathbf{f_A})_\alpha{}^\sigma \quad . \tag{A.205}$$

A.12.16 Metric

The universal metric at the origin was expressed in equation (1.31): $\overset{\circ}{g}_\alpha{}^\beta{}_\mu{}^\nu = \chi\,\delta^\nu_\alpha\,\delta^\beta_\mu + \frac{\psi-\chi}{n}\,\delta^\beta_\alpha\,\delta^\nu_\mu$. Its transport from the origin δ^α_β to an arbitrary point $X^\alpha{}_\beta$ is made using equation (A.205), $g_\alpha{}^\beta{}_\mu{}^\nu = \overset{\circ}{g}_\alpha{}^\rho{}_\mu{}^\sigma\,\overline{X}^\beta{}_\rho\,\overline{X}^\nu{}_\sigma = \chi\,\overline{X}^\nu{}_\alpha\,\overline{X}^\beta{}_\mu + \frac{\psi-\chi}{n}\,\overline{X}^\beta{}_\alpha\,\overline{X}^\nu{}_\mu$.

Property A.19 *In the Lie group manifold* GL(n) *with exponential coordinates* $\{X^\alpha{}_\beta\}$, *the universal metric at an arbitrary point is*

$$\boxed{g_\alpha{}^\beta{}_\mu{}^\nu \;=\; \chi\,\overline{X}^\nu{}_\alpha\,\overline{X}^\beta{}_\mu + \frac{\psi-\chi}{n}\,\overline{X}^\beta{}_\alpha\,\overline{X}^\nu{}_\mu \quad .} \tag{A.206}$$

We shall later see how this universal metric relates to the usual Killing-Cartan metric (the Killing-Cartan 'metric' is the Ricci of our universal metric).

The 'contravariant' metric, denoted $g^\alpha{}_\beta{}^\mu{}_\nu$, is defined by the condition $g_\alpha{}^\beta{}_\rho{}^\sigma\;g^\rho{}_\sigma{}^\mu{}_\nu \;=\; \delta^\mu_\alpha\,\delta^\sigma_\nu$, this giving

$$g^\alpha{}_\beta{}^\mu{}_\nu \;=\; \overline{\chi}\,X^\alpha{}_\nu\,X^\mu{}_\beta + \frac{\overline{\psi}-\overline{\chi}}{n}\,X^\alpha{}_\beta\,X^\mu{}_\nu \quad , \tag{A.207}$$

where $\overline{\chi} = 1/\chi$ and $\overline{\psi} = 1/\psi$.

As a special case, choosing $\chi = \psi = 1$, one obtains

$$g_\alpha{}^\beta{}_\mu{}^\nu \;=\; \overline{X}^\nu{}_\alpha\,\overline{X}^\beta{}_\mu \quad ; \quad g^\alpha{}_\beta{}^\mu{}_\nu \;=\; X^\alpha{}_\nu\,X^\mu{}_\beta \quad . \tag{A.208}$$

This special expression of the metric is sufficient to understand most of the geometric properties of the Lie group manifold GL(n).

A.12.17 Volume Element

Once the metric tensor is defined over a manifold, we can express the volume element (or 'measure'), as the volume density is always given by $\sqrt{-\det \mathbf{g}}$. Here, as we are using as coordinates the $X^\alpha{}_\beta$ the volume element shall have the form

$$dV = \sqrt{-\det \mathbf{g}} \prod_{\substack{1 \leq \alpha \leq n \\ 1 \leq \beta \leq n}} dX^\alpha{}_\beta \quad . \tag{A.209}$$

Given the expression (A.206) for the metric, one obtains[15] $\sqrt{-\det \mathbf{g}} = (\psi\, \chi^{n^2-1})^{1/2}\, (\det \overline{\mathbf{X}})^n$, i.e.,

$$\sqrt{-\det \mathbf{g}} = \frac{(\psi\, \chi^{n^2-1})^{1/2}}{(\det \mathbf{X})^n} \quad . \tag{A.210}$$

Except for our (constant) factor $(\psi\, \chi^{n^2-1})^{1/2}$, this is identical to the well-known *Haar measure* defined over Lie groups (see, for instance, Terras, 1988). Should we choose $\psi = \chi$ (i.e., to give equal weight to homotheties and to isochoric transformations), then $\sqrt{-\det \mathbf{g}} = \chi^{n^2/2} / (\det \mathbf{X})^n$.

A.12.18 Finite Distance Between Points

With the universal metric $g_\alpha{}^\beta{}_\mu{}^\nu$ given in equation (A.206), the squared (infinitesimal) distance between point $\mathbf{X} = \{X^\alpha{}_\beta\}$ and point $\mathbf{X}+d\mathbf{X} = \{X^\alpha{}_\beta + dX^\alpha{}_\beta\}$ $ds^2 = g_\alpha{}^\beta{}_\mu{}^\nu\, dX^\alpha{}_\beta\, dX^\mu{}_\nu$, this giving

$$ds^2 = \chi\, dX^\alpha{}_\beta\, \overline{X}^\beta{}_\mu\, dX^\mu{}_\nu\, \overline{X}^\nu{}_\alpha + \frac{\psi - \chi}{n}\, dX^\alpha{}_\beta\, \overline{X}^\beta{}_\alpha\, dX^\mu{}_\nu\, \overline{X}^\nu{}_\mu \quad . \tag{A.211}$$

It is easy to express the finite distance between two points:

Property A.20 *With the universal metric* (A.206)*, the squared distance between point* $\mathbf{X} = \{X^\alpha{}_\beta\}$ *and point* $\mathbf{X}' = \{X'^\alpha{}_\beta\}$ *is*

$$D^2(\mathbf{X}', \mathbf{X}) = \| \mathbf{t} \|^2 \equiv \chi \operatorname{tr} \tilde{\mathbf{t}}^2 + \psi \operatorname{tr} \bar{\mathbf{t}}^2 \quad , \quad \textit{where} \quad \mathbf{t} = \log(\mathbf{X}'\, \mathbf{X}^{-1}) \; , \tag{A.212}$$

and where $\tilde{\mathbf{t}}$ *and* $\bar{\mathbf{t}}$ *respectively denote the deviatoric and the isotropic parts of* $\mathbf{t}$ *(equations 1.34).*

---

[15]To evaluate $\sqrt{-\det \mathbf{g}}$ we can, for instance, use the definition of determinant given in footnote 37, that is valid when using a single-index notation, and transform the expression (A.206) of the metric into a single-index notation, as was done in equation (A.180) for the expression of the metric at the origin. The coefficients $\Lambda^\alpha{}_{\beta i}$ to be introduced must verify the relation (A.175).

The norm $\| \mathbf{t} \|$ defining this squared distance has already appeared in property 1.3.

To demonstrate the property A.20, one simply sets $\mathbf{X}' = \mathbf{X} + d\mathbf{X}$ in equation (A.212), uses the property $\log(\mathbf{I} + \mathbf{A}) = \mathbf{A} + \dots$, to write the series $D^2(\mathbf{X}+d\mathbf{X}, \mathbf{X}) = \underbrace{g_\alpha{}^\beta{}_\mu{}^\nu \, dX^\alpha{}_\beta \, dX^\mu{}_\nu}_{ds^2} + \dots$ (only the second-order term needs to be evaluated). This produces exactly the expression (A.211) for the ds^2.

A.12.19 Levi-Civita Connection

The transport associated to the metric is defined via the Levi-Civita connection. Should we wish to use vector-like notation, we would write $\{^i{}_{jk}\} = \frac{1}{2} g^{is} (\partial g_{ks}/\partial x^j + \partial g_{js}/\partial x^k - \partial g_{jk}/\partial x^s)$. Using double-index notation, $\{^\alpha{}_{\beta\mu}{}^\nu{}_\rho{}^\sigma\} = \frac{1}{2} g^\alpha{}_\beta{}^\omega{}_\pi (\partial g_\rho{}^\sigma{}_\omega{}^\pi/\partial X^\mu{}_\nu + \partial g_\mu{}^\nu{}_\omega{}^\pi/\partial X^\rho{}_\sigma - \partial g_\mu{}^\nu{}_\rho{}^\sigma/\partial X^\pi{}_\omega)$. The computation is easy to perform,[16] and gives

$$\{^\alpha{}_{\beta\mu}{}^\nu{}_\rho{}^\sigma\} = -\tfrac{1}{2} (\overline{X}^\nu{}_\rho \, \delta^\sigma_\beta \, \delta^\alpha_\mu + \overline{X}^\sigma{}_\mu \, \delta^\alpha_\rho \, \delta^\nu_\beta) \quad . \tag{A.213}$$

A.12.20 Covariant Torsion

The metric can be used to lower the 'contravariant index' of the torsion, according to $T_\alpha{}^\beta{}_\mu{}^\nu{}_\rho{}^\sigma = g_\alpha{}^\beta{}_\epsilon{}^\pi \, T^\epsilon{}_{\pi\mu}{}^\nu{}_\rho{}^\sigma$, to obtain

$$\boxed{T_\alpha{}^\beta{}_\mu{}^\nu{}_\rho{}^\sigma = \chi \left(\overline{X}^\beta{}_\mu \, \overline{X}^\nu{}_\rho \, \overline{X}^\sigma{}_\alpha - \overline{X}^\beta{}_\rho \, \overline{X}^\nu{}_\alpha \, \overline{X}^\sigma{}_\mu \right) \quad .} \tag{A.214}$$

One easily verifies the (anti)symmetries

$$T_\alpha{}^\beta{}_\mu{}^\nu{}_\rho{}^\sigma = -T_\mu{}^\nu{}_\alpha{}^\beta{}_\rho{}^\sigma = -T_\alpha{}^\beta{}_\rho{}^\sigma{}_\mu{}^\nu \quad . \tag{A.215}$$

Property A.21 *The torsion of the Lie group manifold* GL(n)*, endowed with the universal metric is totally antisymmetric.*

As explained in appendix A.11, when the torsion is totally antisymmetric (with respect to a given metric), the autoparallels of the connection and the geodesics of the metric coincide. Therefore, we have the following

Property A.22 *In the Lie group manifold* GL(n)*, the geodesics of the metric are the autoparallels of the connection (and vice versa).*

Therefore, we could have replaced everywhere in this section the term 'autoparallel' by the term 'geodesic'. From now on: when working with Lie group manifolds, the 'autoparallel lines' become 'geodesic lines'.

---

[16] Hint: from $\overline{X}^\alpha{}_\sigma \, X^\sigma{}_\beta = \delta^\alpha_\beta$ it follows that $\partial \overline{X}^\alpha{}_\beta / \partial X^\mu{}_\nu = -\overline{X}^\alpha{}_\mu \, \overline{X}^\nu{}_\beta$.

A.12.21 Curvature and Ricci of the Metric

The curvature ("the Riemann of the metric") is defined as a function of the Levi-Civita connection with the same expression used to define the Riemann as a function of the (total) connection (equation 1.110). Using vector notation this would be $C^i{}_{jk\ell} = \partial\{^i{}_{kj}\}/\partial x^\ell - \partial\{^i{}_{\ell j}\}/\partial x^k + \{^i{}_{\ell s}\}\,\{^s{}_{kj}\} - \{^i{}_{ks}\}\,\{^s{}_{\ell j}\}$, the translation using the present double-index notation being $C^\alpha{}_{\beta\mu}{}^\nu{}_\rho{}^\sigma{}_\epsilon{}^\pi = \partial\{^\alpha{}_{\beta\rho}{}^\sigma{}_\mu{}^\nu\}/\partial X^\epsilon{}_\pi - \partial\{^\alpha{}_{\beta\epsilon}{}^\pi{}_\mu{}^\nu\}/\partial X^\rho{}_\sigma + \{^\alpha{}_{\beta\epsilon}{}^\pi{}_\varphi{}^\phi\}\,\{^\varphi{}_{\phi\rho}{}^\sigma{}_\mu{}^\nu\} - \{^\alpha{}_{\beta\rho}{}^\sigma{}_\varphi{}^\phi\}\,\{^\varphi{}_{\phi\epsilon}{}^\pi{}_\mu{}^\nu\}$. Using equation (A.213) this gives, after several computations,

$$\boxed{C^\alpha{}_{\beta\mu}{}^\nu{}_\rho{}^\sigma{}_\epsilon{}^\pi = \tfrac{1}{4}\, T^\alpha{}_{\beta\mu}{}^\nu{}_\varphi{}^\phi\, T^\varphi{}_{\phi\epsilon}{}^\pi{}_\rho{}^\sigma \quad ,} \tag{A.216}$$

where $T^\alpha{}_{\beta\mu}{}^\nu{}_\rho{}^\sigma$ is the torsion obtained in equation (A.201). Therefore, one has

Property A.23 *In the Lie group manifold* GL(n)*, the curvature of the metric is proportional to the squared of the torsion, i.e., equation* (A.216) *holds.*

The Ricci of the metric is defined as $C_\alpha{}^\beta{}_\mu{}^\nu = C^\rho{}_{\sigma\alpha}{}^\beta{}_\rho{}^\sigma{}_\mu{}^\nu$. In view of equation (A.216), this gives

$$C_\alpha{}^\beta{}_\mu{}^\nu = \tfrac{1}{4}\, T^\rho{}_{\sigma\alpha}{}^\beta{}_\varphi{}^\phi\, T^\varphi{}_{\phi\mu}{}^\nu{}_\rho{}^\sigma \quad , \tag{A.217}$$

i.e., using the expression (A.201) for the torsion,

$$C_\alpha{}^\beta{}_\mu{}^\nu = \tfrac{n}{2}\left(\overline{X}^\nu{}_\alpha\,\overline{X}^\beta{}_\mu - \tfrac{1}{n}\,\overline{X}^\beta{}_\alpha\,\overline{X}^\nu{}_\mu\right) \quad . \tag{A.218}$$

At this point, we may remark that the one-index version of equation (A.217) would be

$$C_{ij} = \tfrac{1}{4}\, T^r{}_{is}\, T^s{}_{jr} \quad . \tag{A.219}$$

Up to a numerical factor, this expression corresponds to the usual definition of the "Cartan metric" of a Lie group (Goldberg, 1998): the usual 'structure coefficients' are nothing but the components of the torsion at the origin. Here, we obtain directly the the Ricci of the Lie group manifold at an arbitrary point with coordinates $X^\alpha{}_\beta$, while the 'Cartan metric' (or 'Killing form') is usually introduced at the origin only (i.e., for the linear tangent space at the origin), but there is no problem in the standard presentation of the theory, to "drag" it to an arbitrary point (see, for instance, Choquet-Bruhat *et al.*, 1977). We have thus arrived at the following

Property A.24 *The so-called Cartan metric is the Ricci of the Lie group manifold* GL(n) *(up to a numerical factor).*

Many properties of Lie groups are traditionally attached to the properties of the Cartan metric of the group.[17] The present discussion suggests that

[17]For instance, a Lie group is 'semi-simple' if its Cartan metric is nonsingular.

the the wording of these properties could be changed, replacing everywhere 'Cartan metric' by 'Ricci of the (universal) metric'.

One obvious question, now, concerns the relation that the Cartan metric bears with the actual metric of the Lie group manifold (the universal metric). The proper question, of course, is about the relation between the universal metric and its Ricci. Expression (A.218) can be compared with the expression (A.206) of the universal metric when one sets $\psi = 0$ (i.e., when one gives zero weight to the homotheties):

$$g_\alpha{}^\beta{}_\mu{}^\nu = \chi \left(\overline{X}^\nu{}_\alpha \, \overline{X}^\beta{}_\mu - \tfrac{1}{n} \, \overline{X}^\beta{}_\alpha \, \overline{X}^\nu{}_\mu \right) \quad ; \quad (\psi = 0) \quad . \tag{A.220}$$

We thus obtain

Property A.25 *If in the universal metric one gives zero weight to the homotheties ($\psi = 0$), then, the Ricci of the (universal) metric is proportional to the (universal) metric:*

$$\boxed{C_\alpha{}^\beta{}_\mu{}^\nu = \tfrac{n}{2\chi} \, g_\alpha{}^\beta{}_\mu{}^\nu \quad ; \quad (\psi = 0) \quad .} \tag{A.221}$$

This suggests that the 'Cartan metric' fails to properly take into account the homotheties.

A.12.22 Connection (again)

Given the torsion and the Levi-Civita connection, the (total) connection is expressed as (equation (A.156) with a totally antisymmetric torsion)

$$\Gamma^\alpha{}_{\beta\mu}{}^\nu{}_\rho{}^\sigma = \tfrac{1}{2} \, T^\alpha{}_{\beta\mu}{}^\nu{}_\rho{}^\sigma + \{^\alpha{}_{\beta\mu}{}^\nu{}_\rho{}^\sigma\} \quad . \tag{A.222}$$

With the torsion in equation (A.201) and the Levi-Civita connection obtained in equation (A.213), this gives $\Gamma^\alpha{}_{\beta\mu}{}^\nu{}_\rho{}^\sigma = -\overline{X}^\sigma{}_\mu \, \delta^\alpha_\rho \, \delta^\nu_\beta$, i.e., the expression found in equation (A.183).

A.12.23 Expressions in Arbitrary Coordinates

By definition, the exponential coordinates cover the whole GL(n) manifold (as every matrix of GL(n) corresponds to a point, and vice versa). The analysis of the subgroups of GL(n) is better made using coordinates $\{x^i\}$ that, when taking independent values, cover the submanifold. Let us then consider a system $\{x^1, x^2 \dots x^p\}$ of p coordinates ($1 \leq p \leq n^2$), and assume given the functions

$$X^\alpha{}_\beta = X^\alpha{}_\beta(x^i) \tag{A.223}$$

and the partial derivatives

$$\Lambda^\alpha{}_{\beta i} = \frac{\partial X^\alpha{}_\beta}{\partial x^i} \quad . \tag{A.224}$$

Example A.11 *The Lie group manifold* SL(2) *is covered by the three coordinates* $\{x^1, x^2, x^3\} = \{e, \alpha, \varphi\}$ *, that are related to the exponential coordinates* $X^\alpha{}_\beta$ *of* GL(2) *through (see example A.12 for details)*

$$\mathbf{X} = \cosh e \begin{pmatrix} \cos\alpha & \sin\alpha \\ -\sin\alpha & \cos\alpha \end{pmatrix} + \sinh e \begin{pmatrix} \sin\varphi & \cos\varphi \\ \cos\varphi & -\sin\varphi \end{pmatrix} \quad . \tag{A.225}$$

Note that if the functions $X^\alpha{}_\beta(x^i)$ are given, by inversion of the matrix $\mathbf{X} = \{X^\alpha{}_\beta(x^i)\}$ we can also consider the functions $\overline{X}^\alpha{}_\beta(x^i)$ are given.

As there may be less than n^2 coordinates x^i , the relations (A.223) cannot be solved to give the inverse functions $x^i = x^i(X^\alpha{}_\beta)$. Therefore the partial derivatives $\Lambda_\alpha{}^{\beta i} = \partial x^i / \partial X^\alpha{}_\beta$ cannot, in general, be computed. But given the partial derivatives in equation (A.224), it is possible to define the reciprocal coefficients $\Lambda_\alpha{}^{\beta i}$ as was done in equations (A.171) and (A.172). Then, $P^\alpha{}_{\beta\mu}{}^\nu = \Lambda^\alpha{}_{\beta i}\,\Lambda_\mu{}^{\nu i}$ is a projector over the p-dimensional linear subspace $\mathfrak{gl}(n)_p$ locally defined by the p coordinates x^i .

The components of the tensors in the new coordinates are obtained using the standard rules associated to the change of variables. For the torsion one has $T^i{}_{jk} = \Lambda_\alpha{}^{\beta i}\,\Lambda^\mu{}_{\nu j}\,\Lambda^\rho{}_{\sigma k}\,T^\alpha{}_{\beta\mu}{}^\nu{}_\rho{}^\sigma$, and using $T^\alpha{}_{\beta\mu}{}^\nu{}_\rho{}^\sigma = \overline{X}^\nu{}_\rho\,\delta^\sigma_\beta\,\delta^\alpha_\mu - \overline{X}^\sigma{}_\mu\,\delta^\alpha_\rho\,\delta^\nu_\beta$ (expression (A.201)) this gives

$$T^i{}_{jk} = \overline{X}^\mu{}_\nu\,\Lambda_\alpha{}^{\beta i}\,(\Lambda^\alpha{}_{\mu j}\,\Lambda^\nu{}_{\beta k} - \Lambda^\alpha{}_{\mu k}\,\Lambda^\nu{}_{\beta j}) \quad , \tag{A.226}$$

an expression that reduces to (A.178) at the origin. For the metric, $g_{ij} = \Lambda^\alpha{}_{\beta i}\,\Lambda^\mu{}_{\nu j}\,g_\alpha{}^\beta{}_\mu{}^\nu$. Using the expression $g_\alpha{}^\beta{}_\mu{}^\nu = \chi\,\overline{X}^\nu{}_\alpha\,\overline{X}^\beta{}_\mu + \frac{\psi-\chi}{n}\,\overline{X}^\beta{}_\alpha\,\overline{X}^\nu{}_\mu$ (equation A.206) this gives

$$g_{ij} = \overline{X}^\nu{}_\alpha\,\overline{X}^\beta{}_\mu\,(\chi\,\Lambda^\alpha{}_{\beta i}\,\Lambda^\mu{}_{\nu j} + \frac{\psi-\chi}{n}\,\Lambda^\mu{}_{\beta i}\,\Lambda^\alpha{}_{\nu j}) \quad . \tag{A.227}$$

an expression that reduces to (A.180) at the origin. Finally, the totally covariant expression for the torsion, $T_{ijk} \equiv g_{is}\,T^s{}_{jk}$, can be obtained, for instance, using equation (A.214):

$$T_{ijk} = \chi\,\overline{X}^\beta{}_\mu\,\overline{X}^\nu{}_\rho\,\overline{X}^\sigma{}_\alpha\,(\Lambda^\alpha{}_{\beta i}\,\Lambda^\mu{}_{\nu j}\,\Lambda^\rho{}_{\sigma k} - \Lambda^\rho{}_{\beta i}\,\Lambda^\alpha{}_{\nu j}\,\Lambda^\mu{}_{\sigma k}) \quad , \tag{A.228}$$

an expression that reduces to (A.181) at the origin. One clearly has

$$T_{ijk} = -T_{ikj} = -T_{jik} \quad . \tag{A.229}$$

As the metric on a submanifold is the metric induced by the metric on the manifold, equation (A.227) can, in fact, be used to obtain the metric on any submanifold of the Lie group manifold: this equation makes perfect sense. For instance, we can use this formula to obtain the metric on the SL(n) and the SO(n) submanifolds of GL(n) . This property does not extend to the formulas (A.226) and (A.228) expressing the torsion on arbitrary coordinates.

Example A.12 Coordinates over $\mathrm{GL}^+(2)$**.** *In section 1.4.6, where the manifold of the Lie group* $\mathrm{GL}^+(2)$ *is studied, a matrix* $\mathbf{X} \in \mathrm{GL}^+(2)$ *is represented (see equation* (1.181)*) using four parameters* $\{\kappa, e, \alpha, \varphi\}$*,*

$$\mathbf{X} = \exp\kappa \left[\cosh e \begin{pmatrix} \cos\alpha & \sin\alpha \\ -\sin\alpha & \cos\alpha \end{pmatrix} + \sinh e \begin{pmatrix} \sin\varphi & \cos\varphi \\ \cos\varphi & -\sin\varphi \end{pmatrix} \right] \quad , \tag{A.230}$$

which, in fact, are four coordinates $\{x^0, x^1, x^2, x^3\}$ *over the Lie group manifold. The partial derivatives* $\Lambda^\alpha{}_{\beta i}$*, defined in equations* (A.224)*, are easily obtained, and the components of the metric tensor in these coordinates are then obtained using equation* (A.227) *(the inverse matrix* $\mathbf{X}^{-1}$ *is given in equation* (1.183)*). The metric so obtained (that happens to be diagonal in these coordinates) gives to the expression* $ds^2 = g_{ij}\, dx^i\, dx^j$ *the form*[18]

$$ds^2 = 2\,\psi\, d\kappa^2 + 2\,\chi\,(de^2 - \cosh^2 e\; d\alpha^2 + \sinh^2 e\; d\varphi^2) \quad . \tag{A.231}$$

The torsion is directly obtained using (A.228)*:*

$$T_{ijk} = \frac{1}{\sqrt{\psi\,\chi}}\, \epsilon_{0ijk} \quad , \tag{A.232}$$

where $\epsilon_{ijk\ell}$ *is the Levi-Civita tensor of the space.*[19] *In particular, all the components of the torsion* T_{ijk} *with an index 0 vanish. One should note that the three coordinates* $\{e, \alpha, \varphi\}$*, are 'cylindrical-like', so they are singular along* $e = 0$.

A.12.24 SL(n)

The two obvious subgroups of $\mathrm{GL}^+(n)$ (of dimension n^2), are $\mathrm{SL}(n)$ (of dimension $n^2 - 1$) and $\mathrm{H}(n)$ (of dimension 1). As this partition of $\mathrm{GL}^+(n)$ into $\mathrm{SL}(n)$ and $\mathrm{H}(n)$ corresponds to the fundamental geometric structure of the $\mathrm{GL}^+(n)$ manifold, it is important that we introduce a coordinate system adapted to this partition.

Let us first decompose the matrices $\mathbf{X}$ (representing the coordinates of a point) in an appropriate way, writing

$$\mathbf{X} = \lambda\, \mathbf{Y} \quad , \quad \text{with} \quad \lambda = (\det \mathbf{X})^{1/n} \quad \text{and} \quad \mathbf{Y} = \frac{1}{(\det \mathbf{X})^{1/n}}\, \mathbf{X} \tag{A.233}$$

so that one has[20] $\det \mathbf{Y} = 1$.

---

[18]Choosing, for instance, $\psi = \chi = 1/2$, this simplifies to $ds^2 = d\kappa^2 + de^2 - \cosh^2 e\; d\alpha^2 + \sinh^2 e\; d\varphi^2$.

[19]I.e., the totally antisymmetric tensor defined by the condition $\epsilon_{0123} = \sqrt{-\det \mathbf{g}} = 2\,\psi^{1/2}\,\chi^{3/2}\,\sinh 2e$.

[20]Note that $\log\lambda = \frac{1}{n}\log(\det\mathbf{X}) = \frac{1}{n}\,\mathrm{tr}\,(\log\mathbf{X})$.

The n^2 parameters $\{x^0,\dots,x^{(n^2-1)}\}$ can be separated into two sets, the parameter x^0 used to parameterize the scalar λ, and the parameters $\{x^1,\dots,x^{(n^2-1)}\}$ used to parameterize $\mathbf{Y}$:

$$\lambda = \lambda(x^0) \quad ; \quad \mathbf{Y} = \mathbf{Y}(x^1,\dots,x^{(n^2-1)}) \tag{A.234}$$

(one may choose for instance the parameter $x^0 = \lambda$ or $x^0 = \log\lambda$).

In what follows, the indices $a,b,\dots$ shall be used for the range $\{1,\dots,(n^2-1)\}$. With this decomposition, the expression (A.227) for the metric g_{ij} separates into[21]

$$\boxed{g_{00} = \frac{\psi\, n}{\lambda^2}\left(\frac{\partial\lambda}{\partial x^0}\right)^2 \quad ; \quad g_{ab} = \chi\,\frac{\partial Y^\alpha{}_\beta}{\partial x^a}\,\overline{Y}^\beta{}_\mu\,\frac{\partial Y^\mu{}_\nu}{\partial x^b}\,\overline{Y}^\nu{}_\alpha} \tag{A.235}$$

and $g_{0a} = g_{a0} = 0$. As one could have expected, the metric separates into an H(n) part, depending on ψ, and one SL(n) part, depending on χ. For the contravariant metric, one obtains $g^{00} = 1/g_{00}$, $g^{0a} = g^{a0} = 0$ and $g^{ab} = \overline{\chi}\,(\partial x^a/\partial Y^\alpha{}_\beta)\,Y^\alpha{}_\nu\,(\partial x^b/\partial Y^\mu{}_\nu)\,Y^\mu{}_\beta$.

The metric Ricci of H(n) is zero, as the manifold is one-dimensional. The metric Ricci of SL(n) has to be computed from g_{ab}. But, as H(n) and SL(n) are orthogonal subspaces (i.e., as $g_{0a} = g_{a0} = 0$), the metric Ricci of H(n) and that of SL(n) can, more simply, be obtained as the $\{00\}$ and the $\{ab\}$ components of the metric Ricci C_{ij} of $GL^+(n)$ (as given, for instance, by equation (A.219)). One obtains

$$\boxed{C_{ab} = \frac{n}{2}\,\frac{\partial Y^\alpha{}_\beta}{\partial x^a}\,\overline{Y}^\beta{}_\mu\,\frac{\partial Y^\mu{}_\nu}{\partial x^b}\,\overline{Y}^\nu{}_\alpha} \tag{A.236}$$

and $C_{ij} = 0$ if any index is 0. The part of the Ricci associated to H(n) vanishes, and the part associated to SL(n), which is independent of χ, is proportional to the metric:

$$C_{ab} = \frac{n}{2\chi}\,g_{ab} \quad . \tag{A.237}$$

Therefore, one has

Property A.26 *In* SL(n), *the Ricci of the metric is proportional to the metric.*

As already mentioned in section A.12.21, what is known in the literature as the Cartan metric (or Killing form) of a Lie group corresponds to the expressions in equation (A.236). This as an unfortunate confusion.

The metric of a subspace $\mathbb{F}$ of a space $\mathbb{E}$ is the metric induced (in the tensorial sense of the term) on $\mathbb{F}$ by the metric of $\mathbb{E}$. This means that, given

[21]For the demonstration, use the property $(\partial Y^i{}_j/\partial x^a)\,\overline{Y}^j{}_i = 0$, that follows from the condition $\det\mathbf{Y} = 1$.

a covariant tensor of $\mathbb{E}$, and a coordinate system adapted to the subspace $\mathbb{F} \subset \mathbb{E}$, the components of the tensor that depend only on the subspace coordinates "induce" on the subspace a tensor, called the induced tensor. Because of this, the expression (A.235) of the metric for SL(n) can directly be used for any subgroup of SL(n) —for instance, for SO(n),— using adapted coordinates.[22] This property does not extend to the Ricci: the tensor induced on a subspace $\mathbb{F} \subset \mathbb{E}$ by the Ricci tensor of the metric of $\mathbb{E}$ is generally not the Ricci tensor of the metric induced on $\mathbb{F}$ by the metric of $\mathbb{E}$. Briefly put, expression (A.235) can be used to compute the metric of any subgroup of SL(n) if adapted coordinates are used. The expression (A.236) of the metric Ricci *cannot* be used to compute the metric Ricci of a subgroup of SL(n). I have not tried to develop an explicit expression for the metric Ricci of SO(n).

For the (all-covariant) torsion, one easily obtains, using equation (A.228),

$$T_{abc} = \chi\, \overline{Y}^{\beta}{}_{\mu}\, \overline{Y}^{\nu}{}_{\rho}\, \overline{Y}^{\sigma}{}_{\alpha} \left(\frac{\partial Y^{\alpha}{}_{\beta}}{\partial x^{a}} \frac{\partial Y^{\mu}{}_{\nu}}{\partial x^{b}} \frac{\partial Y^{\rho}{}_{\sigma}}{\partial x^{c}} - \frac{\partial Y^{\rho}{}_{\beta}}{\partial x^{a}} \frac{\partial Y^{\alpha}{}_{\nu}}{\partial x^{b}} \frac{\partial Y^{\mu}{}_{\sigma}}{\partial x^{c}} \right) \tag{A.238}$$

and $T_{ijk} = 0$ if any index is 0. As one could have expected, the torsion only affects the SL(n) subgroup of GL$^+$(n).

A.12.25 Geometrical Structure of the GL$^+$(n) Group Manifold

We have seen, using coordinates adapted to SL(n), that the components g_{0a} of the metric vanish. This means that, in fact, the GL$^+$(n) manifold is a continuous "orthogonal stack" of many copies of SL(n).[23] Equation (A.212), for instance, shows that, concerning the distances between points, we can treat independently the SL(n) part and the (one-dimensional) H(n) part. As the torsion and the metric are adapted (the torsion is totally antisymmetric), this has an immediate translation in terms of torsion: not only the one-dimensional subgroup H(n) has zero torsion (as any one-dimensional subspace), but all the components of the torsion containing a zero index also vanish (equations A.238).

This is to say that all interesting geometrical features of GL$^+$(n) come from SL(n), nothing remarkable happening with the addition of H(n).

So, the SL(n) manifold has the metric g_{ab} given in the third of equations (A.235) and the torsion T_{abc} given in the second of equations (A.238).

[22] By the same token, the expression (A.227) for the metric in GL$^+$(n) can also be used for SL(n), instead of (A.235).

[23] As if one stacks many copies of a geographical map, the squared distance between two arbitrary points of the stack being defined as the sum of the squared vertical distance between the two maps that contain each one of the points (weighted by a constant ψ) plus the squared of the actual geographical distance (in one of the maps) between the projections of the two points (weighted by a constant χ).

The torsion is constant over the manifold[24] and the Riemann (of the connection) vanishes. This last property means that over SL(n) (in fact, over GL(n)) there exists a notion of absolute parallelism (when transporting a vector between two points, the transport path doesn't matter). The space has curvature and has torsion, but they balance to give the property of absolute parallelism, a property that is usually only found in linear manifolds.

We have made some effort above to introduce the notion of near neutral subset. The points of the group manifold that are outside this subset cannot be joined from the origin using a geodesic line. Rather than trying to develop the general theory here, it is better to make a detailed analysis in the case of the simplest group presenting this behavior, the four-dimensional group $GL^+(2)$. This is done in section 1.4.6.

A.13 Lie Groups as Groups of Transformations

We have seen that the points of the Lie group manifold associated to the set of matrices in GL(n) are the matrices themselves. In addition to the matrices $\mathbf{A}, \mathbf{B} \dots$ we have also recognized the importance of the oriented geodesic segments connecting two points of the manifold (i.e., connecting two matrices).

It is important, when working with the set of linear transformations over a linear space, to not mistake these linear transformations for points of the GL(n) manifold: it is better to interpret the (matrices representing the) points of the GL(n) manifold as representing the set of all possible *bases* of a linear space, and to interpret the set of all linear transformation over the linear space as the geodesic segments connecting two points of the manifold, i.e., connecting two bases. For although a linear transformation is usually seen as transforming one vector into another vector, it can perfectly well be seen as transforming one basis into another basis, and it is this second point of view that helps one understand the geometry behind a group of linear transformations.

A.13.1 Reference Basis

Let $\mathbb{E}_n$ be an n-dimensional linear space, and let $\{\mathbf{e}_\alpha\}$ (for $\alpha = 1, \dots, n$) be a basis of $\mathbb{E}_n$. Different bases of $\mathbb{E}_n$ are introduced below, and changes of bases considered, but this particular basis $\{\mathbf{e}_\alpha\}$ plays a special role, so let us call it the *reference basis*. Let also $\mathbb{E}^*_n$ be the dual of $\mathbb{E}_n$, and $\{\mathbf{e}^\alpha\}$ the dual of the reference basis. Then, $\langle\, \mathbf{e}^\alpha \,,\, \mathbf{e}_\beta \,\rangle \;=\; \delta^\alpha_\beta$. Finally, let $\mathbb{E}_n \otimes \mathbb{E}^*_n$ be the tensor product of $\mathbb{E}_n$ by $\mathbb{E}^*_n$, with the induced reference basis $\{\mathbf{e}^\alpha \otimes \mathbf{e}_\beta\}$.

[24]We have seen that the covariant derivative of the torsion of a Lie group necessarily vanishes.

A.13.2 Other Bases

Consider now a set of n linearly independent vectors $\{\mathbf{u}_1,\dots,\mathbf{u}_n\}$ of $\mathbb{E}_n$, i.e., a basis of $\mathbb{E}_n$. Denoting by $\{\mathbf{u}^1,\dots,\mathbf{u}^n\}$ the dual basis, then, by definition $\langle\, \mathbf{u}^\alpha \,,\, \mathbf{u}_\beta \,\rangle = \delta^\alpha_\beta$. Let us associate to the bases $\{\mathbf{u}_\alpha\}$ and $\{\mathbf{u}^\alpha\}$ the two matrices $\mathbf{U}$ and $\overline{\mathbf{U}}$ with entries

$$U^\alpha{}_\beta \;=\; \langle\, \mathbf{e}^\alpha \,,\, \mathbf{u}_\beta \,\rangle \qquad ; \qquad \overline{U}^\beta{}_\alpha \;=\; \langle\, \mathbf{u}^\beta \,,\, \mathbf{e}_\alpha \,\rangle \quad . \tag{A.239}$$

Then, $\mathbf{u}_\beta = U^\alpha{}_\beta\, \mathbf{e}_\alpha$ and $\mathbf{u}^\beta = \overline{U}^\beta{}_\alpha\, \mathbf{e}^\alpha$, so one has

Property A.27 *$U^\alpha{}_\beta$ is the ith component, on the reference basis, of the vector $\mathbf{u}_\beta$, while $\overline{U}^\beta{}_\alpha$ is the ith component, on the reference dual basis, of the form $\mathbf{u}^\beta$.*

The duality condition $\langle\, \mathbf{u}^\alpha \,,\, \mathbf{u}_\beta \,\rangle = \delta^\alpha_\beta$ gives $\overline{U}^\alpha{}_k\, U^k{}_\beta \;=\; \delta^\alpha_\beta$, i.e., $\overline{\mathbf{U}}\,\mathbf{U} = \mathbf{I}$, an expression that is consistent with the notation

$$\overline{\mathbf{U}} \;=\; \mathbf{U}^{-1} \quad , \tag{A.240}$$

used everywhere in this book: the matrix $\{U^\alpha{}_\beta\}$ is the inverse of the matrix $\{\overline{U}^\alpha{}_\beta\}$.

One has

Property A.28 *The reference basis is, by definition, represented by the identity matrix $\mathbf{I}$. Other bases are represented by matrices $\mathbf{U}\,,\,\mathbf{V}\dots$ of the set GL(n). The inverse matrices $\mathbf{U}^{-1}\,,\,\mathbf{V}^{-1}\dots$ can be either interpreted as just other bases or as the duals of the bases $\mathbf{U}\,,\,\mathbf{V}\dots$.*

A change of reference basis

$$\mathbf{e}_\alpha \;=\; \Lambda^\beta{}_\alpha\, \widetilde{\mathbf{e}}_\beta \tag{A.241}$$

changes the matrix $U^\alpha{}_\beta$ into $\widetilde{U}^\alpha{}_\beta = \langle \widetilde{\mathbf{e}}^\alpha \,,\, \mathbf{u}_\beta \,\rangle = \langle \Lambda^\alpha{}_\mu\, \mathbf{e}^\mu \,,\, \mathbf{u}_\beta \,\rangle = \Lambda^\alpha{}_\mu \,\langle\, \mathbf{e}^\mu \,,\, \mathbf{u}_\beta \,\rangle$, i.e.,

$$\widetilde{U}^\alpha{}_\beta \;=\; \Lambda^\alpha{}_\mu\, U^\mu{}_\beta \quad . \tag{A.242}$$

We see, in particular, that the coefficients $U^\alpha{}_\beta$ do not transform like the components of a contravariant–covariant tensor.

A.13.3 Transformation of Vectors and of Bases

Definition A.9 *Any ordered pair of vector bases $\{\{\mathbf{u}_\alpha\}\,,\,\{\mathbf{v}_\alpha\}\}$ of $\mathbb{E}_n$ defines a linear transformation for the vectors of $\mathbb{E}_n$, the transformation that to any vector $\mathbf{a}$ associates the vector*

$$\mathbf{b} \;=\; \mathbf{v}_\beta \,\langle\, \mathbf{u}^\beta \,,\, \mathbf{a} \,\rangle \quad . \tag{A.243}$$

Introducing the reference basis $\{\mathbf{e}_\alpha\}$, this equation can be transformed into $\langle\, \mathbf{e}^\alpha \,,\, \mathbf{b} \,\rangle = \langle\, \mathbf{e}^\alpha \,,\, \mathbf{v}_\beta \,\rangle \langle\, \mathbf{u}^\beta \,,\, \mathbf{a} \,\rangle = \langle\, \mathbf{e}^\alpha \,,\, \mathbf{v}_\beta \,\rangle \langle\, \mathbf{u}^\beta \,,\, \mathbf{e}_\sigma \,\rangle \langle\, \mathbf{e}^\sigma \,,\, \mathbf{a} \,\rangle$, i.e.,

$$b^\alpha = T^\alpha{}_\sigma \, a^\sigma \tag{A.244}$$

where the coefficients $T^\alpha{}_\sigma$ are defined as

$$T^\alpha{}_\sigma = \langle\, \mathbf{e}^\alpha \,,\, \mathbf{v}_\beta \,\rangle \langle\, \mathbf{u}^\beta \,,\, \mathbf{e}_\sigma \,\rangle = V^\alpha{}_\beta \, \overline{U}^\beta{}_\sigma \quad . \tag{A.245}$$

For short, equations (A.244) and (A.245) can be written

$$\mathbf{b} = \mathbf{T}\,\mathbf{a} \quad , \quad \text{where} \qquad \mathbf{T} = \mathbf{V}\,\mathbf{U}^{-1} \quad , \tag{A.246}$$

or, alternatively,

$$\mathbf{b} = (\exp \mathbf{t})\,\mathbf{a} \quad , \quad \text{where} \qquad \mathbf{t} = \log(\mathbf{V}\,\mathbf{U}^{-1}) \quad . \tag{A.247}$$

We have seen that the matrix coefficients $V^\alpha{}_\beta$ and $\overline{U}^\alpha{}_\beta$ do not transform like the components of a tensor. It is easy to see[25] that the combinations $V^\alpha{}_\beta \, \overline{U}^\beta{}_\gamma$ do, and, therefore also their logarithms. We then have the following

Property A.29 *Both, the $\{T^\alpha{}_\beta\}$ and the $\{t^\alpha{}_\beta\}$ are the components of tensors.*

Definition A.10 *Via equation* (A.243), *any ordered pair of vector bases $\{\{\mathbf{u}_\alpha\}, \{\mathbf{v}_\alpha\}\}$ of $\mathbb{E}_n$ defines a linear transformation for the vectors of $\mathbb{E}_n$. Therefore it also defines a linear transformation for the vector bases of $\mathbb{E}_n$ that to any vector basis $\{\mathbf{a}_\alpha\}$ associates the vector basis*

$$\boxed{\mathbf{b}_\alpha = \mathbf{v}_\beta \langle\, \mathbf{u}^\beta \,,\, \mathbf{a}_\alpha \,\rangle \quad .} \tag{A.248}$$

As done above, we can use the reference basis $\{\mathbf{e}_\alpha\}$ to transform this equation into $\langle\, \mathbf{e}^\sigma \,,\, \mathbf{b}_\alpha \,\rangle = \langle\, \mathbf{e}^\sigma \,,\, \mathbf{v}_\beta \,\rangle \langle\, \mathbf{u}^\beta \,,\, \mathbf{a}_\alpha \,\rangle = \langle\, \mathbf{e}^\sigma \,,\, \mathbf{v}_\beta \,\rangle \langle\, \mathbf{u}^\beta \,,\, \mathbf{e}_\rho \,\rangle \langle\, \mathbf{e}^\rho \,,\, \mathbf{a}_\alpha \,\rangle$, i.e.,

$$B^\sigma{}_\alpha = T^\sigma{}_\rho \, A^\rho{}_\alpha \tag{A.249}$$

where the components $T^\alpha{}_\beta$ have been defined in equation (A.245). For short, the transformation of vector bases (defined by the two bases $\mathbf{U}$ and $\mathbf{V}$) is the transformation that to any vector basis $\mathbf{A}$ associates the vector basis

$$\boxed{\mathbf{B} = \mathbf{T}\,\mathbf{A} \quad , \quad \text{where} \qquad \mathbf{T} = \mathbf{V}\,\mathbf{U}^{-1} \quad ,} \tag{A.250}$$

[25] Keeping the two bases $\{\mathbf{u}_\alpha\}$ and $\{\mathbf{v}_\alpha\}$ fixed, the change (A.241) in the reference basis transforms $T^\alpha{}_\beta$ into $\widetilde{T}^\alpha{}_\beta = \widetilde{V}^\alpha{}_\mu \, \overline{\widetilde{U}}^\mu{}_\beta = \langle\, \widetilde{\mathbf{e}}^\alpha \,,\, \mathbf{v}_\mu \,\rangle \langle\, \mathbf{u}^\mu \,,\, \widetilde{\mathbf{e}}_\beta \,\rangle = \Lambda^\alpha{}_\rho \langle\, \mathbf{e}_\rho \,,\, \mathbf{v}_\mu \,\rangle \langle\, \mathbf{u}^\mu \,,\, \mathbf{e}_\sigma \,\rangle \overline{\Lambda}^\sigma{}_\beta$, i.e., $\widetilde{T}^\alpha{}_\beta = \Lambda^\alpha{}_\mu \, T^\mu{}_\rho \, \overline{\Lambda}^\rho{}_\beta$, this being the standard transformation for the components of a contravariant–covariant tensor of $\mathbb{E}_n \otimes \mathbb{E}_n^*$ under a change of basis.

or, alternatively,

$$\boxed{\mathbf{B} = (\exp \mathbf{t})\, \mathbf{A} \quad , \quad \text{where} \quad \mathbf{t} = \log(\mathbf{V}\,\mathbf{U}^{-1}) \quad .} \tag{A.251}$$

A linear transformation is, therefore, equivalently characterized when one gives

- some basis $\mathbf{U}$ and the transformed basis $\mathbf{V}$, i.e., an ordered pair of bases $\mathbf{U}$ and $\mathbf{V}$;
- the components $T^\alpha{}_\beta$ of the tensor $\mathbf{T} = \exp \mathbf{t}$;
- the components $t^\alpha{}_\beta$ of the tensor $\mathbf{t} = \log \mathbf{T}$.

We now have an interpretation of the points of the $\mathrm{GL}(n)$ manifold:

Property A.30 *The points of the Lie group manifold* $\mathrm{GL}(n)$ *can be interpreted, via equation* (A.239), *as bases of a linear space* $\mathbb{E}_n$. *The exponential coordinates of the manifold are the "components" of the matrices. A linear transformation (of bases) is characterized as soon as an ordered pair of points of the manifold, say* $\{\mathbf{U}, \mathbf{V}\}$ *has been chosen. An ordered pair of points defines an oriented geodesic segment (from point* $\mathbf{U}$ *to point* $\mathbf{V}$*). When transporting this geodesic segment to the origin* $\mathbf{I}$ *one obtains the autovector whose components are* $\mathbf{t} = \log(\mathbf{V}\,\mathbf{U}^{-1})$. *Therefore, that transformation can be written as* $\mathbf{V} = (\exp \mathbf{t})\, \mathbf{U}$. *In particular, this transformation transforms the origin into the point* $\mathbf{T} = (\exp \mathbf{t})\, \mathbf{I} = \exp \mathbf{t}$ *so the transformation that is characterized by the pair of points* $\{\mathbf{U}, \mathbf{V}\}$ *is also characterized by the pair of points* $\{\mathbf{I}, \mathbf{T}\}$, *with* $\mathbf{T} = \exp \mathbf{t} = \mathbf{V}\,\mathbf{U}^{-1}$. *If* $\mathbf{U}$ *and* $\mathbf{V}$ *belong to the set of matrices* $\mathrm{GL}(n)$, *the matrices of the form* $\mathbf{T} = \mathbf{V}\,\mathbf{U}^{-1}$ *also belong to* $\mathrm{GL}(n)$.

The following terminologies are unambiguous: (i) 'the transformation $\{\mathbf{U}, \mathbf{V}\}$'; (ii) 'the transformation $\{\mathbf{I}, \mathbf{T}\}$', with $\mathbf{T} = \mathbf{V}\,\mathbf{U}^{-1}$; (iii) 'the transformation $\mathbf{t}$', with $\mathbf{t} = \log \mathbf{T} = \log(\mathbf{V}\,\mathbf{U}^{-1})$. By language abuse, one may also say 'the transformation $\mathbf{T}$'.

It is important to understand that the points of a Lie group manifold do not represent transformations, but bases of a linear space, the transformations being the oriented geodesic segments joining two points (when they can be geodesically connected). These oriented geodesic segments can be transported to the origin, and the set of all oriented geodesic segments at the origin forms the associative autovector space that is a local Lie group.

The composition of two transformations $\mathbf{t}_1$ and $\mathbf{t}_2$ is the geometric sum

$$\mathbf{t}_3 = \mathbf{t}_2 \oplus \mathbf{t}_1 = \log(\exp \mathbf{t}_2 \, \exp \mathbf{t}_1) \quad , \tag{A.252}$$

that can equivalently be expressed as

$$\mathbf{T}_3 = \mathbf{T}_2\, \mathbf{T}_1 \tag{A.253}$$

(where $\mathbf{T}_n = \exp \mathbf{t}_n$), this last expression being the coordinate representation of the geometric sum.

A.14 SO(3) – 3D Euclidean Rotations

A.14.1 Introduction

At a point $\mathcal{P}$ of the physical 3D space $\mathfrak{E}$ (that can be assumed Euclidean or not) consider a solid with a given 'attitude' or 'orientation' that can *rotate*, around its center of mass, so its orientation in space may change. Let us now introduce an abstract manifold $\mathfrak{O}$ each point of which represents one possible attitude of the solid. This manifold is three-dimensional (to represent the attitude of a solid one uses three angles, for instance Euler angles). Below, we identify this manifold as that associated to the Lie group SO(3), so this manifold is a metric manifold (with torsion). Two points of this manifold $\mathcal{O}_1$ and $\mathcal{O}_2$ are connected by a geodesic line, that represents the rotation transforming the orientation $\mathcal{O}_1$ into point $\mathcal{O}_2$.

Clearly, the set of all possible attitudes of a solid situated at point $\mathcal{P}$ of the physical space $\mathfrak{E}$ is identical to the set of all possible orthonormal basis (of the linear tangent space) that can be considered at point $\mathcal{P}$. From now on, then, instead of different attitudes of a solid, we may just consider different orthonormal bases of the Euclidean 3D space $\mathbb{E}_3$.

The transformation that transforms an orthonormal basis into another orthonormal basis is, by definition, a rotation. We know that a rotation has two different standard representations: (i) as a real special[26] orthogonal matrix $\mathbf{R}$, or as its logarithm, $\mathbf{r} = \log \mathbf{R}$, that, except when the rotation angle equals π (see appendix A.7 for details) is a real antisymmetric matrix.

I leave it to the reader to verify that if $\mathbf{R}$ is an orthogonal rotation operator, and if $\mathbf{r}$ is the antisymmetric tensor

$$\mathbf{r} = \log \mathbf{R} \quad , \tag{A.254}$$

then, the dual of $\mathbf{r}$,

$$\rho_i = \tfrac{1}{2}\, \epsilon_{ijk}\, r^{jk} \tag{A.255}$$

is the usual "rotation vector" (in fact, a pseudo-vector). This is easily seen by considering the eigenvalues and eigenvectors of both, $\mathbf{R}$ and $\mathbf{r}$. Let us call $\mathbf{r}$ the *rotation tensor.*

Example A.13 *In an Euclidean space with Cartesian coordinates, let ρ_i be the components of the rotation (pseudo)vector, and let r^{ij} be the components of its dual, the (antisymmetric) rotation tensor. They are related by*

$$\rho_i = \tfrac{1}{2}\, \epsilon_{ijk}\, r^{jk} \quad ; \qquad r^{ij} = \epsilon^{ijk}\, \rho_k \quad . \tag{A.256}$$

Explicitly, in an orthonormal referential,

---

[26]The determinant of an orthogonal matrix is ± 1. Special here means that only the matrices with determinant equal to $+1$ are considered.

$$\begin{pmatrix} r^{xx} & r^{xy} & r^{xz} \\ r^{yx} & r^{yy} & r^{yz} \\ r^{zx} & r^{zy} & r^{zz} \end{pmatrix} = \begin{pmatrix} 0 & \rho_z & -\rho_y \\ -\rho_z & 0 & \rho_x \\ \rho_y & -\rho_x & 0 \end{pmatrix} \quad . \tag{A.257}$$

Example A.14 *Let r^{ij} be a 3D antisymmetric tensor, and $\rho_i = \frac{1}{2!}\,\epsilon_{ijk}\,r^{jk}$ its dual. We have*

$$\|\,\mathbf{r}\,\| = \sqrt{\tfrac{1}{2}\,r_{ij}\,r^{ji}} = \sqrt{\tfrac{1}{2}\,\epsilon_{ijk}\,\epsilon^{ji\ell}\,\rho^k\,\rho_\ell} = \sqrt{-\,\rho^k\,\rho_k} = i\,\|\,\boldsymbol{\rho}\,\| \quad , \tag{A.258}$$

where $\|\,\boldsymbol{\rho}\,\|$ is the ordinary vectorial norm.[27]

Example A.15 *Using a system of Cartesian coordinates in the Euclidean space, let $\mathbf{R}$ be the orthogonal matrix $\mathbf{R} = \begin{pmatrix} \cos\theta & \sin\theta & 0 \\ -\sin\theta & \cos\theta & 0 \\ 0 & 0 & 1 \end{pmatrix}$ and let be $\mathbf{r} = \log\mathbf{R} = \begin{pmatrix} 0 & \theta & 0 \\ -\theta & 0 & 0 \\ 0 & 0 & 0 \end{pmatrix}$. Both matrices represent a rotation of angle θ "around the z axis". The angle θ may take negative values. Defining $r = |\theta|$, the eigenvalues of $\mathbf{r}$ are $\{0, -ir, +ir\}$, and the norm of $\mathbf{r}$ is $\|\,\mathbf{r}\,\| = \sqrt{\frac{1}{2}\,\mathrm{trace}\,\mathbf{r}^2} = i\,r$.*

Example A.16 *The three eigenvalues of a 3D rotation (antisymmetric) tensor $\mathbf{r}$, logarithm of the associated rotation operator $\mathbf{R}$, are $\{\lambda_1, \lambda_2, \lambda_3\} = \{0, +i\alpha, -i\alpha\}$. Then, $\|\,\mathbf{r}\,\| = \frac{1}{2}(\lambda_1^2 + \lambda_2^2 + \lambda_3^2) = i\,\alpha$. A rotation tensor $\mathbf{r}$ is a "time-like" tensor.*

It is well-known that the composition of two rotations corresponds to the product of the orthogonal operators,

$$\mathbf{R} = \mathbf{R}_2\,\mathbf{R}_1 \quad . \tag{A.259}$$

In terms of the rotation tensors, the composition of rotations clearly corresponds to the o-sum

$$\mathbf{r} = \mathbf{r}_2 \oplus \mathbf{r}_1 \equiv \log(\,\exp\mathbf{r}_2\,\exp\mathbf{r}_1\,) \quad . \tag{A.260}$$

It is only for small rotations that

$$\mathbf{r}_2 \oplus \mathbf{r}_1 \approx \mathbf{r}_2 + \mathbf{r}_1 \quad , \tag{A.261}$$

i.e., in terms of the dual (pseudo)vectors, "for small rotations, the composition of rotations is approximately equal to the sum of the rotation (pseudo)vectors". According to the terminology proposed in section 1.5, $\mathbf{r} = \log\mathbf{R}$ is a geotensor.

[27] The norm of a vector $\mathbf{v}$, denoted $\|\,\mathbf{v}\,\|$, is defined through $\|\,\mathbf{t}\,\|^2 = t_i\,t^i = t_i\,t^i = g_{ij}\,t^i\,t^j = g^{ij}\,t_i\,t_j$. This is an actual norm if the metric is elliptic, and it is a pseudo-norm if the metric is hyperbolic (like the space-time Minkowski metric).

A.14.2 Exponential of a Matrix of $\mathfrak{so}(3)$

A matrix $\mathbf{r}$ in $\mathfrak{so}(3)$, is a 3×3 antisymmetric matrix. Then,

$$\operatorname{tr} \mathbf{r} = 0 \quad ; \quad \det \mathbf{r} = 0 \ . \tag{A.262}$$

It follows from the Cayley-Hamilton theorem (see appendix A.4) that such a matrix satisfies

$$\mathbf{r}^3 = r^2 \, \mathbf{r} \qquad \text{with} \qquad r = \| \mathbf{r} \| = \sqrt{\frac{\operatorname{tr} \mathbf{r}^2}{2}} \tag{A.263}$$

(the value $\operatorname{tr} \mathbf{r}^2$ is negative, and $r = \| \mathbf{r} \|$ is imaginary). Then, one has, for any odd and any even power of $\mathbf{r}$,

$$\mathbf{r}^{2i+1} = r^{2i} \, \mathbf{r} \quad ; \quad \mathbf{r}^{2i} = r^{2i-2} \, \mathbf{r}^2 \ . \tag{A.264}$$

The exponential of $\mathbf{r}$ is $\exp \mathbf{r} = \sum_{i=0}^{\infty} \frac{1}{n!} \mathbf{r}^i$. Separating the even from the odd powers, and using equation (A.264), the exponential series can equivalently be written, for the considered matrices, as

$$\exp \mathbf{r} = \mathbf{I} + \frac{1}{r} \left(\sum_{i=0}^{\infty} \frac{r^{2i+1}}{(2i+1)!} \right) \mathbf{r} + \frac{1}{r^2} \left(\left(\sum_{i=0}^{\infty} \frac{r^{2i}}{(2i)!} \right) - 1 \right) \mathbf{r}^2 \ , \tag{A.265}$$

i.e.,[28]

$$\boxed{\exp \mathbf{r} = \mathbf{I} + \frac{\sinh r}{r} \, \mathbf{r} + \frac{\cosh r - 1}{r^2} \, \mathbf{r}^2 \quad ; \quad r = \sqrt{\frac{\operatorname{tr} \mathbf{r}^2}{2}} \ .} \tag{A.266}$$

As r is imaginary, one may introduce the (positive) real number α through

$$r = \| \mathbf{r} \| = i \alpha \ , \tag{A.267}$$

in which case one may write[29]

$$\boxed{\exp \mathbf{r} = \mathbf{I} + \frac{\sin \alpha}{\alpha} \, \mathbf{r} + \frac{1 - \cos \alpha}{\alpha^2} \, \mathbf{r}^2 \quad ; \quad \alpha = \sqrt{-\frac{\operatorname{tr} \mathbf{r}^2}{2}} \ .} \tag{A.268}$$

This result for the exponential of a "rotation vector" is known as the *Rodrigues' formula,* and seems to be more than 150 years old (Rodrigues, 1840). As it is not widely known, it is rediscovered from time to time (see, for instance, Neutsch, 1996). Observe that this exponential function is a periodic function of α, with period 2π.

---

[28]This demonstration could be simplified by remarking that $\exp \mathbf{r} = \cosh \mathbf{r} + \sinh \mathbf{r}$, and showing that $\cosh \mathbf{r} = \mathbf{I} + \frac{1-\cos\alpha}{\alpha^2} \, \mathbf{r}^2$ and $\sinh \mathbf{r} = \frac{\sin\alpha}{\alpha} \, \mathbf{r}$, this separating the exponential of a rotation vector into its symmetric and its antisymmetric part.

[29]Using $\sinh i\alpha = i \sin \alpha$ and $\cosh i\alpha = \cos \alpha$.

A.14.3 Logarithm of a Matrix of SO(3)

The expression for the logarithm $\mathbf{r} = \log \mathbf{R}$ is easily obtained solving for $\mathbf{r}$ in the expression above,[30] and gives (the principal determination of) the logarithm of an orthogonal matrix,

$$\boxed{\mathbf{r} = \log \mathbf{R} = \frac{\alpha}{\sin \alpha} \frac{1}{2} (\mathbf{R} - \mathbf{R}^*) \quad ; \quad \cos \alpha = \frac{\operatorname{trace} \mathbf{R} - 1}{2} \ .} \tag{A.269}$$

As $\mathbf{R}$ is an orthogonal matrix, $\mathbf{r} = \log \mathbf{R}$ is an antisymmetric matrix. Equivalently, using the imaginary quantity r,

$$\boxed{\mathbf{r} = \log \mathbf{R} = \frac{r}{\sinh r} \frac{1}{2} (\mathbf{R} - \mathbf{R}^*) \quad ; \quad \cosh r = \frac{\operatorname{trace} \mathbf{R} - 1}{2} \ .} \tag{A.270}$$

A.14.4 Geometric Sum

Let $\mathbf{R}$ be a rotation (i.e. an orthogonal) operator, and $\mathbf{r} = \log \mathbf{R}$, the associated (antisymmetric) geotensor. With the geometric sum defined as

$$\boxed{\mathbf{r}_2 \oplus \mathbf{r}_1 \equiv \log(\exp \mathbf{r}_2 \ \exp \mathbf{r}_1) \ ,} \tag{A.271}$$

the group operation (composition of rotations) has the two equivalent expressions

$$\mathbf{R} = \mathbf{R}_2 \, \mathbf{R}_1 \quad \Longleftrightarrow \quad \mathbf{r} = \mathbf{r}_2 \oplus \mathbf{r}_1 \ . \tag{A.272}$$

Using the expressions just obtained for the logarithm and the exponential, this gives, after some easy simplifications,

$$\boxed{\begin{aligned} \mathbf{r}_2 \oplus \mathbf{r}_1 = \frac{\alpha}{\sin(\alpha/2)} \Big(& \frac{\sin(\alpha_2/2)}{\alpha_2} \cos(\alpha_1/2) \, \mathbf{r}_2 + \cos(\alpha_2/2) \frac{\sin(\alpha_1/2)}{\alpha_1} \, \mathbf{r}_1 \\ & + \frac{\sin(\alpha_2/2)}{\alpha_2} \frac{\sin(\alpha_1/2)}{\alpha_1} \, (\mathbf{r}_2 \, \mathbf{r}_1 - \mathbf{r}_1 \, \mathbf{r}_2) \Big) \ , \end{aligned}} \tag{A.273}$$

where the norms of $\mathbf{r}_1$ and $\mathbf{r}_2$ have been written $\| \mathbf{r}_1 \| = i \, \alpha_1$ and $\| \mathbf{r}_2 \| = i \, \alpha_2$ (so α_1 and α_2 are the two rotation angles), and where the positive scalar α is given through

$$\boxed{\cos(\alpha/2) = \cos(\alpha_2/2) \, \cos(\alpha_1/2) + \frac{1}{2} \frac{\sin(\alpha_2/2)}{\alpha_2} \frac{\sin(\alpha_1/2)}{\alpha_1} \operatorname{tr}(\mathbf{r}_2 \, \mathbf{r}_1) \ .} \tag{A.274}$$

We see that the geometric sum for rotations depends on the half-angle of rotation, this being reminiscent of what happens when using quaternions

[30]Note that $\mathbf{r}^2$ is symmetric.

to represent rotations: the composition of quaternions corresponds in fact to the geometric sum for SO(3). The geometric sum operation is a more general concept, valid for any Lie group.

One could, of course, use a different definition of rotation vector, $\sigma = \log \mathbf{R}^{1/2} = \frac{1}{2}\,\mathbf{r}$, that would absorb the one-half factors in the geometric sum operation (see footnote[31]). I rather choose to stick to the rule that the o-sum operation has to be identical to the group operation, without any factors.

The two formulas (A.273)–(A.274), although fundamental for the theory of 3D rotations, are not popular. They can be found in Engø (2001) and Coll and San José (2002).

As, in a group, $\mathbf{r}_2 \ominus \mathbf{r}_1 = \mathbf{r}_2 \oplus (\text{-}\mathbf{r}_1)$, we immediately obtain the equivalent of formulas (A.273) and (A.274) for the o-difference:

$$\begin{aligned}\mathbf{r}_2 \ominus \mathbf{r}_1 \;=\; \frac{\alpha}{\sin(\alpha/2)} \Big(&\frac{\sin(\alpha_2/2)}{\alpha_2}\, \cos(\alpha_1/2)\, \mathbf{r}_2 - \cos(\alpha_2/2)\, \frac{\sin(\alpha_1/2)}{\alpha_1}\, \mathbf{r}_1 \\ &- \frac{\sin(\alpha_2/2)}{\alpha_2}\, \frac{\sin(\alpha_1/2)}{\alpha_1}\, (\mathbf{r}_2\, \mathbf{r}_1 - \mathbf{r}_1\, \mathbf{r}_2) \Big) \quad ,\end{aligned} \tag{A.275}$$

with

$$\cos(\alpha/2) \;=\; \cos(\alpha_2/2)\, \cos(\alpha_1/2) - \frac{1}{2}\, \frac{\sin(\alpha_2/2)}{\alpha_2}\, \frac{\sin(\alpha_1/2)}{\alpha_1}\, \mathrm{tr}\,(\mathbf{r}_2\, \mathbf{r}_1) \quad . \tag{A.276}$$

A.14.5 Small Rotations

From equation (A.273), valid for the composition of any two finite rotations, one easily obtains, when one of the two rotations is small, the first-order approximation

$$\mathbf{r} \oplus d\mathbf{r} \;=\; \left(1 + \left(\cos r/2\, \frac{r/2}{\sin r/2} - 1\right) \frac{\mathbf{r} \cdot d\mathbf{r}}{r^2}\right) \mathbf{r} + \cos r/2\, \frac{r/2}{\sin r/2}\, d\mathbf{r} + \tfrac{1}{2}\, \mathbf{r} \times d\mathbf{r} + \dots \tag{A.277}$$

When both rotations are small,

$$d\mathbf{r}_2 \oplus d\mathbf{r}_1 = (d\mathbf{r}_2 + d\mathbf{r}_1) + \tfrac{1}{2}\, d\mathbf{r}_2 \times d\mathbf{r}_1 + \dots \quad . \tag{A.278}$$

[31]When introducing the half-rotation geotensor $\sigma = \log \mathbf{R}^{1/2} = (1/2)\,\mathbf{r}$, whose norm $\|\, \sigma \,\| = i\beta$ is i times the half-rotation angle, $\beta = \alpha/2$, then, the group operation (composition of rotations) would correspond to the definition $\sigma_2 \oplus \sigma_1 \equiv \log(\, (\exp \sigma_2)^2\, (\exp \sigma_1)^2\,)^{1/2} = \frac{1}{2} \log(\exp 2\sigma_2\ \exp 2\sigma_1\,)$, this giving, using obvious definitions, $\sigma_2 \oplus \sigma_1 = (\beta/\sin\beta)(\,(\sin\beta_2/\beta_2)\, \cos\beta_1\, \sigma_2 + \cos\beta_2\, (\sin\beta_1/\beta_1)\, \sigma_1 + (\sin\beta_2/\beta_2)\, (\sin\beta_1/\beta_1)\, (\sigma_2\, \sigma_1 - \sigma_1\, \sigma_2)\,)$, β being characterized by $\cos\beta = \cos\beta_2\, \cos\beta_1 + (1/2)\, (\sin\beta_2/\beta_2)\, (\sin\beta_1/\beta_1)\, \mathrm{tr}\,(\sigma_2\, \sigma_1)$. The norm of $\sigma = \sigma_2 \oplus \sigma_1$ is $i\beta$.

A.14.6 Coordinates over SO(3)

The coordinates $\{x, y, z\}$ defined as

$$\mathbf{r} = \begin{pmatrix} 0 & z & -y \\ -z & 0 & x \\ y & -x & 0 \end{pmatrix} , \tag{A.279}$$

define a system of geodesic coordinates (locally Cartesian at the origin). Passing to a coordinate system $\{r, \vartheta, \varphi\}$ that is locally spherical[32] at the origin gives

$$x = r \cos\vartheta \cos\varphi \tag{A.280}$$

$$y = r \cos\vartheta \sin\varphi \tag{A.281}$$

$$z = r \sin\vartheta \quad ; \tag{A.282}$$

this shows that $\{r, \vartheta, \varphi\}$ are spherical (in fact, geographical) coordinates. The coordinates $\{x, y, z\}$ take any real value, while the spherical coordinates have the range

$$0 < r < \infty \tag{A.283}$$

$$-\pi/2 < \vartheta < \pi/2 \tag{A.284}$$

$$-\pi < \varphi < \pi \quad . \tag{A.285}$$

In spherical coordinates, the norm of $\mathbf{r}$ is

$$\| \mathbf{r} \| = \sqrt{\frac{\operatorname{tr} \mathbf{r}^2}{2}} = i\, r \quad , \tag{A.286}$$

and the eigenvalues are $\{0, \pm i\, r\}$.

One obtains

$$\mathbf{R} = \exp \mathbf{r} = \cos r\, \mathbf{U} + \sin r\, \mathbf{V} + \mathbf{W} \quad , \tag{A.287}$$

where

$$\mathbf{U} = \begin{pmatrix} \cos^2\vartheta \sin^2\varphi + \sin^2\vartheta & -\cos^2\vartheta \cos\varphi \sin\varphi & -\cos\vartheta \cos\varphi \sin\vartheta \\ -\cos^2\vartheta \cos\varphi \sin\varphi & \cos^2\vartheta \cos^2\varphi + \sin^2\vartheta & -\cos\vartheta \sin\vartheta \sin\varphi \\ -\cos\vartheta \cos\varphi \sin\vartheta & -\cos\vartheta \sin\vartheta \sin\varphi & \cos^2\vartheta \end{pmatrix} , \tag{A.288}$$

$$\mathbf{V} = \begin{pmatrix} 0 & \sin\vartheta & -\cos\vartheta \sin\varphi \\ -\sin\vartheta & 0 & \cos\vartheta \cos\varphi \\ \cos\vartheta \sin\varphi & -\cos\vartheta \cos\varphi & 0 \end{pmatrix} \tag{A.289}$$

[32]Note that I choose the latitude ϑ rather than the colatitude (i.e., spherical coordinate) θ, that would correspond to the choice $x = r \sin\theta \cos\varphi$, $y = r \sin\theta \sin\varphi$ and $z = r \cos\theta$.

and

$$\mathbf{W} = \begin{pmatrix} \cos^2\vartheta\cos^2\varphi & \cos^2\vartheta\cos\varphi\sin\varphi & \cos\vartheta\cos\varphi\sin\vartheta \\ \cos^2\vartheta\cos\varphi\sin\varphi & \cos^2\vartheta\sin^2\varphi & \cos\vartheta\sin\vartheta\sin\varphi \\ \cos\vartheta\cos\varphi\sin\vartheta & \cos\vartheta\sin\vartheta\sin\varphi & \sin^2\vartheta \end{pmatrix} \quad . \tag{A.290}$$

To obtain the inverse operator (which, in this case, equals the transpose operator), one may make the replacement $(\vartheta,\varphi) \to (-\vartheta,\varphi+\pi)$ or, equivalently, write

$$\mathbf{R}^{-1} = \mathbf{R}^* = \cos r\,\mathbf{U} - \sin r\,\mathbf{V} + \mathbf{W} \quad . \tag{A.291}$$

A.14.7 Metric

As SO(3) is a subgroup of SL(3) we can use expression (A.235) to obtain the metric. Using the parameters $\{r,\vartheta,\varphi\}$, we obtain[33]

$$-ds^2 = dr^2 + \left(\frac{\sin r/2}{r/2}\right)^2 r^2\left(d\vartheta^2 + \cos^2\vartheta\,d\varphi^2\right) \quad . \tag{A.292}$$

Note that the metric is negative definite. The associated volume density is

$$\sqrt{\det \mathbf{g}} = i\left(\frac{\sin r/2}{r/2}\right)^2 r^2\cos\vartheta \quad . \tag{A.293}$$

For small r,

$$-ds^2 \approx dr^2 + r^2(d\vartheta^2 + \cos^2\vartheta\,d\varphi^2) = dx^2 + dy^2 + dz^2 \quad . \tag{A.294}$$

The reader may easily demonstrate that the *distance* between two rotations $\mathbf{r}_1$ and $\mathbf{r}_2$ satisfies the following properties:

- Property 1: The distance between two rotations is the angle of the relative rotation.[34]
- Property 2: The distance between two rotations $\mathbf{r}_1$ and $\mathbf{r}_2$ is $D = \| \mathbf{r}_2 \ominus \mathbf{r}_1 \|$.

A.14.8 Ricci

A direct computation of the Ricci from the expression (A.292) for the metric gives[35]

$$C_{ij} = \tfrac{1}{2}\,g_{ij} \quad . \tag{A.295}$$

Note that the formula (A.237) does not apply here, as we are not in SL(3), but in the subgroup SO(3).

[33] Or, using the more general expression for the metric, with the arbitrary constants χ and ψ, $ds^2 = -2\chi\,(dr^2 + (\sin r/2/r/2)^2\,r^2\,(d\vartheta^2 + \cos^2\vartheta\,d\varphi^2))$.

[34] The angle of rotation is, by definition, a positive quantity, because of the screw-driver rule.

[35] Note: say somewhere that, as SO(3) is three-dimensional, the indices $\{A,B,C,\dots\}$ can be identified to the indices $\{i,j,k,\dots\}$.

A.14.9 Torsion

Using equation (A.238) one obtains

$$T_{ijk} = \frac{i}{2}\,\epsilon_{ijk} \quad , \tag{A.296}$$

with the definition $\epsilon_{123} = \sqrt{\det \mathbf{g}}$ (the volume density is given in equation (A.293)).

A.14.10 Geodesics

The general equation of a geodesic is

$$\frac{d^2x^i}{ds^2} + \{^i{}_{jk}\}\,\frac{dx^j}{ds}\,\frac{dx^k}{ds} = 0 \quad , \tag{A.297}$$

and this gives

$$\begin{aligned}
\frac{d^2r}{ds^2} - \sin r\left(\left(\frac{d\vartheta}{ds}\right)^2 + \cos^2\vartheta\left(\frac{d\varphi}{ds}\right)^2\right) &= 0 \\
\frac{d^2\vartheta}{ds^2} + \operatorname{cotg}\frac{r}{2}\,\frac{dr}{ds}\,\frac{d\vartheta}{ds} + \sin\vartheta\,\cos\vartheta\left(\frac{d\varphi}{ds}\right)^2 &= 0 \\
\frac{d^2\varphi}{ds^2} + \operatorname{cotg}\frac{r}{2}\,\frac{dr}{ds}\,\frac{d\varphi}{ds} - 2\,\tan\vartheta\,\frac{d\vartheta}{ds}\,\frac{d\varphi}{ds} &= 0 \quad .
\end{aligned} \tag{A.298}$$

Figure A.5 displays some of the geodesics defined by this differential system.

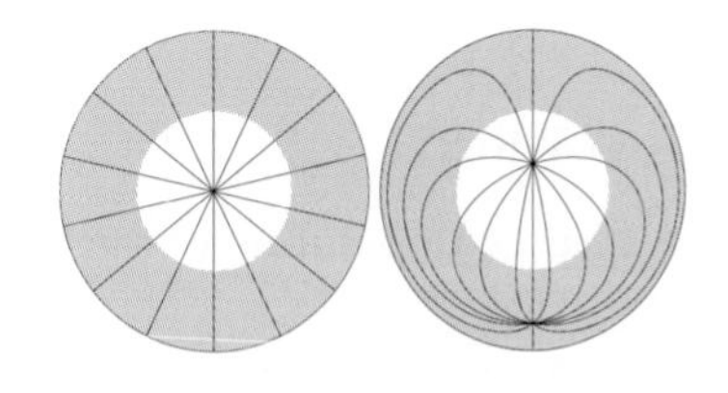

Fig. A.5. *The geodesics defined by the differential system (A.298) give, in the coordinates $\{r, \vartheta\}$ (plotted as polar coordinates, as in figure A.7) curves that are the meridians of an azimuthal equidistant geographical projection. The geodesics at the right are obtained when starting with $\varphi = 0$ and $d\varphi/ds = 0$, so that φ identically vanishes. The shadowed region correspond to the half of the spherical surface not belonging to the SO(3) manifold.*

A.14.11 Pictorial Representation

What is, geometrically, a 3D space of constant, positive curvature, with radius of curvature $R = 2$? As our immediate intuition easily grasps the notion of a 2D curved surface inside a 3D Euclidean space, we may just remark that any 2D geodesic section of our abstract space of orientations will be geometrically equivalent to the 2D surface of an ordinary sphere of radius $R = 2$ in an Euclidean 3D space (see figure A.6).

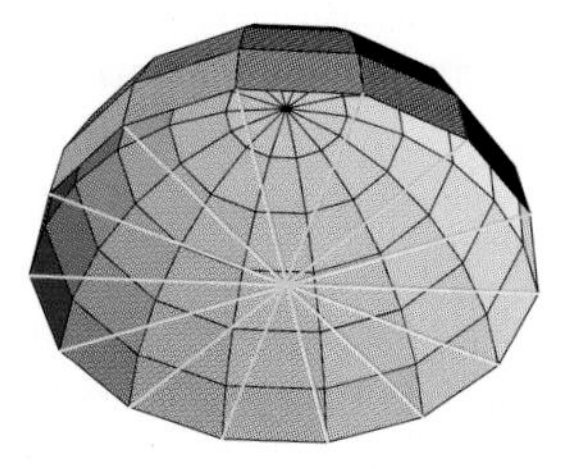

Fig. A.6. *Any two-dimensional (geodesic) section of the three-dimensional manifold SO(3) is, geometrically, one-half of the surface of an ordinary 3D sphere, with antipodal points in the "equator" identified two by two. The figure sketches a bottom view of such an object, the identification of points being suggested by some diameters.*

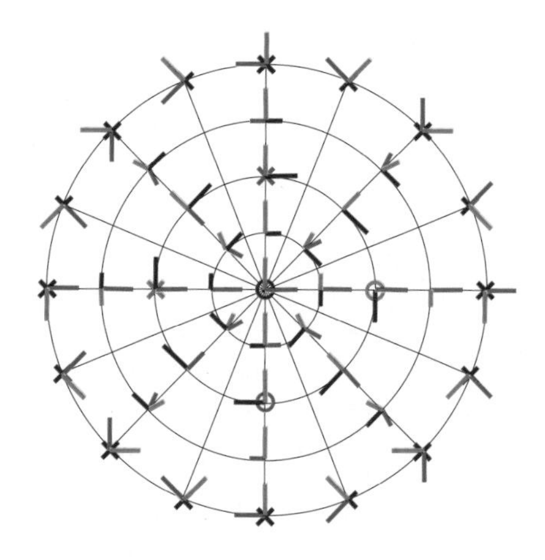

Fig. A.7. *A 2D geodesic section of the 3D (curved) space of the possible orientations of a referential. This is a 2D space of constant curvature, with radius of curvature $R = 2$, geometrically equivalent to one-half the surface of an ordinary sphere (illustrated in figure A.6). The "flat representation" used here is analogous to an azimuthal equidistant projection (see figure A.5). Any two points of the surface may be connected by a geodesic (the rotation leading from one orientation to the other), and the composition of rotations corresponds to the sum of geodesics.*

A flat view of such a 2D surface is represented in figure A.7. As each point of our abstract space corresponds to a possible orientation of a referential, I have suggested, in the figure, using a perspective view, the orientation associated to each point. As this is a 2D section of our space, one degree of freedom has been blocked: all the orientations of the figure can be obtained from the orientation at the center by a rotation "with horizontal axis". The border of the disk represented corresponds to the point antipodal to that at the center of the representation: as the radius of the sphere is $R = 2$, to travel from one point to the antipodal point one must travel a distance πR, i.e., 2π. This corresponds to the fact that rotating a referential round any axis by the angle 2π gives the original orientation.

The space of orientations is a space of constant curvature, with radius of curvature $R = 2$. The geodesic joining two orientations represents the (progressive) rotation around a given axis that transforms one orientation into another, and the sum of two geodesics corresponds to the composition of rotations. Such a sum of geodesics can be performed, geometrically, using spherical triangles. More practically, the sum of geodesics can be performed algebraically: if the two rotations are represented by two rotation operators $\mathbf{R}_1$ and $\mathbf{R}_2$, by the product $\mathbf{R}_2 \cdot \mathbf{R}_1$, and, if the two rotations are represented by the two rotation 'vectors' $\mathbf{r}_1$ and $\mathbf{r}_2$ (logarithms of $\mathbf{R}_1$ and $\mathbf{R}_2$), by the noncommutative sum $\mathbf{r}_2 \oplus \mathbf{r}_1$ defined above.

This remark unveils the true nature of a "rotation vector": it is not an element of a linear space, but a geodesic of a curved space. This explains, in particular, why it does not make any sense to define a commutative sum of two rotation "vectors", as the sum is only commutative in flat spaces. Of course, for small rotations, we have small geodesics, and the sum can approximately be performed in the tangent linear space: this is why the composition of small rotation is, approximately, commutative. In fact, in the limit when $\alpha \to 0$, the metric (A.292) becomes (A.294), that is the expression for an ordinary vector.

A.14.12 The Cardan-Brauer Angles

Although the Euler angles are quite universally used, it is sometimes better to choose an X-Y-Z basis than an Z-X-Z basis. When a 3D rotation is defined by rotating around the axis X first, by an angle θ_x, then around the axis Y, by an angle θ_y and, finally, around the axis Z, by an angle θ_z, the three angles $\{\theta_x, \theta_y, \theta_z\}$ are sometimes called the Cardan angles. Srinivasa Rao, in his book about the representation of the rotation and the Lorentz groups mentions that this "resolution" of a rotation is due to Brauer. Let us call these angles the Cardan-Brauer angles.

Example A.17 SO(3) *When parameterizing a rotation using the three Cardan-Brauer angles, defined by performing a rotation around each of three orthogonal axes,* $\mathbf{R} = \mathbf{R}_z(\gamma)\,\mathbf{R}_y(\beta)\,\mathbf{R}_x(\alpha)$, *one obtains the distance element*

$$-ds^2 \;=\; d\alpha^2 + d\beta^2 + d\gamma^2 - 2\,\sin\beta\, d\alpha\, d\gamma \quad . \tag{A.299}$$

When parameterizing a rotation using the three Euler angles, $\mathbf{R} = \mathbf{R}_x(\gamma)\,\mathbf{R}_y(\beta)\,\mathbf{R}_x(\alpha)$ *one obtains*

$$-ds^2 \;=\; d\alpha^2 + d\beta^2 + d\gamma^2 + 2\,\cos\beta\, d\alpha\, d\gamma \quad . \tag{A.300}$$

As a final example, writing $\mathbf{R} = \begin{pmatrix} 0 & z & -y \\ -z & 0 & x \\ y & -x & 0 \end{pmatrix}$ *with* $x = a\cos\chi\cos\varphi$, $y = a\cos\chi\sin\varphi$ *and* $z = a\sin\chi$, *gives*

$$-ds^2 \;=\; da^2 + \left(\frac{\sin a/2}{a/2}\right)^2 a^2\,(d\chi^2 + \cos^2\chi\, d\varphi^2) \quad . \tag{A.301}$$

A.15 SO(3,1) – Lorentz Transformations

In this appendix a few basic considerations are made on the Lorentz group SO(3,1). The expansions exposed here are much less complete that those presented for the Lie group GL(2) (section 1.4.6) of for the rotation group SO(3) (section A.14).

A.15.1 Preliminaries

In the four-dimensional space-time of special relativity, assume for the metric $g_{\alpha\beta}$ the signature $(-,+,+,+)$. As usual, we shall denote by $\epsilon_{\alpha\beta\gamma\delta}$ the Levi-Civita totally antisymmetric tensor, with $\epsilon_{0123} = \sqrt{-\det \mathbf{g}}$. The dual of a tensor $\mathbf{t}$ is defined, for instance, through $t^*_{\alpha\beta} = \frac{1}{2}\,\epsilon_{\alpha\beta\gamma\delta}\,t^{\gamma\delta}$.

To fix ideas, let us start by considering a system of Minkowskian coordinates $\{x^\alpha\} = \{x^0, x^1, x^2, x^3\}$ (i.e., one Newtonian time coordinate and three spatial Cartesian coordinates). Then $g_{\alpha\beta} = \text{diagonal}(-1,+1,+1,+1)$, and $\epsilon_{0123} = 1$.

As usual in special relativity, consider that from this referential we observe another referential, and that the two referentials have coincident space-time origins. The second referential may then be described by its velocity and by the rotation necessary to make coincident the two spatial referentials. The rotation is characterized by the rotation "vector"

$$\mathbf{r} = \{r_x, r_y, r_z\} \quad . \tag{A.302}$$

Pure space rotations have been analyzed in section A.14, where we have seen in which sense $\mathbf{r}$ is an autovector. To characterize the velocity of the referential we can use any of the three colinear vectors $\mathbf{v} = \{v_x, v_y, v_z\}$, $\boldsymbol{\beta} = \{\beta_x, \beta_y, \beta_z\}$ or $\boldsymbol{\psi} = \{\psi_x, \psi_y, \psi_z\}$, where, v, β and ψ, the norms of the three vectors, are related by

$$\tanh \psi = \beta = \frac{v}{c} \ . \tag{A.303}$$

The *celerity* vector $\boldsymbol{\psi}$ is of special interest for us, as the 'relativistic sum' of colinear velocities, $\beta = \frac{\beta_1+\beta_2}{1+\beta_2\,\beta_2}$ simply corresponds to $\psi = \psi_1 + \psi_2$: the "vector"

$$\boldsymbol{\psi} = \{\psi_x, \psi_y, \psi_z\} \tag{A.304}$$

is in fact, an autovector (recall that the geometric sum of two colinear autovectors equals their sum).

From the 3D rotation vector $\mathbf{r}$ and the 3D velocity vector $\boldsymbol{\psi}$ we can form the 4D antisymmetric tensor $\boldsymbol{\Lambda}$ whose covariant and mixed components are, respectively,

$$\{\lambda_{\alpha\beta}\} = \begin{pmatrix} 0 & -\psi_x & -\psi_y & -\psi_z \\ \psi_x & 0 & r_z & -r_y \\ \psi_y & -r_z & 0 & r_x \\ \psi_z & r_y & -r_x & 0 \end{pmatrix} \quad ; \qquad \{\lambda^\alpha{}_\beta\} = \begin{pmatrix} 0 & \psi_x & \psi_y & \psi_z \\ \psi_x & 0 & r_z & -r_y \\ \psi_y & -r_z & 0 & r_x \\ \psi_z & r_y & -r_x & 0 \end{pmatrix} \quad . \tag{A.305}$$

The *Lorentz transformation* associated to this *Lorentz autovector* simply is

$$\Lambda = \exp \lambda \quad . \tag{A.306}$$

Remember that it is the contravariant–covariant version $\lambda^\alpha{}_\beta$ that must appear in the series expansion defining the exponential of the autovector λ.

A.15.2 The Exponential of a Lorentz Geotensor

In a series of papers, Coll and San José (1990, 2002) and Coll (2002), give the exponential of a tensor in $\mathfrak{so}(3,1)$ the logarithm of a tensor in SO(3,1) and a finite expression for the BCH operation (the geometric sum of two autovectors of $\mathfrak{so}(3,1)_0$, in our terminology). This work is a good example of seriously taking into account the log-exp mapping in a Lie group of fundamental importance for physics.

The exponential of an element λ in the algebra of the Lorentz group is found to be, using arbitrary space-time coordinates (Coll and San José, 1990),

$$\exp \lambda = p\,\mathbf{I} + q\,\lambda + r\,\lambda^* + s\,\mathbf{T} \quad , \tag{A.307}$$

where $\mathbf{I}$ is the identity tensor (the metric, if covariant–covariant components are used), λ^* is the dual of λ, $\lambda^*_{\alpha\beta} = \frac{1}{2}\,\epsilon_{\alpha\beta\gamma\delta}\,\lambda^{\gamma\delta}$, $\mathbf{T}$ is the stress-energy tensor

$$\mathbf{T} = \tfrac{1}{2}\,(\lambda^2 + (\lambda^*)^2) \quad , \tag{A.308}$$

and where the four real numbers p, q, r, s are defined as

$$p = \frac{\cosh\alpha + \cos\beta}{2} \quad ; \quad q = \frac{\alpha\sinh\alpha + \beta\sin\beta}{\alpha^2+\beta^2}$$
$$r = \frac{\alpha\sin\beta - \beta\sinh\alpha}{\alpha^2+\beta^2} \quad ; \quad s = \frac{\cosh\alpha - \cos\beta}{\alpha^2+\beta^2} \quad , \tag{A.309}$$

where the two nonnegative real numbers α, β are defined by writing the four eigenvalues of λ under the form $\{\pm\alpha, \pm i\beta\}$. Alternatively, these two real numbers can be obtained by solving the system $2(\alpha^2 - \beta^2) = \operatorname{tr}\lambda^2$, $-4\,\alpha\,\beta = \operatorname{tr}(\lambda\,\lambda^*)$.

Example A.18 Special Lorentz Transformation. *When using Minkowskian coordinates, the Lorentz autovector is that in equation* (A.305), *the dual is easy to* *obtain,*[36] *as is the stress energy tensor* $\mathbf{T}$. *When the velocity* $\boldsymbol{\psi}$ *is aligned along the x axis,* $\boldsymbol{\psi} = \{\psi_x, 0, 0\}$, *the Lorentz transformation* $\Lambda = \exp \boldsymbol{\lambda}$, *as given by equation* (A.307), *is*

$$\{\Lambda^\alpha{}_\beta\} = \begin{pmatrix} \cosh\psi & \sinh\psi & 0 & 0 \\ \sinh\psi & \cosh\psi & 0 & 0 \\ 0 & 0 & 1 & 0 \\ 0 & 0 & 0 & 1 \end{pmatrix} \quad . \tag{A.310}$$

Example A.19 Space Rotation. *When the two referentials are relatively at rest, they only may differ by a relative rotation. Taking the z axis as axis of rotation, the Lorentz transformation* $\Lambda = \exp \boldsymbol{\lambda}$, *as given by equation* (A.307), *is the 4D version of a standard 3D rotation operator:*

$$\{\Lambda^\alpha{}_\beta\} = \exp\begin{pmatrix} 0 & 0 & 0 & 0 \\ 0 & 0 & \varphi & 0 \\ 0 & -\varphi & 0 & 0 \\ 0 & 0 & 0 & 0 \end{pmatrix} = \begin{pmatrix} 1 & 0 & 0 & 0 \\ 0 & \cos\varphi & \sin\varphi & 0 \\ 0 & -\sin\varphi & \cos\varphi & 0 \\ 0 & 0 & 0 & 1 \end{pmatrix} \quad . \tag{A.311}$$

A.15.3 The Logarithm of a Lorentz Transformation

Reciprocally, let Λ be a Lorentz transformation. Its logarithm is found to be (Coll and San José, 1990)

$$\log \Lambda = p\,\widetilde{\Lambda} + \epsilon\, q\, \widetilde{\Lambda}^* \quad , \tag{A.312}$$

where the antisymmetric part of Λ, is introduced by $\widetilde{\Lambda} = \frac{1}{2}(\Lambda - \Lambda^t)$, and where

$$p = \frac{\sqrt{\nu_+^2 - 1}\,\operatorname{arccosh}\nu_+ + \sqrt{1-\nu_-^2}\,\arccos\nu_-}{\nu_+^2 - \nu_-^2}$$
$$q = \frac{\sqrt{\nu_+^2 - 1}\,\operatorname{arccosh}\nu_+ + \sqrt{1-\nu_-^2}\,\arccos\nu_-}{\nu_+^2 - \nu_-^2} \tag{A.313}$$

where the scalars $\nu_\pm$ are the invariants

$$\nu_\pm = \tfrac{1}{4}\left(\operatorname{tr}\Lambda \pm \sqrt{2\operatorname{tr}\Lambda^2 - \operatorname{tr}^2\Lambda + 8}\right) \quad . \tag{A.314}$$

[36]One has $\lambda^*_{\alpha\beta} = \begin{pmatrix} 0 & r_x & r_y & r_z \\ -r_x & 0 & -\psi_z & \psi_y \\ -r_y & \psi_z & 0 & -\psi_x \\ -r_z & -\psi_y & \psi_x & 0 \end{pmatrix}$.

A.15.4 The Geometric Sum

Let $\boldsymbol{\lambda}$ and $\boldsymbol{\mu}$ be two Lorentz geotensors. We have called geometric sum the operation

$$\boldsymbol{\nu} = \boldsymbol{\mu} \oplus \boldsymbol{\lambda} \equiv \log(\exp \boldsymbol{\mu} \, \exp \boldsymbol{\lambda}) \quad , \tag{A.315}$$

that is, at least locally, a representation of the group operation (i.e., the composition of the two Lorentz transformations $\boldsymbol{\Lambda} = \exp \boldsymbol{\lambda}$ and $\mathbf{M} = \exp \boldsymbol{\mu}$). Coll and San José (2002) analyze this operation exactly. Its result is better expressed through a 'complexification' of the Lorentz group. Instead of the Lorentz geotensors $\boldsymbol{\lambda}$ and $\boldsymbol{\mu}$ consider

$$\mathbf{a} = \boldsymbol{\lambda} - i\,\boldsymbol{\lambda}^* \quad ; \quad \mathbf{b} = \boldsymbol{\mu} - i\,\boldsymbol{\mu}^* \quad ; \quad \mathbf{c} = \boldsymbol{\nu} - i\,\boldsymbol{\nu}^* \quad . \tag{A.316}$$

Then, for $\mathbf{c} = \mathbf{b} \oplus \mathbf{a}$ one obtains

$$\mathbf{c} = \frac{\sinh c}{c} \left(\frac{\sinh b}{b} \cosh a \; \mathbf{b} + \cosh b \, \frac{\sinh a}{a} \, \mathbf{a} + \frac{\sinh b}{b} \, \frac{\sinh a}{a} \, (\mathbf{b}\,\mathbf{a} - \mathbf{a}\,\mathbf{b}) \right) \quad , \tag{A.317}$$

where the scalar c is defined through

$$\cosh c = \cosh b \cosh a + \frac{1}{2} \frac{\sinh b}{b} \, \frac{\sinh a}{a} \, \mathrm{tr}\,(\mathbf{b}\,\mathbf{a}) \quad . \tag{A.318}$$

The reader may note the formal identity between these two equations and equations (1.178) and (1.179) expressing the o-sum in $\mathfrak{sl}(2)$.

A.15.5 Metric in the Group Manifold

In the 6D manifold $SO(3,1)$, let us choose the coordinates

$$\{x^1, x^2, x^3, x^4, x^5, x^6\} = \{\psi_x, \psi_y, \psi_z, r_x, r_y, r_z\} \quad . \tag{A.319}$$

The goal of this section is to obtain an expression for the metric tensor in these coordinates. First, note that as as a Lorentz geotensor is traceless, its norm, as defined by the universal metric (see the main text), simplifies here to (choosing $\chi = 1/2$)

$$\| \boldsymbol{\lambda} \| = \sqrt{\frac{\mathrm{tr}\, \boldsymbol{\lambda}^2}{2}} = \sqrt{\frac{\lambda^\alpha{}_\beta \, \lambda^\beta{}_\alpha}{2}} \quad . \tag{A.320}$$

Obtaining the expression of the metric at the origin is trivial, as the ds^2 at the origin simply corresponds to the squared norm of the infinitesimal autovector

$$d\boldsymbol{\lambda} = \{d\lambda^\alpha{}_\beta\} = \begin{pmatrix} 0 & d\psi_x & d\psi_y & d\psi_z \\ d\psi_x & 0 & dr_z & \text{-}dr_y \\ d\psi_y & \text{-}dr_z & 0 & dr_x \\ d\psi_z & dr_y & \text{-}dr_x & 0 \end{pmatrix} \quad . \tag{A.321}$$

This gives

$$ds^2 = d\lambda_x^2 + d\lambda_y^2 + d\lambda_z^2 - dr_x^2 - dr_y^2 - dr_z^2 \quad . \tag{A.322}$$

We see that we have a six-dimensional Minkowskian space, with three space-like dimensions and three time-like dimensions. The coordinates $\{\psi_x, \psi_y, \psi_z, r_x, r_y, r_z\}$ are, in an infinitesimal neighborhood of the origin, Cartesian-like.

A.15.6 The Fundamental Operations

Consider three Galilean referentials $\mathcal{G}_1$, $\mathcal{G}_2$ and $\mathcal{G}_3$. We know that if Λ_{21} is the space-time rotation (i.e., Lorentz transformation) transforming $\mathcal{G}_1$ into $\mathcal{G}_2$, and if Λ_{32} is the space-time rotation transforming $\mathcal{G}_2$ into $\mathcal{G}_3$, the space-time rotation transforming $\mathcal{G}_1$ into $\mathcal{G}_3$ is

$$\Lambda_{31} = \Lambda_{32} \cdot \Lambda_{21} \; . \tag{A.323}$$

Equivalently, we have

$$\Lambda_{32} = \Lambda_{31} \, / \, \Lambda_{21} \; , \tag{A.324}$$

where, as usual, $\mathbf{A}/\mathbf{B}$ means $\mathbf{A} \cdot \mathbf{B}^{-1}$.

So much for the Lorentz operators. What about the relative velocities and the relative rotations between the referentials? As velocities and rotations are described by the (antisymmetric) tensor $\lambda = \log \Lambda$, we just need to rewrite equations (A.323)–(A.324) using the logarithms of the Lorentz transformation. This gives

$$\lambda_{31} = \lambda_{32} \oplus \lambda_{21} \qquad ; \qquad \lambda_{32} = \lambda_{31} \ominus \lambda_{21} \; , \tag{A.325}$$

where the operations $\oplus$ and $\ominus$ are defined, as usual, by

$$\lambda_A \oplus \lambda_B = \log(\exp \lambda_A \cdot \exp \lambda_B) \tag{A.326}$$

and

$$\lambda_A \ominus \lambda_B = \log(\exp \lambda_A \, / \, \exp \lambda_B) \; . \tag{A.327}$$

A numerical implementation of these formulas may simply use the series expansion of the logarithm and of the exponential of a tensor, of the Jordan decomposition. Analytic expansions may use the results of Coll and San José (1990) for the exponential of a 4D antisymmetric tensor.

A.15.7 The Metric in the Velocity Space

Let us focus in the special Lorentz transformation, i.e., in the case where the rotation vector is zero:

$$\boldsymbol{\lambda} = \{\lambda^\alpha{}_\beta\} = \begin{pmatrix} 0 & \psi_x & \psi_y & \psi_z \\ \psi_x & 0 & 0 & 0 \\ \psi_y & 0 & 0 & 0 \\ \psi_z & 0 & 0 & 0 \end{pmatrix} . \tag{A.328}$$

Let, with respect to a given referential, denoted '0', be a first referential with celerity $\boldsymbol{\lambda}_{10}$ and a second referential with celerity $\boldsymbol{\lambda}_{20}$. The relative celerity of the second referential with respect to the first, $\boldsymbol{\lambda}_{21}$, is

$$\boldsymbol{\lambda}_{21} = \boldsymbol{\lambda}_{20} \ominus \boldsymbol{\lambda}_{10} . \tag{A.329}$$

Taking the norm in equation (A.329) defines the distance between two celerities, and that distance is the unique one that is invariant under Lorentz transformations:

$$D(\boldsymbol{\psi}_2, \boldsymbol{\psi}_1) = \| \boldsymbol{\lambda}_{21} \| = \| \boldsymbol{\lambda}_{20} \ominus \boldsymbol{\lambda}_{10} \| . \tag{A.330}$$

Here,

$$\| \boldsymbol{\lambda} \| = \sqrt{\frac{\operatorname{tr} \boldsymbol{\lambda}^2}{2}} = \sqrt{\frac{\lambda^\alpha{}_\beta \, \lambda^\beta{}_\alpha}{2}} . \tag{A.331}$$

Let us now parameterize the 'celerity vector' $\boldsymbol{\psi}$ not by its components $\{\psi_x, \psi_y, \psi_z\}$, but by its modulus ψ and two spherical angles θ and φ defining its orientation. We write $\boldsymbol{\psi} = \{\psi, \theta, \varphi\}$. The distance element ds between the celerity $\{\psi, \theta, \varphi\}$ and the celerity $\{\psi + d\psi, \theta + d\theta, \varphi + d\varphi\}$ is obtained by developing expression (A.330) (using the definition (A.327)) up to the second order:

$$ds^2 = d\psi^2 + \sinh^2\psi \, (d\theta^2 + \sin^2\theta \, d\varphi^2) . \tag{A.332}$$

Evrard (1995) was interested in applying to cosmology some of the conceptual tools of Bayesian probability theory. He, first, demonstrated that the probability distribution represented by the probability density $f(\beta, \theta, \varphi) = \beta^2 \sin\theta \,/\, (1 - \beta^2)^2$ is 'noninformative' (homogeneous, we would say), and, second, he demonstrated that the metric (A.332) (with the change of variables $\tanh\psi = \beta$) is the only (isotropic) one leading to the volume element $dV = (\beta^2 \sin\theta / (1 - \beta^2)^2) \, d\beta \, d\theta \, d\varphi$, from which follows the homogeneous property of the probability distribution $f(\beta, \theta, \varphi)$. To my knowledge, this was the first instance when the metric defined by equation (A.332) was considered.

A.16 Coordinates over SL(2)

A matrix of SL(2) can always be written

$$\mathbf{M} = \begin{pmatrix} a + b & c - d \\ c + d & a - b \end{pmatrix} \tag{A.333}$$

with the constraint

$$\det \mathbf{M} = (a^2 + d^2) - (b^2 + c^2) = 1 \quad . \tag{A.334}$$

As, necessarily, $(a^2 + d^2) \geq 1$, one can always introduce a positive real number e such that one has

$$a^2 + d^2 = \cosh^2 e \quad ; \quad b^2 + c^2 = \sinh^2 e \quad . \tag{A.335}$$

The condition $\det \mathbf{M} = 1$ is then automatically satisfied. Given the two equations (A.335), one can always introduce a circular angle α such that

$$a = \cosh e \cos \alpha \quad ; \quad d = \cosh e \sin \alpha \quad , \tag{A.336}$$

and a circular angle φ such that

$$b = \sinh e \sin \varphi \quad ; \quad c = \sinh e \cos \varphi \quad . \tag{A.337}$$

This is equation (1.181), except for an overall factor $\exp \kappa$ passing from a matrix of SL(2) to a matrix of $GL^+(2)$. It is easy to solve the equations above, to obtain the parameters $\{e, \varphi, \alpha\}$ as a function of the parameters $\{a, b, c, d\}$:

$$e = \operatorname{arccosh} \sqrt{a^2 + d^2} = \operatorname{arcsinh} \sqrt{b^2 + c^2} \tag{A.338}$$

and

$$\alpha = \arcsin \frac{d}{\sqrt{a^2 + d^2}} \quad ; \quad \varphi = \arcsin \frac{b}{\sqrt{b^2 + c^2}} \quad . \tag{A.339}$$

When passing from a matrix of SL(2) to a matrix of $GL^+(2)$, one needs to account for the determinant of the matrix. As the determinant is positive, one can introduce

$$\kappa = \tfrac{1}{2} \log \det \mathbf{M} \quad , \tag{A.340}$$

This now gives exactly equation (1.181).

When introducing $\mathbf{m} = \log \mathbf{M}$ (equation 1.184), one can also write

$$\kappa = \tfrac{1}{2} \operatorname{tr} \log \mathbf{M} = \tfrac{1}{2} \operatorname{tr} \mathbf{m} \quad . \tag{A.341}$$

A.17 Autoparallel Interpolation Between Two Points

A musician remarks that the pitch of a given key of her/his piano depends on the fact that the weather is cold or hot. She/he measures the pitch on a very cold day and on a very hot day, and wishes to interpolate to obtain the pitch on another day. How is the interpolation to be done?

In the 'pitch space' or 'grave–acute space' $\mathfrak{P}$, the frequency ν or the period $\tau = 1/\nu$ can equivalently be used as a coordinate to position a musical note. There is no physical argument suggesting we define over the grave–acute space any distance other than the usual musical distance (in octaves) that is (proportional to)

$$D_{\mathfrak{P}} = | \log \frac{\nu_2}{\nu_1} | = | \log \frac{\tau_2}{\tau_1} | \quad . \tag{A.342}$$

In the cold–hot space $\mathfrak{C}/\mathfrak{H}$ one may choose to use the temperature T of the thermodynamic parameter $\beta = 1/kT$. The distance between two points is (equation 3.30)

$$D_{\mathfrak{C}/\mathfrak{H}} = | \log \frac{T_2}{T_1} | = | \log \frac{\beta_2}{\beta_1} | \quad . \tag{A.343}$$

Let us first solve the problem using frequency ν and temperature T. It is not difficult to see[37] that the autoparallel mapping passing through the two points $\{T_1, \nu_1\}$ and $\{T_2, \nu_2\}$ is the mapping $T \mapsto \nu(T)$ defined by the expression

$$\nu / \overline{\nu} = (T / \overline{T})^\alpha \quad , \tag{A.344}$$

where $\alpha = \log(\nu_2/\nu_1) / \log(T_2/T_1)$, where $\overline{T}$ is the temperature coordinate of the point at the center of the interval $\{T_1, T_2\}$, $\overline{T} = \sqrt{T_1\, T_2}$, and where $\overline{\nu}$ is the frequency coordinate of the point at the center of the interval $\{\nu_1, \nu_2\}$, $\overline{\nu} = \sqrt{\nu_1\, \nu_2}$.

If, for instance, instead of frequency one had used the period as coordinate over the grave–acute space, the solution would have been

$$\tau / \overline{\tau} = (T / \overline{T})^\gamma \quad , \tag{A.345}$$

with $\overline{\tau} = \sqrt{\tau_1\, \tau_2} = 1/\overline{\nu}$ and $\gamma = \log(\tau_2/\tau_1) / \log(T_2/T_1) = -\alpha$.

Of course, the two equations (A.344) and (A.345) define exactly the same (geodesic) mapping between the cold–hot space and the grave–acute space. Calling this relation "geodesic" rather than "linear" is just to avoid misunderstandings with the usual relations called "linear", which are just formally linear in the coordinates being used.

A.18 Trajectory on a Lie Group Manifold

A.18.1 Declinative

Consider a one-dimensional metric manifold, with a coordinate t that is assumed to be metric (the distance between point t_1 and point t_2 is $|t_2 - t_1|$).

[37]For instance, one may introduce the logarithmic frequency and the logarithmic temperature, in which case the geodesic interpolation is just the formally linear interpolation.

Also consider a multiplicative group of matrices $\mathbf{M}_1, \mathbf{M}_2 \ldots$ We know that the matrix $\mathbf{m} = \log \mathbf{M}$ can be interpreted as the oriented geodesic segment from point $\mathbf{I}$ to point $\mathbf{M}$. The group operation can equivalently be represented by the matrix product $\mathbf{M}_2 \mathbf{M}_1$ or by the geometric sum $\mathbf{m}_2 \oplus \mathbf{m}_1 = \log(\exp \mathbf{m}_2 \ \exp \mathbf{m}_1)$. A 'trajectory' on the Lie group manifold is a mapping that can equivalently be represented by the mapping

$$t \mapsto \mathbf{M}(t) \tag{A.346}$$

or the mapping

$$t \mapsto \mathbf{m}(t) \quad . \tag{A.347}$$

As explained in example 2.5 (page 96) the declinative of such a mapping is given by any of the two equivalent expressions

$$\mu(t) = \lim_{t' \to t} \frac{\mathbf{m}(t') \ominus \mathbf{m}(t)}{t' - t} = \lim_{t' \to t} \frac{\log(\mathbf{M}(t')\, \mathbf{M}(t)^{-1})}{t' - t} \quad . \tag{A.348}$$

The declinative belongs to the linear space tangent to the group at its origin (the point $\mathbf{I}$).

A.18.2 Geometric Integral

Let $\mathbf{w}(t)$ be a "time dependent" vector of the linear space tangent to the group at its origin. For any value Δt, the vector $\mathbf{w}(t)\, \Delta t$ can either be interpreted as a vector of the linear tangent space or as an oriented geodesic segment of the manifold (with origin at the origin of the manifold). For any t and any t', both of the expressions

$$\mathbf{w}(t)\, \Delta t + \mathbf{w}(t')\, \Delta t = (\mathbf{w}(t) + \mathbf{w}(t'))\, \Delta t \tag{A.349}$$

and

$$\mathbf{w}(t)\, \Delta t \oplus \mathbf{w}(t')\, \Delta t = \log(\ \exp(\mathbf{w}(t)\, \Delta t)\ \exp(\mathbf{w}(t')\, \Delta t)\ ) \tag{A.350}$$

make sense. Using the geometric sum $\oplus$, let us introduce the *geometric integral*

$$\begin{aligned} &\int_{t_1}^{t_2} dt\ \mathbf{w}(t) = \\ &\lim_{\Delta t \to 0} \mathbf{w}(t_2)\, \Delta t \oplus \mathbf{w}(t_2 - \Delta t)\, \Delta t \oplus \cdots \oplus \mathbf{w}(t_1 + \Delta t)\, \Delta t \oplus \mathbf{w}(t_1)\, \Delta t \quad . \end{aligned} \tag{A.351}$$

Because of the geometric interpretation of the operation $\oplus$, this expression defines an oriented geodesic segment on the Lie group manifold, having as origin the origin of the group. We do not need to group the terms of the sum using parentheses because the operation $\oplus$ is associative in a group.

A.18.3 Basic Property

We have a fundamental theorem linking declinative to geometric sum, that we stated as follows.

Property A.31 *Consider a mapping from "the real line" into a Lie group, as expressed, for instance, by equations* (A.346) *and* (A.347), *and let* $\boldsymbol{\mu}(t)$ *be the declinative of the mapping (that is given, for instance, by any of the two expressions in equation* (A.348). *Then,*

$$\fint_{t_0}^{t_1} dt\, \boldsymbol{\mu}(t) \;=\; \log(\,\mathbf{M}(t_1)\,\mathbf{M}(t_0)^{-1}\,) \;=\; \mathbf{m}(t_1) \ominus \mathbf{m}(t_0) \quad , \tag{A.352}$$

this showing that the geodesic integration is an operation inverse to the declination.

The demonstration of the property is quite simple, and is given as a footnote.[38]

This property is the equivalent —in our context— of Barrow's fundamental theorem of calculus.

Example A.20 *If a body is rotating with (instantaneous) angular velocity* $\boldsymbol{\omega}(t)$, *the exponential of the geometric integral of* $\boldsymbol{\omega}(t)$ *between instants* t_1 *and* t_2, *gives the relative rotation between these two instants,*

$$\exp\Big(\fint_{t_0}^{t_1} dt\, \boldsymbol{\omega}(t)\Big) \;=\; \mathbf{R}(t_1)\,\mathbf{R}(t_0)^{-1} \quad . \tag{A.353}$$

A.18.4 Propagator

From equation (A.352) it follows that

$$\mathbf{R}(t) \;=\; \exp\Big(\fint_{t_1}^{t_2} dt\, \boldsymbol{\omega}(t)\Big)\,\mathbf{R}(t_0) \quad . \tag{A.354}$$

Equivalently, defining the *propagator*

$$\mathbf{P}(t,t_0) \;=\; \exp\Big(\fint_{t_0}^{t} dt'\, \boldsymbol{\omega}(t')\Big) \quad . \tag{A.355}$$

---

[38] One has $\mathbf{r}(t_2) \ominus \mathbf{r}(t_1) = (\mathbf{r}(t_2) \ominus \mathbf{r}(t_2 - \Delta t)) \oplus (\mathbf{r}(t_2 - \Delta t) \ominus \mathbf{r}(t_2 - 2\Delta t)) \oplus \cdots \oplus (\mathbf{r}(t_1 + \Delta t) \ominus \mathbf{r}(t_1))$. Using the definition of declinative (first of expressions (A.348)), we can equivalently write $\mathbf{r}(t_2) \ominus \mathbf{r}(t_1) = (\mathbf{v}(t_2 - \frac{\Delta t}{2})\, \Delta t) \oplus (\mathbf{v}(t_2 - \frac{3\Delta t}{2})\, \Delta t) \oplus (\mathbf{v}(t_2 - \frac{5\Delta t}{2})\, \Delta t) \oplus \cdots \oplus (\mathbf{v}(t_1 + \frac{5\Delta t}{2})\, \Delta t) \oplus (\mathbf{v}(t_1 + \frac{3\Delta t}{2})\, \Delta t) \oplus (\mathbf{v}(t_1 + \frac{\Delta t}{2})\, \Delta t)$, where the expression for the geometric integral appears. The points used in this footnote, while clearly equivalent, in the limit, to those in equation (A.351), are better adapted to discrete approximations.

one has

$$\mathbf{R}(t) \;=\; \mathbf{P}(t,t_0)\,\mathbf{R}(t_0) \quad . \tag{A.356}$$

These equations show that "the exponential of the geotensor representing the transformation is the propagator of the transformation operator".

There are many ways for evaluating the propagator $\mathbf{P}(t,t_0)$. First, of course, using the series expansion of the exponential gives, using equation (A.355),

$$\exp \int_{t_0}^{t} dt'\, \mathbf{v}(t') \;=\; \mathbf{I} + \Big(\int_{t_0}^{t} dt'\, \mathbf{v}(t') \Big) + \frac{1}{2!} \Big(\int_{t_0}^{t} dt'\, \mathbf{v}(t') \Big)^2 + \dots \tag{A.357}$$

It is also easy to see[39] that the propagator can be evaluated as

$$\exp \int_{t_0}^{t} dt\, \mathbf{v}(t) \;=\; \mathbf{I} + \int_{t_0}^{t} dt'\, \mathbf{v}(t') + \int_{t_0}^{t} dt'\, \mathbf{v}(t') \int_{t_0}^{t'} dt''\, \mathbf{v}(t'') + \dots \quad , \tag{A.358}$$

the expression on the right corresponding to what is usually named the 'matrizant' or 'matricant' (Gantmacher, 1967). Finally, from the definition of noncommutative integral, it follows[40]

$$\begin{aligned} &\exp\Big(\int_{t_0}^{t} dt\, \mathbf{v}(t) \Big) \\ &= \lim_{\Delta t \to 0} (\,\mathbf{I} + \mathbf{v}(t)\,\Delta t\,)\,(\,\mathbf{I} + \mathbf{v}(t-\Delta t)\,\Delta t\,) \;\cdots\; (\,\mathbf{I} + \mathbf{v}(t_0)\,\Delta t\,) \quad . \end{aligned} \tag{A.359}$$

The infinite product on the right-hand side was introduced by Volterra in 1887, with the name 'multiplicative integral' (see, for instance, Gantmacher, 1967). We see that it corresponds to the exponential of the noncommutative integral (sum) defined here. Volterra also introduced the 'multiplicative derivative', inverse of its 'multiplicative integral'. Volterra's 'multiplicative derivative' is exactly equivalent to the declinative of a trajectory on a Lie group, as defined in this text.

[39] For from equation (A.358) follows the two properties $\frac{d\mathbf{P}}{dt}(t,t_0) = \mathbf{v}(t)\,\mathbf{P}(t,t_0)$ and $\mathbf{P}(t_0,t_0) = \mathbf{I}$. If we define $\mathbf{S}(t) = \mathbf{P}(t,t_0)\,\mathbf{U}(t_0)$ we immediately obtain $\frac{d\mathbf{S}}{dt}(t) = \mathbf{v}(t)\,\mathbf{S}(t)$. As this is identical to the defining equation for $\mathbf{v}(t)$, $\frac{d\mathbf{U}}{dt}(t) \;=\; \mathbf{v}(t)\,\mathbf{U}(R)$, we see that $\mathbf{S}(t)$ and $\mathbf{U}(t)$ are identical up to a multiplicative constant. But the equations above imply that $\mathbf{S}(t_0) = \mathbf{U}(t_0)$, so $\mathbf{S}(t)$ and $\mathbf{U}(t)$ are, in fact, identical. The equation $\mathbf{S}(t) = \mathbf{P}(t,t_0)\,\mathbf{U}(t_0)$ then becomes $\mathbf{U}(t) \;=\; \mathbf{P}(t,t_0)\,\mathbf{U}(t_0)$, that is identical to (A.356), so we have the same propagator, and the identity of the two expressions is demonstrated.

[40] This is true because one has, using obvious notation, $\exp(\int_{t1}^{t_2} dt\;\, \mathbf{v}(t)\;) \;=\; \exp(\lim_{\Delta t\to 0}(\mathbf{v}_n\,\Delta t) \oplus (\mathbf{v}_{n-1}\,\Delta t) \oplus \dots \oplus (\mathbf{v}_1\,\Delta t)\,) \;=\; \exp(\lim_{\Delta t\to 0} \log \prod_{i=1}^{n} \exp(\mathbf{v}_i\,\Delta t)\;) \;=\; \lim_{\Delta t\to 0} \prod_{i=1}^{n} (\,\mathbf{I} + \mathbf{v}_i\,\Delta t + \dots) = \lim_{\Delta t\to 0} \prod_{i=1}^{n} (\,\mathbf{I} + \mathbf{v}_i\,\Delta t\,)$.

A.19 Geometry of the Concentration–Dilution Manifold

There are different definitions of the *concentration* in chemistry. For instance, when one considers the mass concentration of a product i in a mixing of n products, one defines

$$c^i = \frac{\text{mass of } i}{\text{total mass}} \quad , \tag{A.360}$$

and one has the constraint

$$\sum_{i=1}^{n} c^i = 1 \quad , \tag{A.361}$$

the range of variation of the concentration being

$$0 \leq c^i \leq 1 \quad . \tag{A.362}$$

To have a Jeffreys quantity (that should have a range of variation between zero and infinity) we can introduce the *eigenconcentration*

$$K^i = \frac{\text{mass of } i}{\text{mass of not } i} \quad . \tag{A.363}$$

Then,

$$0 \leq K^i \leq \infty \quad . \tag{A.364}$$

The inverse parameter $1/K^i$ having an obvious meaning, we clearly now face a Jeffreys quantity. The relations between concentration and eigenconcentration are easy to obtain:

$$K^i = \frac{c^i}{1 - c^i} \quad ; \quad c^i = \frac{K^i}{1 + K^i} \quad . \tag{A.365}$$

The constraint in equation (A.361) now becomes

$$\sum_{i=1}^{n} \frac{K^i}{1 + K^i} = 1 \quad . \tag{A.366}$$

From the Jeffreys quantities K^i we can introduce the *logarithmic eigenconcentrations*

$$k^i = \log K^i \quad , \tag{A.367}$$

that are Cartesian quantities, with the range of variation

$$-\infty \leq k^i \leq +\infty \quad , \tag{A.368}$$

subjected to the constraint

$$\sum_{i=1}^{n} \frac{e^{k^i}}{1 + e^{k^i}} = 1 \quad . \tag{A.369}$$

Should we not have the constraint expressed by the equations (A.361), (A.366) and (A.369), we would face an n-dimensional manifold, with different choices of coordinates, the coordinates $\{c^i\}$, the coordinates $\{K^i\}$, or the coordinates $\{k^i\}$. As the quantities k^i, logarithm of the Jeffreys quantities K^i, play the role of Cartesian coordinates, the distance between a point k^i_a and a point k^i_b is

$$D = \sqrt{\sum_{i=1}^{n} (k^i_b - k^i_a)^2} \quad . \tag{A.370}$$

Replacing here the different definition of the different quantities, we can express the distance by any of the three expressions

$$D_n = \sqrt{\sum_{i=1}^{n} \Big(\log \frac{c^i_b\,(1-c^i_a)}{c^i_a\,(1-c^i_b)} \Big)^2} = \sqrt{\sum_{i=1}^{n} \Big(\log \frac{K^i_b}{K^i_a} \Big)^2} = \sqrt{\sum_{i=1}^{n} (k^i_b - k^i_a)^2} \quad . \tag{A.371}$$

The associated distance elements are easy to obtain (by direct differentiation):

$$ds_n^2 = \sum_{i=1}^{n} \left(\frac{dc^i}{c^i\,(1-c^i)} \right)^2 = \sum_{i=1}^{n} \left(\frac{dK^i}{K^i} \right)^2 = \sum_{i=1}^{n} (dk^i)^2 \quad . \tag{A.372}$$

To express the volume element of the manifold in these different coordinates we just need to evaluate the metric determinant $\sqrt{\mathrm{g}}$, to obtain

$$dv_n = \frac{dc^1}{c^1\,(1-c^1)} \, \frac{dc^2}{c^2\,(1-c^2)} \cdots = \frac{dK^1}{K^1} \, \frac{dK^2}{K^2} \cdots = dk^1\, dk^2 \cdots \quad . \tag{A.373}$$

In reality, we do not work in this n-dimensional manifold. As we have n quantities and one constraint (that expressed by the equations (A.361), (A.366) and (A.369)), we face a manifold with dimension $n-1$. While the n-dimensional manifold can se seen as a Euclidean manifold (that accepts the Cartesian coordinates $\{k^i\}$), this $(n-1)$-dimensional manifold is not Euclidean, as the constraint (A.369) is not a linear constraint in the Cartesian coordinates. Of course, under the form (A.361) the constraint is formally linear, but the coordinates $\{c^i\}$ are not Cartesian.

The metric over the $(n-1)$-dimensional manifold is that induced by the metric over the n-dimensional manifold. It is easy to evaluate this induced metric, and we use now one of the possible methods.

Because the simplicity of the metric may be obscured when addressing the general case, let us make the derivation when we have only three chemical elements, i.e., when $n = 3$. From this special case, the general formulas for the n-dimensional case will be easy to write. Also, in what follows, let us consider only the quantities c^i (the ordinary concentrations), leaving as an exercise for the reader to obtain equivalent results for the eigenconcentrations K^i or the eigenconcentrations k^i.

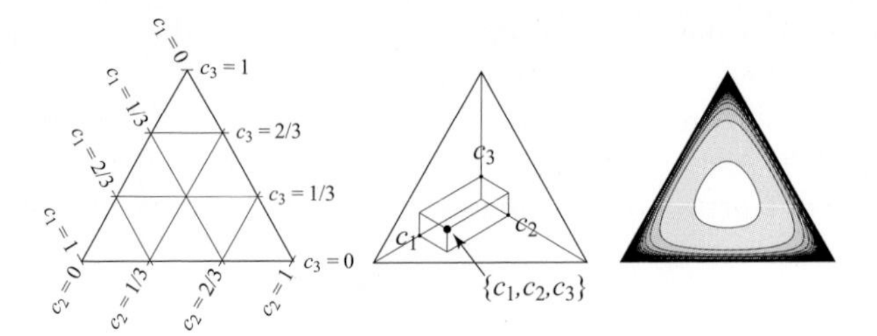

Fig. A.8. *Top left, when one has three quantities* $\{c^2, c^2, c^3\}$ *related by the constraint* $c^1 + c^2 + c^3 = 1$ *one may use any of the two equivalent representations the usual one (left) or a "cube corner" representation (middle). At the right, the volume density* $\sqrt{\mathrm{g}}$, *as expressed by equation* (A.378) *(here, in fact, we have a surface density). Dark grays correspond to large values of the volume density.*

When we have only three chemical elements, the constraint in equation (A.361), becomes, explicitly,

$$c^1 + c^2 + c^3 = 1 \quad , \tag{A.374}$$

and the distance element (equation A.372) becomes

$$ds_3^2 = \left(\frac{dc^1}{c^1\,(1-c^1)} \right)^2 + \left(\frac{dc^2}{c^2\,(1-c^2)} \right)^2 + \left(\frac{dc^3}{c^3\,(1-c^3)} \right)^2 \quad . \tag{A.375}$$

As coordinates over the two-dimensional manifold defined by the constraint, let us arbitrarily choose the first two coordinates $\{c^1, c^2\}$, dropping c^3. Differentiating the constraint (A.374) gives $dc^3 = -dc^1 - dc^2$, expression that we can insert in (A.375), to obtain the following expressions for the distance element over the two-dimensional manifold:

$$ds_2^2 = \Big(\frac{1}{Q^1} + \frac{1}{Q^3} \Big) (dc^1)^2 + \Big(\frac{1}{Q^2} + \frac{1}{Q^3} \Big) (dc^2)^2 + \frac{2\,dc^1\,dc^2}{Q^3} \quad , \tag{A.376}$$

where

$$\begin{aligned} Q^1 &= (c^1)^2\,(1-c^1)^2 \\ Q^2 &= (c^2)^2\,(1-c^2)^2 \\ Q^3 &= (c^3)^2\,(1-c^3)^2 \quad , \end{aligned} \tag{A.377}$$

and where $c^3 = 1 - c^1 - c^2$. From this expression we evaluate the metric determinant $\sqrt{\mathrm{g}}$, to obtain the volume element (here, in fact, surface element):

$$dv_2 = \frac{\sqrt{1 + (Q^1 + Q^2)/Q^3}}{\sqrt{Q^1\,Q^2}}\; dc^1\,dc^2 \quad . \tag{A.378}$$

This volume density (in fact, surface density) is represented in figure A.8.

A.20 Dynamics of a Particle

The objective of this section is just to show how the (second) Newton's law of dynamics of a particle can be written with adherence to the generalized tensor formulation developed in this text (allowed by the introduction of a connection or a metric in all relevant quality manifolds). While the space variables are always treated tensorially, this is generally not the case for the time variable. So this section serves as an introduction to the tensor notation for the time space, to pave the way for the other theory to be developed below —where, for instance, the cold–hot space is treated tensorially.—

The *physical space*, denoted $\mathfrak{E}$, is a three-dimensional manifold (Euclidean or not), endowed with some coordinates $\{y^i\} = \{y^1, y^2, y^3\}$, and with a metric

$$ds^2_{\mathfrak{E}} = g_{ij}\, dx^i\, dx^j \quad ; \quad (\, i, j, \ldots \in \{1, 2, 3\}\,) \quad . \tag{A.379}$$

The *time manifold*, denoted $\mathfrak{T}$, is a one-dimensional manifold, endowed with an arbitrary coordinate $\{\tau^a\} = \{\tau^1\}$, and with a metric

$$ds^2_{\mathfrak{T}} = G_{ab}\, d\tau^a\, d\tau^b \quad ; \quad (\, a, b, \ldots \in \{1\}\,) \quad . \tag{A.380}$$

The existence of the $ds_{\mathfrak{E}}$ in equation (A.379) implies the existence of the notion of *length* of a line on $\mathfrak{E}$, while the $ds_{\mathfrak{T}}$ in equation (A.380) implies the existence of the notion of *duration* associated to a segment of $\mathfrak{T}$, this corresponding to the postulate of existence of Newtonian time in mechanics. When using a Newtonian time t as coordinate, $ds^2_{\mathfrak{T}} = dt^2$. Then, when using some arbitrary coordinate τ^1, we write

$$\begin{aligned} ds^2_{\mathfrak{T}} &= dt^2 = \left(\frac{dt}{d\tau^1}\right)^2 (d\tau^1)^2 \\ ds^2_{\mathfrak{T}} &= G_{ab}\, d\tau^a\, d\tau^b \quad , \end{aligned} \tag{A.381}$$

from where it follows that the unique component of the 1×1 metric tensor G_{ab} is

$$G_{11} = \left(\frac{dt}{d\tau^1}\right)^2 \quad , \tag{A.382}$$

and, therefore, one has

$$\frac{dt}{d\tau^1} = \pm\sqrt{G_{11}} \quad ; \quad \frac{d\tau^1}{dt} = \pm\frac{1}{\sqrt{G_{11}}} \quad , \tag{A.383}$$

the sign depending of the orientation defined in the time manifold by the arbitrary coordinate τ^1.

Consider now a trajectory, i.e., a mapping from $\mathfrak{T}$ into $\mathfrak{E}$. Using coordinates, a trajectory is defined by the three functions

$$\tau^1 \mapsto \begin{cases} y^1(\tau^1) \\ y^2(\tau^1) \\ y^3(\tau^1) \end{cases} \tag{A.384}$$

The *velocity tensor* along the trajectory is defined as the derivative of the mapping:

$$V_a{}^i = \frac{\partial y^i}{\partial \tau^a} \quad . \tag{A.385}$$

Although there is only one coordinate τ^a, it is better to use general notation, and use ∂/∂ instead of d/d.

The particle describing the trajectory may be submitted, at each point, to a force f^i, that, as usual, must be defined independently of the dynamics of the particle (for instance, using linear springs). The question, then is that of relating the force vector f^i to the velocity tensor $V_a{}^i$. What we need here is a formulation that allows us to work with arbitrary coordinates both on the physical space and on the time manifold, that is tensorial, and that reduces to the standard Newton's law when Newtonian time is used on the time manifold. There is not much freedom in selecting the appropriate equations: except for minor details, we arrive at the following mathematical model,

$$V_a{}^i = \frac{\partial y^i}{\partial \tau^a} \quad ; \quad P^i = p^a\, V_a{}^i \quad ; \quad Q_a{}^i = \frac{\partial P^i}{\partial \tau^a} \quad ; \quad f^i = q^a\, Q_a{}^i \quad , \tag{A.386}$$

where p^a and q^a are two (one-dimensional) vectors of the time manifold[41] $\mathfrak{T}$. In (A.386), the first three equations can be considered as mere definitions. The fourth equation is a postulate, relating two objects f^i and $q^a\, Q_a{}^i$, that have been defined independently.

The norm of the two tensors $V_a{}^i$ and $Q_a{}^i$ is, respectively,

$$\|\, \mathbf{V} \,\| = \sqrt{G^{ab}\, g_{ij}\, V_a{}^i\, V_b{}^j} \quad ; \quad \|\, \mathbf{Q} \,\| = \sqrt{G^{ab}\, g_{ij}\, Q_a{}^i\, Q_b{}^j} \quad , \tag{A.387}$$

the norm of the two vectors P^i and f^i is given by the usual formulas for space vectors, $\|\, \mathbf{P} \,\| = (\, g_{ij}\, P^i\, P^j\,)^{1/2}$, and $\|\, \mathbf{f} \,\| = (\, g_{ij}\, f^i\, f^j\,)^{1/2}$, and, finally, the norm of the two one-dimensional vectors p^a and q^a is, respectively,

$$\|\, \mathbf{p} \,\| = \sqrt{G_{ab}\, p^a\, p^b} \quad ; \quad \|\, \mathbf{q} \,\| = \sqrt{G_{ab}\, q^a\, q^b} \quad . \tag{A.388}$$

Introducing the two scalars $p = \|\, \mathbf{p} \,\|$ and $q = \|\, \mathbf{q} \,\|$, one easily obtains

$$p = \sqrt{G_{11}}\; |\, p^1\, | \quad ; \quad q = \sqrt{G_{11}}\; |\, q^1\, | \quad . \tag{A.389}$$

Our basic system of equations (A.386) can be written as a single equation,

$$f^i = q^a\, \frac{\partial}{\partial \tau^a} \left(p^b\, \frac{\partial y^i}{\partial \tau^b} \right) \quad , \tag{A.390}$$

---

[41] Or, to speak properly, two vectors belonging to the linear space tangent to $\mathfrak{T}$ at the given point.

an expression that, using the different results just obtained, leads to[42]

$$f^i = m \frac{d^2 y^i}{dt^2} \quad , \tag{A.391}$$

where $m = p\,q$ is to be interpreted as the *mass* of the particle. This, of course, is the traditional form of Newton's second law of dynamics, valid only when using a Newtonian time coordinate on the time manifold $\mathfrak{T}$.

To be complete, let us relate the velocity tensor $V_a{}^i$ to the usual velocity vector v^i. Letting t be a Newtonian time coordinate, running from past to future (while τ^1 is still an arbitrary coordinate, with arbitrary orientation), the usual velocity vector is defined as

$$v^i = dy^i/dt \quad , \tag{A.392}$$

with norm $\| \mathbf{v} \| = (g_{ij}\, v^i\, v^j)^{1/2}$. Evaluating $V_1{}^i$ successively gives $V_1{}^i = dy^i/d\tau^1 = (dt/d\tau^1)\,(dy^i/dt)$, i.e., using the first of equations (A.383),

$$V_1{}^i = \pm \sqrt{G_{11}}\; v^i \quad . \tag{A.393}$$

We can now evaluate the norm of the velocity tensor $V_a{}^i$, using the first of equations (A.387). Taking into account (A.393), one immediately obtains

$$\| \mathbf{V} \| = \| \mathbf{v} \| \quad . \tag{A.394}$$

The norm of the velocity tensor $V_a{}^i$ is identical to the norm of the ordinary velocity vector v^i.

There is no simple identification between the tensor $P_a{}^i$ introduced in (A.386) and the ordinary linear momentum $p^i = m\, v^i$.

With this example, we have learned here that the requirement of using an arbitrary coordinate on the time manifold has slightly altered the writing of our dynamical tensor equations, by adding to the usual index set $\{i, j, \dots\}$ a new set $\{a, b, \dots\}$, corresponding to the one-dimensional time manifold.

A.21 Basic Notation for Deformation Theory

A.21.1 Transpose and Adjoint of a Tensor

In appendix A.1 the definition of the adjoint of an operator mapping one vector space into another vector space has been examined. We need to particularize here to the case where the considered mapping maps one space into itself.

---

[42]We start writing $f^i = q^a \frac{\partial}{\partial \tau^a}(p^b \frac{\partial y^i}{\partial \tau^b}) = q^1 \frac{d}{d\tau^1}(p^1 \frac{dy^i}{d\tau^1})$. Using equation (A.389), this can be written $f^i = pq\,(1/\sqrt{G_{11}})\,\frac{d}{d\tau^1}((1/\sqrt{G_{11}})\,\frac{dy^i}{d\tau^1})$, i.e., using equation (A.383), $f^i = pq\,\frac{d\tau^1}{dt}\,\frac{d}{d\tau^1}(\frac{d\tau^1}{dt}\,\frac{dy^i}{d\tau^1})$, from which equation (A.391) immediately follows.

Consider a manifold with some coordinate system, a given point of the manifold and the natural basis for the local linear space. Consider also, as usual, the dual space at the given point, as well as the dual basis. If $\mathbf{f} = \{f_i\}$ is a form and $\mathbf{v} = \{v^i\}$ a vector, the duality product is, by definition,

$$\langle \mathbf{f} \, , \, \mathbf{v} \rangle = f_i v^i \quad . \tag{A.395}$$

Any (real) tensor $\mathbf{Z} = \{Z^i{}_j\}$ can be considered as a linear mapping that to every vector v^i associates the vector $w^i = Z^i{}_j v^j$. The transpose $\mathbf{Z}^t$ of $\mathbf{Z}$ is the mapping with components $(\mathbf{Z}^t)_i{}^j$ that to every form f_i associates a form $h_i = (\mathbf{Z}^t)_i{}^j f_j$ with the property

$$\langle \mathbf{f} \, , \, \mathbf{Z}\mathbf{v} \rangle = \langle \mathbf{Z}^t \mathbf{f} \, , \, \mathbf{v} \rangle \quad , \tag{A.396}$$

i.e., $f_i (\mathbf{Z}\mathbf{v})^i = (\mathbf{Z}^t \mathbf{f})_j v^j$, or, more explicitly $f_i Z^i{}_j v^j = (\mathbf{Z}^t)_j{}^i f_i v^j$. This leads to

$$(\mathbf{Z}^t)_j{}^i = Z^i{}_j \quad . \tag{A.397}$$

Assume now that the manifold is metric, let g_{ij} be the covariant components of the metric at the given point, in the local natural basis, and g^{ij} the contravariant components. The scalar product of two vectors is

$$(\, \mathbf{w} \, , \, \mathbf{v} \,) = w^i g_{ij} v^j \quad . \tag{A.398}$$

The *adjoint* of the linear operator $\mathbf{Z}$, denoted $\mathbf{Z}^*$, also maps vectors into vectors, and we write an equation like $\mathbf{w} = \mathbf{Z}^* \mathbf{v}$ as $w^i = (\mathbf{Z}^*)^i{}_j v^j$. We say that $\mathbf{Z}^*$ is the adjoint of $\mathbf{Z}$ if for any vectors $\mathbf{v}$ and $\mathbf{w}$, one has

$$(\, \mathbf{w} \, , \, \mathbf{Z}\mathbf{v} \,) = (\, \mathbf{Z}^* \mathbf{w} \, , \, \mathbf{v} \,) \quad , \tag{A.399}$$

i.e., $w^j g_{ji} (\mathbf{Z}\mathbf{v})^i = (\mathbf{Z}^* \mathbf{w})^i g_{ik} w^k$, or, more explicitly $w^j g_{ji} Z^i{}_k w^k = (\mathbf{Z}^*)^i{}_j w^j g_{ik}$ w^k. This leads to $g_{ji} Z^i{}_k = (\mathbf{Z}^*)^i{}_j g_{ik}$, i.e.,

$$(\mathbf{Z}^*)^i{}_j = g^{ik} Z^\ell{}_k g_{\ell j} \quad , \tag{A.400}$$

This can also be written $(\mathbf{Z}^*)^i{}_j = g^{ik} (\mathbf{Z}^t)_k{}^\ell g_{\ell j}$ or, more formally,

$$\mathbf{Z}^* = \mathbf{g}^{-1} \mathbf{Z}^t \mathbf{g} \quad , \tag{A.401}$$

an equation that can also be interpreted as involving matrix products (see section A.21.3 below for details on matrix notation).

Definition A.11 *A tensor* $\mathbf{Z} = \{Z_i{}^j\}$ *is called* orthogonal *if its adjoint equals its inverse:*

$$\mathbf{Z}^* = \mathbf{Z}^{-1} \quad . \tag{A.402}$$

Then, $\mathbf{Z}^* \mathbf{Z} = \mathbf{Z} \mathbf{Z}^* = \mathbf{I}$, or using equation (A.401),

$$\mathbf{Z}^t \mathbf{g} \mathbf{Z} = \mathbf{g} \quad ; \quad \mathbf{Z} \mathbf{g}^{-1} \mathbf{Z}^t = \mathbf{g}^{-1} \quad , \tag{A.403}$$

i.e., $(Z^t)_i{}^k g_{k\ell} Z^\ell{}_j = g_{ij}$, $Z^{ik} g^{k\ell} (Z^t)_\ell{}^j = g^{ij}$, or using expression (A.397) for the transpose,

$$Z^k{}_i \, g_{k\ell} \, Z^\ell{}_j = g_{ij} \quad ; \quad Z^i{}_k \, g^{k\ell} \, Z^j{}_\ell = g^{ij} \quad . \tag{A.404}$$

Example A.21 *Let* $\mathbf{R} = \{R^i{}_j\}$ *be a rotation tensor. Rotation tensors are orthogonal:* $\mathbf{R}\,\mathbf{R}^* = \mathbf{I}$, $R^k{}_i \, g_{k\ell} \, R^\ell{}_j = g_{ij}$.

Definition A.12 *A tensor* $\mathbf{Q} = \{Q_i{}^j\}$ *is called* symmetric *(or* self-adjoint*) if it equals its adjoint:*

$$\mathbf{Q}^* = \mathbf{Q} \quad . \tag{A.405}$$

Using equation (A.401) this condition can also be written

$$\mathbf{g}\mathbf{Q} = \mathbf{Q}^t \mathbf{g} \quad , \tag{A.406}$$

i.e., $g_{ik} \, Q^k{}_j = (\mathbf{Q}^t)_i{}^k \, g_{kj}$, or using the expression (A.397) for the transpose, $g_{ik} \, Q^k{}_j = Q^k{}_i \, g_{kj}$. When using the metric to lower indices, of course,

$$Q_{ij} = Q_{ji} \quad . \tag{A.407}$$

Writing the symmetry condition as $\mathbf{Q}^t = \mathbf{Q}$, instead of the more correct expressions (A.405) or (A.406), may lead to misunderstandings (except when using Cartesian coordinates in Euclidean spaces).

Example A.22 *Let* $\mathbf{D} = \{D^i{}_j\}$ *represent a pure shear deformation (defined below). Such a tensor is self-adjoint (or symmetric):* $\mathbf{D} = \mathbf{D}^*$, $\mathbf{g}\mathbf{D} = \mathbf{D}^t \mathbf{g}$, $g_{ik} \, D^k{}_j = D^k{}_i \, g_{kj}$.

A.21.2 Polar Decomposition

A transformation $\mathbf{T} = \{T^i{}_j\}$ can uniquely[43] be decomposed as

$$\mathbf{T} = \mathbf{R}\mathbf{E} = \mathbf{F}\mathbf{R} \quad , \tag{A.408}$$

where $\mathbf{R}$ is a special orthogonal operator (a rotation), $\det \mathbf{R} = 1$, $\mathbf{R}^* = \mathbf{R}^{-1}$, and where $\mathbf{E}$ and $\mathbf{F}$ are positive definite symmetric tensors (that we shall call *deformations*), $\mathbf{E}^* = \mathbf{E}$, $\mathbf{F}^* = \mathbf{F}$. One easily arrives at

$$\mathbf{E} = (\mathbf{T}^* \mathbf{T})^{1/2} \quad ; \quad \mathbf{F} = (\mathbf{T}\,\mathbf{T}^*)^{1/2} \quad ; \quad \mathbf{R} = \mathbf{T}\mathbf{E}^{-1} = \mathbf{F}^{-1}\mathbf{T} \quad , \tag{A.409}$$

[43] For a demonstration of the uniqueness of the decomposition, see, for instance, Ogden, 1984.

and one has

$$\mathbf{E} = \mathbf{R}^{-1}\,\mathbf{F}\,\mathbf{R} \quad ; \quad \mathbf{F} = \mathbf{R}\,\mathbf{E}\,\mathbf{R}^{-1} \ . \tag{A.410}$$

Using the expression for the adjoint in terms of the transpose and the metric (equation A.401) the solutions for $\mathbf{E}$ and $\mathbf{F}$ (at left in equation A.409) are written

$$\mathbf{E} = (\mathbf{g}^{-1}\,\mathbf{T}^t\,\mathbf{g}\,\mathbf{T})^{1/2} \quad ; \quad \mathbf{F} = (\mathbf{T}\,\mathbf{g}^{-1}\,\mathbf{T}^t\,\mathbf{g})^{1/2} \ , \tag{A.411}$$

expressions that can directly be interpreted as matrix equations.

A.21.3 Rules of Matrix Representation

The equations written above are simultaneously valid in three possible representations, as intrinsic tensor equations (i.e., tensor equations written without indices), as equations involving (abstract) operators, and, finally, as equations representing matrices. For the matrix representation, the usual rule or matrix multiplication imposes that the first index always corresponds to the rows, and the the second index to columns, and this irrespectively of their upper or lower position. For instance

$$\mathbf{g} = \{g_{ij}\} = \begin{pmatrix} g_{11} & g_{12} & \cdots \\ g_{21} & g_{22} & \cdots \\ \vdots & \vdots & \ddots \end{pmatrix} \quad ; \quad \mathbf{g}^{-1} = \{g^{ij}\} = \begin{pmatrix} g^{11} & g^{12} & \cdots \\ g^{21} & g^{22} & \cdots \\ \vdots & \vdots & \ddots \end{pmatrix}$$
$$\mathbf{P} = \{P^i{}_j\} = \begin{pmatrix} P^1{}_1 & P^1{}_2 & \cdots \\ P^2{}_1 & P^2{}_2 & \cdots \\ \vdots & \vdots & \ddots \end{pmatrix} \quad ; \quad \mathbf{Q} = \{Q_i{}^j\} = \begin{pmatrix} Q_1{}^1 & Q_1{}^2 & \cdots \\ Q_2{}^1 & Q_2{}^2 & \cdots \\ \vdots & \vdots & \ddots \end{pmatrix} \ . \tag{A.412}$$

With this convention, the abstract definition of transpose (equation A.397) corresponds to the usual matrix transposition of rows and columns. To pass from an equation written in index notation to the same equation written in the operator-matrix notation, it is sufficient that in the index notation the indices concatenate. This is how, for instance, the index equation $(\mathbf{Z}^*)^i{}_j = g^{ik}\,(\mathbf{Z}^t)_k{}^\ell\,g_{\ell j}$ corresponds to the operator-matrix equation $\mathbf{Z}^* = \mathbf{g}^{-1}\,\mathbf{Z}^t\,\mathbf{g}$ (equation A.401).

No particular rule is needed to represent vectors and forms, as the context usually suggests unambiguous notation. For instance, $ds^2 = g_{ij}\,dx^i\,dx^j = dx^i\,g_{ij}\,dx^j$ can be written, with obvious matrix meaning, $ds^2 = \mathbf{dx}^t\,\mathbf{g}\,\mathbf{dx}$.

For objects with more than two indices (like the torsion tensor or the elastic compliance), it is better to accompany any abstract notation with its explicit meaning in terms of components in a basis (i.e., in terms of indices).

In deformation theory, when using material coordinates, the components of the metric tensor may depend on time, i.e., more than one metric is considered. To clarify the tensor equations of this chapter, all occurrences of the metric tensor are explicitly documented, and only in exceptional situations shall we absorb the metric into a raising or lowering of indices. For instance, the condition that a tensor $\mathbf{Q} = \{Q^i{}_j\}$ is orthogonal will be written as (equation at left in A.404) $Q^k{}_i\, g_{k\ell}\, Q^\ell{}_j = g_{ij}$, instead of $Q_{si}\, Q^{sj} = \delta_i{}^j$. In abstract notation, $\mathbf{Q}^t\, \mathbf{g}\, \mathbf{Q} = \mathbf{g}$ (equation A.403). Similarly, the condition that a tensor $\mathbf{Q} = \{Q^i{}_j\}$ is self-adjoint (symmetric) will be written as $g_{ik}\, Q^k{}_j = Q^k{}_i\, g_{kj}$, instead of $Q_{ij} = Q_{ji}$. In abstract notation, $\mathbf{g}\, \mathbf{Q} = \mathbf{Q}^t\, \mathbf{g}$ (equation A.406).

A.22 Isotropic Four-indices Tensor

In an n-dimensional space, with metric g_{ij}, the three operators $\mathbf{K}$, $\mathbf{M}$ and $\mathbf{A}$ with components

$$\begin{aligned} K_{ij}{}^{k\ell} &= \frac{1}{n}\, g_{ij}\, g^{k\ell} \\ M_{ij}{}^{k\ell} &= \tfrac{1}{2}\, (\delta_i{}^k\, \delta_j{}^\ell + \delta_i{}^\ell\, \delta_j{}^k) - \frac{1}{n}\, g_{ij}\, g^{k\ell} \\ A_{ij}{}^{k\ell} &= \tfrac{1}{2}\, (\delta_i{}^k\, \delta_j{}^\ell - \delta_i{}^\ell\, \delta_j{}^k) \end{aligned} \qquad \text{(A.413)}$$

are projectors ($\mathbf{K}^2 = \mathbf{K}$; $\mathbf{M}^2 = \mathbf{M}$; $\mathbf{A}^2 = \mathbf{A}$), are orthogonal ($\mathbf{K}\,\mathbf{M} = \mathbf{M}\,\mathbf{K} = \mathbf{K}\,\mathbf{A} = \mathbf{A}\,\mathbf{K} = \mathbf{M}\,\mathbf{A} = \mathbf{A}\,\mathbf{M} = \mathbf{0}$) and their sum is the identity ($\mathbf{K}+\mathbf{M}+\mathbf{A} = \mathbf{I}$). It is clear that $K_{ij}{}^{k\ell}$ maps any tensor t_{ij} into its isotropic part

$$K_{ij}{}^{k\ell}\, t_{ij} = \frac{1}{n}\, t^k{}_k\, g_{ij} \equiv \bar{t}_{ij} \quad , \qquad \text{(A.414)}$$

$M_{ij}{}^{k\ell}$ maps any tensor t_{ij} into its symmetric traceless part

$$M_{ij}{}^{k\ell}\, t_{ij} = \tfrac{1}{2}\, (t_{ij} + t_{ji}) - \bar{t}_{ij} \equiv \hat{t}_{ij} \quad , \qquad \text{(A.415)}$$

and $A_{ij}{}^{k\ell}$ maps any tensor t_{ij} into its antisymmetric part

$$A_{ij}{}^{k\ell}\, t_{ij} = \tfrac{1}{2}\, (t_{ij} - t_{ji}) \equiv \check{t}_{ij} \quad . \qquad \text{(A.416)}$$

In the space of tensors $c_{ij}{}^{k\ell}$ with the symmetry

$$c_{ijk\ell} = c_{k\ell ij} \quad , \qquad \text{(A.417)}$$

the most general isotropic[44] tensor has the form

$$c_{ij}{}^{k\ell} = c_\kappa\, K_{ij}{}^{k\ell} + c_\mu\, M_{ij}{}^{k\ell} + c_\theta\, A_{ij}{}^{k\ell} \quad . \qquad \text{(A.418)}$$

Its eigenvalues are λ_κ, with multiplicity one, c_μ, with multiplicity $n\,(n+1)/2 - 1$, and c_θ, with multiplicity $n\,(n-1)/2$. Explicitly, this gives

[44] I.e., such that the mapping $t_{ij} \mapsto c_{ij}{}^{k\ell}\, t_{k\ell}$ preserves the character of t_{ij} of being an isotropic, symmetric traceless or antisymmetric tensor.

$$c_{ij}{}^{k\ell} = \frac{c_\kappa}{n}\, g_{ij}\, g^{k\ell} + c_\mu \left(\tfrac{1}{2}\,(\delta_i{}^k\,\delta_j{}^\ell + \delta_i{}^\ell\,\delta_j{}^k) - \frac{1}{n}\, g_{ij}\, g^{k\ell}\right) + \frac{c_\theta}{2}\,(\delta_i{}^k\,\delta_j{}^\ell - \delta_i{}^\ell\,\delta_j{}^k) \quad . \tag{A.419}$$

Then, for any tensor t_{ij},

$$c_{ij}{}^{k\ell}\, t_{k\ell} = c_\kappa\, \bar{t}_{ij} + c_\mu\, \hat{t}_{ij} + c_\theta\, \check{t}_{ij} \quad . \tag{A.420}$$

The inverse of the tensor $c_{ij}{}^{k\ell}$, as expressed in equation (A.418), is the tensor

$$d_{ij}{}^{k\ell} = \chi_\kappa\, K_{ij}{}^{k\ell} + \chi_\mu\, M_{ij}{}^{k\ell} + \chi_\theta\, A_{ij}{}^{k\ell} \quad , \tag{A.421}$$

with $\chi_\kappa = 1/c_\kappa$, $\chi_\mu = 1/c_\mu$ and $\chi_\theta = 1/c_\theta$.

A.23 9D Representation of 3D Fourth Rank Tensors

The algebra of 3D, fourth rank tensors is underdeveloped.[45] For instance, routines for computing the eigenvalues of a tensor like $c^{ijk\ell}$, or to compute the inverse tensor, the compliance $s^{ijk\ell}$, are not widely available. There are also psychological barriers, as we are more trained to handle matrices than objects with higher dimensions. This is why it is customary to introduce a 6×6 representation of the tensor $c^{ijk\ell}$. Let us see how this is done (here, in fact, a 9×9 representation). The formulas written below generalize the formulas in the literature as they are valid in the case where stress or strain are not necessarily symmetric.

The stress, the strain and the compliance tensors are written, using the usual tensor bases,

$$\boldsymbol{\sigma} = \sigma^{ij}\, \mathbf{e}_i \otimes \mathbf{e}_j \quad ; \quad \boldsymbol{\varepsilon} = \varepsilon^{ij}\, \mathbf{e}_i \otimes \mathbf{e}_j \quad ; \quad \mathbf{c} = c^{ijk\ell}\, \mathbf{e}_i \otimes \mathbf{e}_j \otimes \mathbf{e}_k \otimes \mathbf{e}_\ell \ . \tag{A.422}$$

When working with orthonormed bases, one introduces a new basis, composed of the three "diagonal elements"

$$\mathbf{E}_1 \equiv \mathbf{e}_1 \otimes \mathbf{e}_1 \quad ; \quad \mathbf{E}_2 \equiv \mathbf{e}_2 \otimes \mathbf{e}_2 \quad ; \quad \mathbf{E}_3 \equiv \mathbf{e}_3 \otimes \mathbf{e}_3 \quad , \tag{A.423}$$

the three "symmetric elements"

$$\begin{aligned} \mathbf{E}_4 &\equiv \tfrac{1}{\sqrt{2}}\,(\mathbf{e}_1 \otimes \mathbf{e}_2 + \mathbf{e}_2 \otimes \mathbf{e}_1) \\ \mathbf{E}_5 &\equiv \tfrac{1}{\sqrt{2}}\,(\mathbf{e}_2 \otimes \mathbf{e}_3 + \mathbf{e}_3 \otimes \mathbf{e}_2) \\ \mathbf{E}_6 &\equiv \tfrac{1}{\sqrt{2}}\,(\mathbf{e}_3 \otimes \mathbf{e}_1 + \mathbf{e}_1 \otimes \mathbf{e}_3) \quad , \end{aligned} \tag{A.424}$$

and the three "antisymmetric elements"

[45] See Itskov (2000) for recent developments.

$$
\begin{aligned}
\mathbf{E}_7 &\equiv \tfrac{1}{\sqrt{2}} (\mathbf{e}_1 \otimes \mathbf{e}_2 - \mathbf{e}_2 \otimes \mathbf{e}_1) \\
\mathbf{E}_8 &\equiv \tfrac{1}{\sqrt{2}} (\mathbf{e}_2 \otimes \mathbf{e}_3 - \mathbf{e}_3 \otimes \mathbf{e}_2) \\
\mathbf{E}_9 &\equiv \tfrac{1}{\sqrt{2}} (\mathbf{e}_3 \otimes \mathbf{e}_1 - \mathbf{e}_1 \otimes \mathbf{e}_3) \quad .
\end{aligned}
\tag{A.425}
$$

In this basis, the components of the tensors are defined using general expressions:

$$
\sigma = S^A \, \mathbf{E}_A \quad ; \quad \varepsilon = E^A \, \mathbf{E}_A \quad ; \quad \mathbf{c} = C^{AB} \, \mathbf{e}_A \otimes \mathbf{e}_B \quad , \tag{A.426}
$$

where all the implicit sums concerning the indices $\{A, B, \dots\}$ run from 1 to 9.

For Hooke's law and for the eigenstiffness–eigenstrain equation, one then, respectively, has the equivalences

$$
\begin{aligned}
\sigma^{ij} = c^{ijk\ell} \, \varepsilon_{k\ell} \quad &\Leftrightarrow \quad S^A = C^{AB} \, E_B \\
c^{ijkl} \, \varepsilon_{k\ell} = \lambda \, \varepsilon^{ij} \quad &\Leftrightarrow \quad c^{AB} \, E_B = \lambda \, E^A \quad .
\end{aligned}
\tag{A.427}
$$

Using elementary algebra, one obtains the following relations between the components of the stress and strain in the two bases:

$$
\begin{pmatrix} S^1 \\ S^2 \\ S^3 \\ S^4 \\ S^5 \\ S^6 \\ S^7 \\ S^8 \\ S^9 \end{pmatrix} = \begin{pmatrix} \sigma^{11} \\ \sigma^{22} \\ \sigma^{33} \\ \sigma^{(12)} \\ \sigma^{(23)} \\ \sigma^{(31)} \\ \sigma^{[12]} \\ \sigma^{[23]} \\ \sigma^{[31]} \end{pmatrix} \quad ; \quad \begin{pmatrix} E^1 \\ E^2 \\ E^3 \\ E^4 \\ E^5 \\ E^6 \\ E^7 \\ E^8 \\ E^9 \end{pmatrix} = \begin{pmatrix} \varepsilon^{11} \\ \varepsilon^{22} \\ \varepsilon^{33} \\ \varepsilon^{(12)} \\ \varepsilon^{(23)} \\ \varepsilon^{(31)} \\ \varepsilon^{[12]} \\ \varepsilon^{[23]} \\ \varepsilon^{[31]} \end{pmatrix}
\tag{A.428}
$$

where the following notation is used:

$$
\alpha^{(ij)} \equiv \tfrac{1}{\sqrt{2}} (\alpha^{ij} + \alpha^{ji}) \quad ; \quad \alpha^{[ij]} \equiv \tfrac{1}{\sqrt{2}} (\alpha^{ij} - \alpha^{ji}) \quad . \tag{A.429}
$$

The new components of the stiffness tensor $\mathbf{c}$ are

$$
\begin{pmatrix}
C^{11} & C^{12} & C^{13} & C^{14} & C^{15} & C^{16} & C^{17} & C^{18} & C^{19} \\
C^{21} & C^{22} & C^{23} & C^{24} & C^{25} & C^{26} & C^{27} & C^{28} & C^{29} \\
C^{31} & C^{32} & C^{33} & C^{34} & C^{35} & C^{36} & C^{37} & C^{38} & C^{39} \\
C^{41} & C^{42} & C^{43} & C^{44} & C^{45} & C^{46} & C^{47} & C^{48} & C^{49} \\
C^{51} & C^{52} & C^{53} & C^{54} & C^{55} & C^{56} & C^{57} & C^{58} & C^{59} \\
C^{61} & C^{62} & C^{63} & C^{64} & C^{65} & C^{66} & C^{67} & C^{68} & C^{69} \\
C^{71} & C^{72} & C^{73} & C^{74} & C^{75} & C^{76} & C^{77} & C^{78} & C^{79} \\
C^{81} & C^{82} & C^{83} & C^{84} & C^{85} & C^{86} & C^{87} & C^{88} & C^{89} \\
C^{91} & C^{92} & C^{93} & C^{94} & C^{95} & C^{96} & C^{97} & C^{98} & C^{99}
\end{pmatrix} =
\tag{A.430}
$$

$$\left(\begin{array}{ccc|ccc|ccc}
c^{1111} & c^{1122} & c^{1133} & c^{11(12)} & c^{11(23)} & c^{11(31)} & c^{11[12]} & c^{11[23]} & c^{11[31]} \\
c^{2211} & c^{2222} & c^{2233} & c^{22(12)} & c^{22(23)} & c^{22(31)} & c^{22[12]} & c^{22[23]} & c^{22[31]} \\
c^{3311} & c^{3322} & c^{3333} & c^{33(12)} & c^{33(23)} & c^{33(31)} & c^{33[12]} & c^{33[23]} & c^{33[31]} \\
\hline
c^{(12)11} & c^{(12)22} & c^{(12)33} & c^{(1212)} & c^{(1223)} & c^{(1231)} & c^{(12)[12]} & c^{(12)[23]} & c^{(12)[31]} \\
c^{(23)11} & c^{(23)22} & c^{(23)33} & c^{(2312)} & c^{(2323)} & c^{(2331)} & c^{(23)[12]} & c^{(23)[23]} & c^{(23)[31]} \\
c^{(31)11} & c^{(31)22} & c^{(31)33} & c^{(3112)} & c^{(3123)} & c^{(3131)} & c^{(31)[12]} & c^{(31)[23]} & c^{(31)[31]} \\
\hline
c^{[12]11} & c^{[12]22} & c^{[12]33} & c^{[12](12)} & c^{[12](23)} & c^{[12](31)} & c^{[1212]} & c^{[1223]} & c^{[1231]} \\
c^{[23]11} & c^{[23]22} & c^{[23]33} & c^{[23](12)} & c^{[23](23)} & c^{[23](31)} & c^{[2312]} & c^{[2323]} & c^{[2331]} \\
c^{[31]11} & c^{[31]22} & c^{[31]33} & c^{[31](12)} & c^{[31](23)} & c^{[31](31)} & c^{[3112]} & c^{[3123]} & c^{[3131]}
\end{array}\right) ,$$

where

$$\begin{aligned}
c^{ij(k\ell)} &\equiv \tfrac{1}{\sqrt{2}}(c^{ijk\ell} + c^{ij\ell k}) \quad ; \quad & c^{ij[k\ell]} &\equiv \tfrac{1}{\sqrt{2}}(c^{ijk\ell} - c^{ij\ell k}) \\
c^{(ij)k\ell} &\equiv \tfrac{1}{\sqrt{2}}(c^{ijk\ell} + c^{jik\ell}) \quad ; \quad & c^{[ij]k\ell} &\equiv \tfrac{1}{\sqrt{2}}(c^{ijk\ell} - c^{jik\ell}) \ ,
\end{aligned} \tag{A.431}$$

and

$$\begin{aligned}
c^{(ijk\ell)} &\equiv \tfrac{1}{2}(c^{ijk\ell} + c^{ij\ell k} + c^{jik\ell} + c^{ji\ell k}) \\
c^{(ij)[k\ell]} &\equiv \tfrac{1}{2}(c^{ijk\ell} - c^{ij\ell k} + c^{jik\ell} - c^{ji\ell k}) \\
c^{[ij](k\ell)} &\equiv \tfrac{1}{2}(c^{ijk\ell} + c^{ij\ell k} - c^{jik\ell} - c^{ji\ell k}) \\
c^{[ijk\ell]} &\equiv \tfrac{1}{2}(c^{ijk\ell} - c^{ij\ell k} - c^{jik\ell} + c^{ji\ell k}) \ .
\end{aligned} \tag{A.432}$$

Should energy considerations suggest imposing the symmetry $c^{ijk\ell} = c^{k\ell ij}$, then, the matrix $\{C^{AB}\}$ would be symmetric.

As an example, for an isotropic medium, the stiffness tensor is given by expression (A.419), and one obtains

$$\{C^{AB}\} = \left(\begin{array}{ccc|ccc|ccc}
(c_\kappa + 2c_\mu)/3 & (c_\kappa - c_\mu)/3 & (c_\kappa - c_\mu)/3 & 0 & 0 & 0 & 0 & 0 & 0 \\
(c_\kappa - c_\mu)/3 & (c_\kappa + 2c_\mu)/3 & (c_\kappa - c_\mu)/3 & 0 & 0 & 0 & 0 & 0 & 0 \\
(c_\kappa - c_\mu)/3 & (c_\kappa - c_\mu)/3 & (c_\kappa + 2c_\mu)/3 & 0 & 0 & 0 & 0 & 0 & 0 \\
\hline
0 & 0 & 0 & c_\mu & 0 & 0 & 0 & 0 & 0 \\
0 & 0 & 0 & 0 & c_\mu & 0 & 0 & 0 & 0 \\
0 & 0 & 0 & 0 & 0 & c_\mu & 0 & 0 & 0 \\
\hline
0 & 0 & 0 & 0 & 0 & 0 & c_\theta & 0 & 0 \\
0 & 0 & 0 & 0 & 0 & 0 & 0 & c_\theta & 0 \\
0 & 0 & 0 & 0 & 0 & 0 & 0 & 0 & c_\theta
\end{array}\right) . \tag{A.433}$$

Using a standard mathematical routine to evaluate the nine eigenvalues of this matrix gives $\{c_\kappa, c_\mu, c_\mu, c_\mu, c_\mu, c_\mu, c_\theta, c_\theta, c_\theta\}$ as it should. These are the eigenvalues of the tensor $\mathbf{c} = c^{ijk\ell}\, \mathbf{e}_i \otimes \mathbf{e}_j \otimes \mathbf{e}_k \otimes \mathbf{e}_\ell = C^{AB}\, \mathbf{E}_A \otimes \mathbf{E}_B$ for an isotropic medium.

If the rotational eigenstiffness vanishes, $c_\theta = 0$, then, the stress is symmetric, and the expressions above simplify, as one can work using a six-dimensional basis. One obtains

$$\begin{pmatrix} S^1 \\ S^2 \\ S^3 \\ S^4 \\ S^5 \\ S^6 \end{pmatrix} = \begin{pmatrix} \sigma^{11} \\ \sigma^{22} \\ \sigma^{33} \\ \sqrt{2}\,\sigma^{12} \\ \sqrt{2}\,\sigma^{23} \\ \sqrt{2}\,\sigma^{31} \end{pmatrix} \quad ; \quad \begin{pmatrix} E^1 \\ E^2 \\ E^3 \\ E^4 \\ E^5 \\ E^6 \end{pmatrix} = \begin{pmatrix} \varepsilon^{11} \\ \varepsilon^{22} \\ \varepsilon^{33} \\ \sqrt{2}\,\varepsilon^{12} \\ \sqrt{2}\,\varepsilon^{23} \\ \sqrt{2}\,\varepsilon^{31} \end{pmatrix} \tag{A.434}$$

$$\begin{pmatrix} C^{11} & C^{12} & C^{13} & C^{14} & C^{15} & C^{16} \\ C^{21} & C^{22} & C^{23} & C^{24} & C^{25} & C^{26} \\ C^{31} & C^{32} & C^{33} & C^{34} & C^{35} & C^{36} \\ C^{41} & C^{42} & C^{43} & C^{44} & C^{45} & C^{46} \\ C^{51} & C^{52} & C^{53} & C^{54} & C^{55} & C^{56} \\ C^{61} & C^{62} & C^{63} & C^{64} & C^{65} & C^{66} \end{pmatrix} = \tag{A.435}$$

$$\left(\begin{array}{ccc|ccc} c^{1111} & c^{1122} & c^{3311} & \sqrt{2}\,c^{1112} & \sqrt{2}\,c^{1123} & \sqrt{2}\,c^{1131} \\ c^{1122} & c^{2222} & c^{2233} & \sqrt{2}\,c^{2212} & \sqrt{2}\,c^{2223} & \sqrt{2}\,c^{2231} \\ c^{3311} & c^{2233} & c^{3333} & \sqrt{2}\,c^{3312} & \sqrt{2}\,c^{3323} & \sqrt{2}\,c^{3331} \\ \hline \sqrt{2}\,c^{1112} & \sqrt{2}\,c^{2212} & \sqrt{2}\,c^{3312} & 2\,c^{1212} & 2\,c^{1223} & 2\,c^{1231} \\ \sqrt{2}\,c^{1123} & \sqrt{2}\,c^{2223} & \sqrt{2}\,c^{3323} & 2\,c^{1223} & 2\,c^{2323} & 2\,c^{2331} \\ \sqrt{2}\,c^{1131} & \sqrt{2}\,c^{2231} & \sqrt{2}\,c^{3331} & 2\,c^{1231} & 2\,c^{2331} & 2\,c^{3131} \end{array}\right) .$$

It is unfortunate that passing from four indices to two indices is sometimes done without care, even in modern literature, as in Auld (1990), where old, nontensor definitions are introduced. We have here followed the canonical way, as in Mehrabadi and Cowin (1990), generalizing it to the case where stresses are not necessarily symmetric.

A.24 Rotation of Strain and Stress

In a first thought experiment, we interpret the transformation $\mathbf{T} = \{T^i{}_j\}$ as a deformation followed by a rotation,

$$\mathbf{T} = \mathbf{R}\,\mathbf{E} \quad , \tag{A.436}$$

i.e., $T^i{}_j = R^i{}_k\, E^k{}_j$. To the unrotated deformation $\mathbf{E} = \{E^i{}_j\}$ we associate the *unrotated strain*

$$\varepsilon_{\mathbf{E}} = \log \mathbf{E} \quad , \tag{A.437}$$

and assume that such a strain is produced by the stress (Hooke's law)

$$\sigma_{\mathbf{E}} = \mathbf{c}\,\varepsilon_{\mathbf{E}} \quad , \tag{A.438}$$

or, explicitly, $(\sigma_{\mathrm{E}})_i{}^j = c_i{}^j{}_k{}^\ell\,(\varepsilon_{\mathrm{E}})^k{}_\ell$. Here, $\mathbf{c} = \{c_i{}^j{}_k{}^\ell\}$ represents the *stiffness tensor* of the medium in its initial configuration. One should keep in mind that, as the elastic medium may be anisotropic, the stiffness tensor of a rotated medium would be the rotated version of $\mathbf{c}$ (see below). To conclude the first thought experiment, we now need to apply the rotation $\mathbf{R} = \{R^i{}_j\}$. The medium is assumed to rotate without resistance, so the stress is not actually modified; it only rotates with the medium. Applying the general rule expressing the change of components of a tensor under a rotation gives the final stress associated to the transformation:

$$\sigma = \mathbf{R}^{-t}\, \sigma_{\mathrm{E}}\, \mathbf{R}^t \quad . \tag{A.439}$$

Explicitly, $\sigma_i{}^j = \overline{R}^k{}_i\, (\sigma_{\mathrm{E}})_k{}^\ell\, R^j{}_\ell$. Putting together expressions (A.437), (A.438), and (A.439) gives

$$\sigma = \mathbf{R}^{-t}\, (\, \mathbf{c}\, \log \mathbf{E}\,)\, \mathbf{R}^t \quad , \tag{A.440}$$

or, explicitly,[46] $\sigma_i{}^j = \overline{R}^k{}_i\, R^j{}_\ell\, c_k{}^\ell{}_r{}^s\, \log E^r{}_s$. This is the stress associated to the transformation $\mathbf{T} = \mathbf{R}\,\mathbf{E}$.

In a second thought experiment, one decomposes the transformation as $\mathbf{T} = \mathbf{F}\,\mathbf{R}$, so one starts by rotating the body with $\mathbf{R}$. This produces no stress, but the stiffness tensor is rotated,[47] becoming

$$(\mathbf{c}_{\mathbf{R}})_i{}^j{}_k{}^\ell \;=\; \overline{R}^p{}_i\, R^j{}_q\, \overline{R}^r{}_k\, R^\ell{}_s\, c_p{}^q{}_r{}^s \quad . \tag{A.441}$$

One next applies the deformation $\mathbf{F} = \{F^i{}_j\}$. The associated *rotated strain* is[48]

$$\varepsilon_{\mathbf{F}} \;=\; \log \mathbf{F} \quad , \tag{A.442}$$

with stress (Hooke's law again)

$$\sigma \;=\; \mathbf{c}_{\mathbf{R}}\, \varepsilon_{\mathbf{F}} \quad . \tag{A.443}$$

Putting together equations (A.441), (A.442), and (A.443) gives

$$\sigma_i{}^j \;=\; \overline{R}^p{}_i\, R^j{}_q\, \overline{R}^r{}_k\, R^\ell{}_s\, c_p{}^q{}_r{}^s\, \log F^k{}_\ell \quad . \tag{A.444}$$

To verify that this is identical to the stress obtained in the first though experiment (equation A.440), we can just replace there $\mathbf{E}$ by $\mathbf{R}^{-1}\,\mathbf{F}\,\mathbf{R}$ (equation 5.37), and use the property $\log(\mathbf{R}^{-1}\,\mathbf{F}\,\mathbf{R}) = \mathbf{R}^{-1}\,(\log \mathbf{F})\,\mathbf{R}$ of the logarithm function. We thus see that the two experiments lead to the same state of stress.

A.25 Macro-rotations, Micro-rotations, and Strain

In section 5.3.2, where the configuration space has been introduced, two simplifications have been made: the consideration of homogeneous transformations only, and the absence of macro-rotations. Let us here introduce the strain in the general case.

[46]Using the notation $\log E^r{}_s \equiv (\log \mathbf{E})^r{}_s$.
[47]Should the medium be isotropic, this would simply give $(\mathbf{c}_{\mathbf{R}})_i{}^j{}_k{}^\ell = c_i{}^j{}_k{}^\ell$.
[48]One has $\varepsilon_{\mathbf{F}} = \log \mathbf{F} = \log(\mathbf{R}\,\mathbf{E}\,\mathbf{R}^{-1}) = \mathbf{R}\,(\log \mathbf{E})\,\mathbf{R}^{-1} = \mathbf{R}\,\varepsilon_{\mathbf{E}}\,\mathbf{R}^{-1}$.

So, consider a (possibly heterogeneous) transformation field $\mathbf{T}$, with it polar decomposition

$$\mathbf{T} = \mathbf{R}\,\mathbf{E} = \mathbf{F}\,\mathbf{R} \quad ; \quad \mathbf{F} = \mathbf{R}\,\mathbf{E}\,\mathbf{R}^{-1} \quad , \tag{A.445}$$

and assume that another rotation field (representing the micro-rotations) is given, that may be represented (at every point) by the orthogonal tensor $\mathbf{S_E}$ or, equivalently, by

$$\mathbf{S_F} = \mathbf{R}\,\mathbf{S_E}\,\mathbf{R}^{-1} \quad . \tag{A.446}$$

Using the terminology introduced in appendix A.24, the *unrotated strain* can be defined as

$$\varepsilon_{\mathrm{E}} = \log \mathbf{E} + \log \mathbf{S_E} \quad , \tag{A.447}$$

and the *rotated strain* as

$$\varepsilon_{\mathrm{F}} = \log \mathbf{F} + \log \mathbf{S_F} \quad . \tag{A.448}$$

Using the relation at right in (A.445), equation (A.447), and the property $\log(\mathbf{M}\,\mathbf{A}\,\mathbf{M}^{-1}) = \mathbf{M}\,(\log \mathbf{A})\,\mathbf{M}^{-1}$ of the logarithm function, one obtains

$$\varepsilon_{\mathrm{F}} = \mathbf{R}\,\varepsilon_{\mathrm{E}}\,\mathbf{R}^{-1} \quad . \tag{A.449}$$

The stress can then be computed as explained in appendix A.24.

A.26 Elastic Energy Density

As explained in section 5.3.2, the configuration $\mathbf{C}$ of a body is characterized by a deformation $\mathbf{E}$ and a micro-rotation $\mathbf{S}$. When the configuration changes from $\{\mathbf{E},\mathbf{S}\}$ to $\{\mathbf{E}+d\mathbf{E},\mathbf{S}+d\mathbf{S}\}$, some differential displacements dx^i and some differential micro-rotations ds^{ij} (antisymmetric tensor) are produced, and a differential work dW is associated to each of these.

In order not to get confused with the simultaneous existence of macro- and micro-rotations, let us evaluate the two differential works separately, and make the sum afterwards. We start by assuming that there are no micro-rotations, and evaluate the work associated to the displacements dx^i.

The elementary work produced by the external actions is then

$$dW = \int_{V(\mathbf{C})} dV\ \varphi_i\, dx^i + \int_{S(\mathbf{C})} dS\ \tau_i\, dx^i \quad , \tag{A.450}$$

where φ_i is the force density, and τ_i is the traction at the surface of the body. Introducing the boundary conditions in equation (5.47), and using the divergence theorem, this gives $dW = \int_{V(\mathbf{C})} dV\ (\sigma_i{}^j\, \nabla_j\, dx^i + (\varphi_i + \nabla_j\, \sigma_i{}^j)\, dx^i\,)$, i.e., using the static equilibrium conditions in equation (5.48),

$$dW = \int_{V(\mathbf{C})} dV\, \sigma_i{}^j\, \nabla_j\, dx^i \quad . \tag{A.451}$$

Using[49]

$$\nabla_j dx^i = dE^i{}_k\, \overline{E}^k{}_j \quad , \tag{A.452}$$

the dW can be written

$$dW = \int_{V(\mathbf{C})} dV\, \sigma_i{}^j\, dE^i{}_k\, \overline{E}^k{}_j \quad . \tag{A.453}$$

We parameterize an evolving configuration by a parameter λ, so we write $\mathbf{E} = \mathbf{E}(\lambda)$. The declinative

$$\nu = \dot{\mathbf{E}}\, \mathbf{E}^{-1} \tag{A.454}$$

corresponds to the deformation velocity (or "strain rate"). With this, one arrives at

$$dW = \int_{V(\lambda)} dV\, \sigma_{ij}\, \nu^{ij}\, d\lambda \quad . \tag{A.455}$$

In this (half) computation, we are assuming that there are no micro-rotations, so the stress is symmetric. The equation above remains unchanged if we write, instead,

$$dW = \int_{V(\lambda)} dV\, \hat{\sigma}_{ij}\, \nu^{ij}\, d\lambda \quad , \tag{A.456}$$

where $\hat{\sigma}_{ij} = \frac{1}{2}(\sigma_{ij} + \sigma_{ji})$. This symmetry of the stress makes that the possible macro-rotations (ν needs not to be symmetric) do not contribute to the evaluation of the work. It will be important that we keep expression (A.456) as it is when we also consider micro-rotations (then the stress may not be symmetric, but the antisymmetric part of the stress produces work on the micro-rotations, not the macro-rotations.

Let us now turn to the evaluation of the work associated to the differential rotations ds^{ij}. The elementary work produced by the external actions is then[50]

[49]To understand the relation $\nabla_j\, dx^i = dE^i{}_k\, \overline{E}^k{}_j$ consider the case of an Euclidean space with Cartesian coordinates. When passing from the reference configuration $\mathbf{I}$ to configuration $\mathbf{E}$, the transformation is $\mathbf{T} = \mathbf{E}\,\mathbf{I}^{-1} = \mathbf{E}$, and the new coordinates x^i of a material point are related to the initial coordinates X^i via (we are considering homogeneous transformations) $x^i = E^i{}_j\, X^j$. When the configuration is $\mathbf{E} + d\mathbf{E}$, the coordinates become $x^i = (E^i{}_j + dE^i{}_j)\, X^j$ so the displacements are $dx^i = dE^i{}_j\, X^j$. To express them in the current coordinates, we solve the relation $x^i = E^i{}_j\, X^j$ to obtain $X^i = \overline{E}^i{}_j\, x^j$. This gives $dx^i = dE^i{}_j\, \overline{E}^j{}_k\, x^k$, from where it follows $\partial_j\, dx^i = dE^i{}_k\, \overline{C}^k{}_j$. The relation $\nabla_j\, dx^i = dE^i{}_k\, \overline{E}^k{}_j$ is the covariant expression of this, valid in an arbitrary coordinate system.

[50]Should one write $\chi_{ij} = \epsilon_{ijk}\, \xi^k$, $\mu_{ij} = \epsilon_{ijk}\, m^k$, and $ds^{ij} = \epsilon^{ijk}\, d\Sigma_k$, then $\frac{1}{2}\, \chi_{ij}\, ds^{ij} = \xi^k\, d\Sigma_k$, and $\frac{1}{2}\, \mu_{ij}\, ds^{ij} = m^k\, d\Sigma_k$.

$$dW = \int_{V(\mathbf{C})} dV \, \tfrac{1}{2} \, \chi_{ij} \, ds^{ij} + \int_{S(\mathbf{C})} dS \, \tfrac{1}{2} \, \mu_{ij} \, ds^{ij} \quad , \tag{A.457}$$

where, as explained in section 5.3.3, χ_{ij} is the moment-force density, and μ_{ij} the moment-traction (at the surface). Introducing the boundary conditions in equation (5.47), and using the divergence theorem, this gives $dW = \int_{V(\mathbf{C})} dV \, (\tfrac{1}{2} \, m_{ij}{}^k \, \nabla_k \, ds^{ij} + \tfrac{1}{2} \, (\chi_{ij} + \nabla_k \, m_{ij}{}^k) \, ds^{ij})$, i.e., using the static equilibrium conditions in equation (5.48),

$$dW = \int_{V(\mathbf{C})} dV \left(\tfrac{1}{2} \, m_{ij}{}^k \, \nabla_k \, ds^{ij} + \tfrac{1}{2} \, \chi_{ij} \, ds^{ij} \right) \quad , \tag{A.458}$$

where the moment force density χ_{ij} is

$$\chi_{ij} = \sigma_{ij} - \sigma_{ji} \quad . \tag{A.459}$$

As we assume that our medium cannot support moment-stresses, $m_{ij}{}^k = 0$, and we are left with

$$dW = \int_{V(\mathbf{C})} dV \, \tfrac{1}{2} \, \chi_{ij} \, ds^{ij} \quad . \tag{A.460}$$

Using[51]

$$ds^i{}_j = dS^i{}_k \, \overline{S}^k{}_j \quad , \tag{A.461}$$

the dW can be written

$$dW = \int_{V(\mathbf{C})} dV \, \tfrac{1}{2} \, \chi_i{}^j \, dS^i{}_k \, \overline{S}^k{}_j \quad . \tag{A.462}$$

We parameterize an evolving configuration by a parameter λ, so we write $\mathbf{S} = \mathbf{S}(\lambda)$. The declinative

$$\omega = \dot{\mathbf{S}} \, \mathbf{S}^{-1} \tag{A.463}$$

corresponds to the micro-rotation velocity. With this, one arrives at

$$dW = \int_{V(\lambda)} dV \, \tfrac{1}{2} \, \chi_{ij} \, \omega^{ij} \, d\lambda \quad . \tag{A.464}$$

This equation can also be written

$$dW = \int_{V(\lambda)} dV \, \check{\sigma}_{ij} \, \omega^{ij} \, d\lambda \quad . \tag{A.465}$$

where $\check{\sigma}_{ij} = \tfrac{1}{2}(\sigma_{ij} - \sigma_{ji})$.

We can now sum the two differential works expressed in equation (A.456) and equation (A.465), to obtain

[51] To understand that the differential micro-rotations are given by $ds^i{}_j = dS^i{}_k \, \overline{S}^k{}_j$, just consider that the rotation velocity is the declinative $\dot{\mathbf{S}} \, \mathbf{S}^{-1}$.

$$dW = \int_{V(\lambda)} dV \left(\hat{\sigma}_{ij} \, v^{ij} + \check{\sigma}_{ij} \, \omega^{ij} \right) d\lambda \quad . \tag{A.466}$$

If the transformation is homogeneous, the volume integral can be performed, to give

$$dW = V(\lambda) \left(\hat{\sigma}_{ij}(\lambda) \, v^{ij}(\lambda) + \check{\sigma}_{ij}(\lambda) \, \omega^{ij}(\lambda) \right) d\lambda \quad , \tag{A.467}$$

where $\hat{\sigma}(\lambda)$ and $\check{\sigma}(\lambda)$ represent $\hat{\sigma}(\mathbf{C}(\lambda))$ and $\check{\sigma}(\mathbf{C}(\lambda))$. We can write dW compactly as

$$dW = V(\lambda) \, \mathrm{tr}\left(\hat{\sigma}(\lambda) \, \mathbf{v}(\lambda)^t + \check{\sigma}(\lambda) \, \boldsymbol{\omega}(\lambda)^t \right) d\lambda \quad . \tag{A.468}$$

Let us now transform a body from some initial configuration $\mathbf{C}_0 = \mathbf{C}(\lambda_0)$ to some final configuration $\mathbf{C}_1 = \mathbf{C}(\lambda_1)$, following an arbitrary path Γ in the configuration space, a path that we parameterize using a parameter λ, $(\lambda_0 \leq \lambda \leq \lambda_1)$. At the point λ of the path, the configuration is $\mathbf{C}(\lambda)$. The total work associated to the path Γ is $dW = \oint dW$, i.e.,

$$W(\mathbf{C}_1, \mathbf{C}_0)_\Gamma = \oint_{\lambda_0}^{\lambda_1} d\lambda \; V(\lambda) \, \mathrm{tr}\left(\hat{\sigma}(\lambda) \, \mathbf{v}(\lambda)^t + \check{\sigma}(\lambda) \, \boldsymbol{\omega}(\lambda)^t \right) \quad . \tag{A.469}$$

Denoting $\mathbf{V}_0$ as the volume of the reference configuration $\mathbf{I}$, one has $V(\lambda) = V_0 \det \mathbf{C}(\lambda)$, and we can write

$$\boxed{W(\mathbf{C}_1; \mathbf{C}_0)_\Gamma = V_0 \oint_{\lambda_0}^{\lambda_1} d\lambda \; \det \mathbf{C}(\lambda) \, \mathrm{tr}\left(\hat{\sigma}(\lambda) \, \mathbf{v}(\lambda)^t + \check{\sigma}(\lambda) \, \boldsymbol{\omega}(\lambda)^t \right) \quad .} \tag{A.470}$$

For isochoric transformations, $\det \mathbf{C} = 1$, and in this case, when using Hooke's law (equation 5.55), the evaluation[52] of this expression shows that

---

[52] We have to evaluate the sum of two expressions, each having the form $I = \int_{\lambda_0}^{\lambda_1} d\lambda \; \mathrm{tr}(\boldsymbol{\Sigma}\,(\dot{\mathbf{U}}\,\mathbf{U}^{-1})^t)$, where $\boldsymbol{\Sigma}$ and $\mathbf{U}$ are matrix functions of λ, $\boldsymbol{\Sigma}$ is symmetric or skew-symmetric and is proportional to $\mathbf{u} = \log \mathbf{U}$, and the dot denotes the derivative with respect to λ. We first will simplify the integrand $\mathbf{X} \equiv \mathrm{tr}(\boldsymbol{\Sigma}\,(\dot{\mathbf{U}}\,\mathbf{U}^{-1})^t) = \mathrm{tr}(\dot{\mathbf{U}}\,\mathbf{U}^{-1}\,\boldsymbol{\Sigma}^t) = \pm\mathrm{tr}(\dot{\mathbf{U}}\,\mathbf{U}^{-1}\,\boldsymbol{\Sigma})$, where the sign depends on whether $\boldsymbol{\Sigma}$ is symmetric or skew-symmetric. Using the property $\dot{\mathbf{U}}\,\mathbf{U}^{-1} = \int_0^1 d\mu \; \mathbf{U}^{\mu}\,\dot{\mathbf{u}}\,\mathbf{U}^{-\mu}$ (footnote 9, page 100), the term $\mathrm{tr}(\dot{\mathbf{U}}\,\mathbf{U}^{-1}\,\boldsymbol{\Sigma})$ transforms into $\mathrm{tr}(\dot{\mathbf{U}}\,\mathbf{U}^{-1}\,\boldsymbol{\Sigma}) = \int_0^1 d\mu \; \mathrm{tr}(\mathbf{U}^{\mu}\,\dot{\mathbf{u}}\,\mathbf{U}^{-\mu}\,\boldsymbol{\Sigma}) = \int_0^1 d\mu \; \mathrm{tr}(\dot{\mathbf{u}}\,\mathbf{U}^{-\mu}\,\boldsymbol{\Sigma}\,\mathbf{U}^{\mu})$. Because $\mathbf{U}^{\mu}$ and $\boldsymbol{\Sigma}$ are power series in the same matrix $\mathbf{U}$, they commute. Therefore $\mathbf{X} = \pm \int_0^1 d\mu \; \mathrm{tr}(\dot{\mathbf{u}}\,\boldsymbol{\Sigma}) = \pm\mathrm{tr}(\dot{\mathbf{u}}\,\boldsymbol{\Sigma})$. Using Hooke's law ($\boldsymbol{\Sigma}$ is proportional to $\mathbf{u}$), and an integration by parts, one obtains $I = \pm\frac{1}{2}\,\mathrm{tr}(\boldsymbol{\Sigma}\,\mathbf{u})|_{\lambda_0}^{\lambda_1} = \frac{1}{2}\,\mathrm{tr}(\boldsymbol{\Sigma}^t\,\mathbf{u})|_{\lambda_0}^{\lambda_1}$. Now, making the sum of the two original terms and using the original notations, this gives $W(\mathbf{C}_1; \mathbf{C}_0)_\Gamma = V_0\,(\,\hat{\sigma}_{ij}\,(\log \mathbf{E})^{ij} + \check{\sigma}_{ij}\,(\log \mathbf{S})^{ij}\,)|_{\lambda_0}^{\lambda_1}$, but $\log \mathbf{E}$ is symmetric and $\log \mathbf{S}$ is antisymmetric, so we can simply write $W(\mathbf{C}_1; \mathbf{C}_0)_\Gamma = V_0\,(\,\sigma_{ij}\,(\log \mathbf{E})^{ij} + \sigma_{ij}\,(\log \mathbf{S})^{ij}\,)|_{\lambda_0}^{\lambda_1} = V_0\,\sigma_{ij}\,(\,\log \mathbf{E} + \log \mathbf{S}\,)^{ij}|_{\lambda_0}^{\lambda_1}$, i.e., $W(\mathbf{C}_1; \mathbf{C}_0)_\Gamma = V_0\,\sigma_{ij}\,\varepsilon^{ij}|_{\lambda_0}^{\lambda_1}$, with $\boldsymbol{\varepsilon} = \log \mathbf{E} + \log \mathbf{S} = \log \mathbf{C}$.

the value of the integral does not depend on the particular path chosen in the configuration space, and one obtains

$$W(\mathbf{C}_1;\mathbf{C}_0)_\Gamma = V_0\,(\,U(\mathbf{C}_1) - U(\mathbf{C}_0)\,) \quad , \tag{A.471}$$

where

$$\boxed{U(\mathbf{C}) = \tfrac{1}{2}\,\mathrm{tr}\,\boldsymbol{\sigma}\,\boldsymbol{\varepsilon}^t = \tfrac{1}{2}\,\sigma_{ij}\,\varepsilon^{ij} = \tfrac{1}{2}\,c_{ijk\ell}\,\varepsilon^{ij}\,\varepsilon^{k\ell} \quad ,} \tag{A.472}$$

with $\boldsymbol{\varepsilon} = \log \mathbf{C}$. Therefore, for isochoric transformations the work depends only on the end points of the transformation path, and not on the path itself. This means that the elastic forces are conservative, and that in this theory one can associate to every configuration an elastic energy density (expressed in equation A.472). The elastic energy density is zero for the reference configuration $\mathbf{C} = \mathbf{I}$. As it must be positive for any $\mathbf{C} \neq \mathbf{I}$, the stiffness tensor $\mathbf{c} = \{c_{ijk\ell}\}$ must be a positive definite tensor.

When $\det \mathbf{C} \neq 1$ the result does not hold, and we face a choice: or we just accept that elastic forces are not conservative when there is a change of volume of the body, or we modify Hooke's law. Instead of the relation $\boldsymbol{\sigma} = \mathbf{c}\,\boldsymbol{\varepsilon}$ one may postulate the modified relation

$$\boldsymbol{\sigma} = \frac{1}{\exp \mathrm{tr}\,\boldsymbol{\varepsilon}}\,\mathbf{c}\,\boldsymbol{\varepsilon} \quad . \tag{A.473}$$

With this stress-strain relation, the elastic forces are always conservative,[53] and the energy density is given by expression (A.472).

A.27 Saint-Venant Conditions

Let us again work in the context of section 5.3.7, where a deformation at a point $\mathbf{x}$ of a medium can be described giving the initial metric $\mathbf{g}(\mathbf{x}) \equiv \mathbf{G}(\mathbf{x}, t_0)$ and the final metric $\mathbf{G}(\mathbf{x}) \equiv \mathbf{G}(\mathbf{x}, t)$. These cannot be arbitrary functions, as the associated Riemann tensors must both vanish (let us consider only Euclidean spaces here).

While the two metrics are here denoted g_{ij} and G_{ij}, let us denote $r_{ijk}{}^\ell$ and $R_{ijk}{}^\ell$ the two associated Riemanns. Explicitly,

$$\begin{aligned} r_{ijk}{}^\ell &= \partial_i \gamma_{jk}{}^\ell - \partial_j \gamma_{ik}{}^\ell + \gamma_{is}{}^\ell\,\gamma_{jk}{}^s - \gamma_{js}{}^\ell\,\gamma_{ik}{}^s \\ R_{ijk}{}^\ell &= \partial_i \Gamma_{jk}{}^\ell - \partial_j \Gamma_{ik}{}^\ell + \Gamma_{is}{}^\ell\,\Gamma_{jk}{}^s - \Gamma_{js}{}^\ell\,\Gamma_{ik}{}^s \end{aligned} \tag{A.474}$$

where

$$\begin{aligned} \gamma_{ij}{}^k &= \tfrac{1}{2}\,g^{ks}\,(\,\partial_i g_{js} + \partial_j g_{is} - \partial_s g_{ij}\,) \\ \Gamma_{ij}{}^k &= \tfrac{1}{2}\,G^{ks}\,(\,\partial_i G_{js} + \partial_j G_{is} - \partial_s G_{ij}\,) \quad . \end{aligned} \tag{A.475}$$

[53] As $\exp \mathrm{tr}\,\boldsymbol{\varepsilon} = \det \mathbf{C}$, the term $\det \mathbf{C}(\lambda)$ in equation (A.470) is canceled.

The two conditions that the metrics satisfy are

$$r_{ijk}{}^{\ell} = 0 \quad ; \quad R_{ijk}{}^{\ell} = 0 \quad . \tag{A.476}$$

These two conditions can be rewritten

$$r_{ijk}{}^{\ell} = 0 \quad ; \quad R_{ijk}{}^{\ell} - r_{ijk}{}^{\ell} = 0 \quad . \tag{A.477}$$

The only variable controlling the difference $R_{ijk}{}^{\ell} - r_{ijk}{}^{\ell}$ is the *tensor*[54]

$$Z_{ij}{}^{k} = \Gamma_{ij}{}^{k} - \gamma_{ij}{}^{k} \quad , \tag{A.478}$$

as the Riemanns are linked through

$$R_{ijk}{}^{\ell} - r_{ijk}{}^{\ell} = \overset{\mathbf{g}}{\nabla}_i Z_{jk}{}^{\ell} - \overset{\mathbf{g}}{\nabla}_j Z_{ik}{}^{\ell} + Z_{is}{}^{\ell} Z_{jk}{}^{s} - Z_{js}{}^{\ell} Z_{ik}{}^{s} \quad , \tag{A.479}$$

where

$$Z_{ij}{}^{k} = \tfrac{1}{2} G^{ks} \left(\overset{\mathbf{g}}{\nabla}_i G_{js} + \overset{\mathbf{g}}{\nabla}_j G_{is} - \overset{\mathbf{g}}{\nabla}_s G_{ij} \right) \quad . \tag{A.480}$$

In these equations, by $\overset{\mathbf{g}}{\nabla}$, one should understand the covariant derivative defined using the metric $\mathbf{g}$. The matrix G^{ij} is the inverse of the matrix G_{ij}.

Also, from now on, let us call g_{ij} *the* metric (as it is used to define the covariant differentiation). Then, we can write ∇ instead of $\overset{\mathbf{g}}{\nabla}$. To avoid misunderstandings, it is then better to replace the notation G^{ij} by $\overline{G}^{ij}$ (a bar denoting the inverse of a matrix).

With the new notation, the condition (on G_{ij}) to be satisfied is (equation A.479)

$$\nabla_i Z_{jk}{}^{\ell} - \nabla_j Z_{ik}{}^{\ell} + Z_{is}{}^{\ell} Z_{jk}{}^{s} - Z_{js}{}^{\ell} Z_{ik}{}^{s} = 0 \quad , \tag{A.481}$$

where (equation A.480)

$$Z_{ij}{}^{k} = \tfrac{1}{2} G^{ks} \left(\nabla_i G_{js} + \nabla_j G_{is} - \nabla_s G_{ij} \right) \quad , \tag{A.482}$$

and with the auxiliary condition (on g_{ij})

$$\partial_i \gamma_{jk}{}^{\ell} - \partial_j \gamma_{ik}{}^{\ell} + \gamma_{is}{}^{\ell} \gamma_{jk}{}^{s} - \gamma_{js}{}^{\ell} \gamma_{ik}{}^{s} = 0 \tag{A.483}$$

where

$$\gamma_{ij}{}^{k} = \tfrac{1}{2} g^{ks} \left(\partial_i g_{js} + \partial_j g_{is} - \partial_s g_{ij} \right) \quad . \tag{A.484}$$

Direct computation shows that the conditions (A.481) and (A.482) become, in terms of $\mathbf{G}$,

$$\boxed{\nabla_i \nabla_j G_{k\ell} + \nabla_k \nabla_\ell G_{ij} - \nabla_i \nabla_\ell G_{kj} - \nabla_k \nabla_j G_{i\ell} = \tfrac{1}{2} \overline{G}^{pq} \left(G_{i\ell p} G_{kjq} - G_{k\ell p} G_{ijq} \right) \quad ,} \tag{A.485}$$

[54]The difference between two connections is a tensor.

where

$$\boxed{G_{ijk} = \nabla_i G_{jk} + \nabla_j G_{ik} - \nabla_k G_{ij} \quad ,} \tag{A.486}$$

and where it should be remembered that $\overline{G}^{ij}$ is defined so that $\overline{G}^{ij} G_{jk} = \delta^i{}_k$. We have, therefore, arrived to the

Property A.32 *When using concomitant (i.e., material) coordinates, with an initial metric g_{ij}, a symmetric tensor field G_{ij} can represent the metric at some other time if and only if equations* (A.485) *and* (A.486) *are satisfied, where the covariant derivative is understood to be with respect to the metric g_{ij}.*

Let us see how these equations can be written when the strain is small. In concomitant coordinates, the strain is (see equation 5.87)

$$\varepsilon^i{}_j = \log \sqrt{g^{ik} G_{kj}} \quad ; \quad \varepsilon_{ij} = g_{ik}\, \varepsilon_{kj} \quad . \tag{A.487}$$

From the first of these equations it follows (with the usual notational abuse) $g^{ik} G_{kj} = \exp(\varepsilon^i{}_j)^2 = \exp(2\,\varepsilon^i{}_j) = \delta^i{}_j + 2\,\varepsilon^i{}_j + \dots$, and, using the second equation,

$$G_{ij} = g_{ij} + 2\,\varepsilon_{ij} + \dots \quad . \tag{A.488}$$

Replacing this in equations (A.485) and (A.486), using the property $\nabla_i g_{jk} = 0$, and retaining only the terms that are first-order in the strain gives

$$\nabla_i \nabla_j\, \varepsilon_{k\ell} + \nabla_k \nabla_\ell\, \varepsilon_{ij} - \nabla_i \nabla_\ell\, \varepsilon_{kj} - \nabla_k \nabla_j\, \varepsilon_{i\ell} = 0 \quad . \tag{A.489}$$

If the covariant derivatives are replaced by partial derivatives, these are the well-known Saint-Venant conditions for the strain. A tensor field $\boldsymbol{\varepsilon}(\mathbf{x})$ can be interpreted as a (small) strain field only if it satisfies these conditions.

A.28 Electromagnetism versus Elasticity

There are some well-known analogies between Maxwell's electromagnetism and elasticity (electromagnetic waves were initially interpreted as elastic waves in the ether). Using the standard four-dimensional formalism of relativity, Maxwell equations are written

$$\nabla_\beta G^{\alpha\beta} = J^\alpha \quad ; \quad \partial_\alpha F_{\beta\gamma} + \partial_\beta F_{\gamma\alpha} + \partial_\gamma F_{\alpha\beta} = 0 \quad . \tag{A.490}$$

Here, J^α is the current vector, the tensor $G^{\alpha\beta}$ "contains" the three-dimensional fields $\{D^i, H^i\}$, and the tensor $F_{\alpha\beta}$ "contains" the three-dimensional fields $\{E_i, B_i\}$. In vacuo, the equations are closed by assuming proportionality between $G^{\alpha\beta}$ and $F_{\alpha\beta}$ (via the permittivity and permeability of the vacuum).

Now, in (nonrelativistic) dynamics of continuous media, the Cauchy stress field σ^{ij} is related to the force density inside the medium, φ^i through the condition

$$\nabla_j \sigma^{ij} = \varphi^i \quad . \tag{A.491}$$

For an elastic medium, if the strain is small, it must satisfy the Saint-Venant conditions (equation 5.88) which, if the space is Euclidean and the coordinates Cartesian, are written

$$\partial_i \partial_j \varepsilon_{k\ell} + \partial_k \partial_\ell \varepsilon_{ij} - \partial_i \partial_\ell \varepsilon_{kj} - \partial_k \partial_j \varepsilon_{i\ell} = 0 \quad . \tag{A.492}$$

The two equations (A.491) and (A.492) are very similar to the Maxwell equations (A.490). In addition, for ideal elastic media, there is proportionality between σ^{ij} and ε_{ij} (Hooke's law), as there is proportionality between $G^{\alpha\beta}$ and $F_{\alpha\beta}$ in vacuo. We have seen here that the stress is a bona-fide tensor, and equation (A.491) has been preserved. But we have learned that the strain is, in fact a geotensor (i.e., an oriented geodesic segment on a Lie group manifold), and this has led to a revision of the Saint-Venant conditions that have taken the nonlinear form presented in equation (5.85) (or equation (A.485) in the appendix), expressing that the metric G_{ij} associated to the strain via $\mathbf{G} = \exp \boldsymbol{\varepsilon}$ must have a vanishing Riemann. The Saint-Venant equation (A.492) is just an approximation of (A.485), valid only for small deformations.

If the analogy between electromagnetism and elasticity was to be maintained, one should interpret the antisymmetric tensor $F_{\alpha\beta}$ as the logarithm of a Lorentz transformation. The Maxwell equation on the left in (A.490) would remain unchanged, while the equation on the right should be replaced by a nonlinear (albeit geodesic) equation; in the same way the linearized Saint-Venant equation (5.88) has become the nonlinear condition (A.485). For weak fields, one would recover the standard (linear) Maxwell equations. To my knowledge, such a theory is yet to be developed.

Bibliography

Askar, A., and Cakmak, A.S., 1968, A structural model of a micropolar continuum, Int. J. Eng. Sci., 6, pp. 583–589

Auld, B.A., 1990, Acoustic fields and waves in solids (2nd edn.), Vol. 1, Krieger, Florida, USA.

Baker, H.F., 1905, Alternants and continuous groups, Proc. London Math. Soc., 3, pp. 24–27.

Balakrishnan, A.V., 1976, Applied functional analysis, Springer-Verlag.

Baraff, D., 2001, Physically based modeling, rigid body simulation, on-line at http://www-2.cs.cmu.edu/ baraff/.

Belinfante, J.G.F., and Kolman, B., 1972, A survey of Lie groups and Lie algebras with applications and computational methods, SIAM.

Bender, C.M., and Orszag, S.A., 1978, Advanced mathematical methods for scientists and engineers, McGraw-Hill.

Benford, F., 1938, The law of anomalous numbers, Proc. Amer. Philo. Soc., 78, pp. 551–572.

Bruck, R.H., 1946, Contributions to the theory of loops, Trans. Amer. Math. Soc., 60, pp. 245–354.

Buchheim, A., 1886, An extension of a theorem of Professor Sylvester relating to matrices, Phil. Mag., (5) 22, pp. 173–174.

Callen, H.B., 1985, Thermodynamics and an introduction to thermostatistics, John Wiley and Sons.

Campbell, J.E., 1897, On a law of combination of operators bearing on the theory of continuous transformation groups, Proc. London Math. Soc., 28, pp. 381–390.

Campbell, J.E., 1898, On a law of combination of operators, Proc. London Math. Soc., 29, pp. 14–32.

Cardoso, J., 2004, An explicit formula for the matrix logarithm, arXiv:math. GM/0410556.

Cartan, E., 1952, La théorie des groupes finis et continus et l'analysis situs, Mémorial des Sciences Mathématiques, Fasc. XLII, Gauthier-Villars, Paris.

Cauchy, A.-L., 1841, Mémoire sur les dilatations, les condensations et les rotations produites par un changement de forme dans un système de points matériels, Oeuvres complètes d'Augustin Cauchy, II–XII, pp. 343–377, Gauthier-Villars, Paris.

Choquet-Bruhat, Y., Dewitt-Morette, C., and Dillard-Bleick, M., 1977, Analysis, manifolds and physics, North-Holland.

Ciarlet, P.G., 1988, Mathematical elasticity, North-Holland.

Cohen, E.R., and Taylor, B.N., August 1981, The fundamental physical constants, Physics Today.

Coll, B. and San José, F., 1990, On the exponential of the 2-forms in relativity, Gen. Relat. Gravit., Vol. 22, No. 7, pp. 811–826.

Coll, B. and San José, F., 2002, Composition of Lorentz transformations in terms of their generators, Gen. Relat. Gravit., 34, pp. 1345–1356.

Coll, B., 2003, Concepts for a theory of the electromagnetic field, Meeting on Electromagnetism, Peyresq, France.

Coquereaux R. and Jadczyk, A., 1988, Riemannian geometry, fiber bundles, Kaluza-Klein theories and all that. . . , World Scientific, Singapore.

Cosserat, E. and F. Cosserat, 1909, Théorie des corps déformables, A. Hermann, Paris.

Courant, R., and Hilbert, D., 1953, Methods of mathematical physics, Interscience Publishers.

Cook, A., 1994, The observational foundations of physics, Cambridge University Press.

Dieci, L., 1996, Considerations on computing real logarithms of matrices, Hamiltomian logarithms, and skew-symmetric logarithms, Linear Algebr. Appl., 244, pp. 35–54.

Engø, K., 2001, On the BCH-formula in so(3), Bit Numer. Math., vol. 41, no. 3, pp. 629–632.

Eisenhart, L.P., 1961, Continuous groups of transformations, Dover Publications, New York.

Eringen, A.C., 1962, Nonlinear theory of continuous media, McGraw-Hill, New York.

Evrard, G., 1995, La recherche des paramètres des modèles standard de la cosmologie vue comme un problème inverse, Thèse de Doctorat, Univ. Montpellier.

Evrard, G., 1995, Minimal information in velocity space, Phys. Lett., A 201, pp. 95–102.

Evrard, G., 1996, Objective prior for cosmological parameters, Proc. of the Maximum Entropy and Bayesian Methods, K. Hanson and R. Silver (eds), Kluwer.

Evrard, G. and P. Coles, 1995, Getting the measure of the flatness problem, Class. Quantum Gravity, Vol. 12, No. 10, pp. L93–L97.

Fourier, J., 1822, Théorie analytique de la chaleur, Firmin Didot, Paris.

Fung, Y.C., 1965, Foundation of solid mechanics, Prentice-Hall.

Gantmacher, F.R., 1967, Teorija matrits, Nauka, Moscow. English translation at Chelsea Pub Co, Matrix Theory, 1990.

Garrigues, J., 2002a, Cours de mécanique des milieux continus, web address http://esm2.imt-mrs.fr/gar.

Garrigues, J., 2002b, Grandes déformations et lois de comportement, web address http://esm2.imt-mrs.fr/gar.

Goldberg, S.I., 1998, Curvature and homology, Dover Publications.

Goldstein, H., 1983, Classical mechanics, Addison-Wesley.

Golovina, L.I., 1974, Lineal algebra and some of its applications, Mir Editions.

Golub, G.H. and Van Loan, C.F., 1983, Matrix computations, The John Hopkins University Press.

Gradshteyn, I.S. and Ryzhik, I.M., 1980, Table of integrals, series and products, Academic Press.

Hall, M., 1976, The theory of groups, Chelsea Publishing.

Hausdorff, F., 1906, Die symbolische Exponential Formel in der Gruppen Theorie, Berichte Über die Verhandlungen, Leipzig, pp. 19–48.

Haskell, N.A., 1953, The dispersion of surface waves in multilayered media, Bull. Seismol. Soc. Am., 43, pp. 17–34.

Hehl, F.W., 1973, Spin and torsion in general relativity: I. Foundations, Gen. Relativ. Gravit., Vol. 4, No. 4, pp. 333–349.

Hehl, F.W., 1974, Spin and torsion in general relativity: II. Geometry and field equations, Gen. Relativ. Gravit., Vol. 5, No. 5, pp. 491–516.

Hencky, H., 1928, Über die Form des Elastizitätsgesetzes bei ideal elastischen Stoffen, Z. Physik, 55, pp. 215–220.

Hencky, H., 1928, Uber die Form des Elastizitatsgesetzes bei ideal elastischen Stoffen, Zeit. Tech. Phys., 9, pp. 215–220.

Hencky, H., 1929, Welche Umstände bedingen die Verfestigung bei der bildsamen Verformung von festen isotropen Körpern?, Zeitschrift fur Physik, 55, pp. 145–155.

Herranz, F.J., Ortega, R., and Santander, M., 2000, Trigonometry of spacetimes: a new self-dual approach to a curvature/signature (in) dependent trigonometry, J. Phys. A: Math. Gen., 33, pp. 4525–4551.

Hildebrand, F.B., 1952, Methods of applied mathematics (2nd edn.), Prentice-Hall, Englewood Cliffs.

Horn, R.A., and Johnson, C.R., 1999, Topics in matrix analysis, Cambridge University Press.

Iserles, A., Munthe-Kaas, H.Z., Nørsett, S.P., and Zanna, A., 2000, Lie-group methods, Acta Numerica, 9, pp. 215–365.

Itskov, M., 2000, On the theory of fourth-order tensors and their applications in computational mechanics, Comput. Methods Appl. Mech. Engrg., 189, pp. 419–438.

Jaynes, E.T., 1968, Prior probabilities, IEEE Trans. Syst. Sci. Cybern., Vol. SSC–4, No. 3, pp. 227–241.

Jaynes, E.T., 2003, Probability theory: the logic of science, Cambridge University Press.

Jaynes, E.T., 1985, Where do we go from here?, in Smith, C. R., and Grandy, W. T., Jr., eds., Maximum-entropy and Bayesian methods in inverse problems, Reidel.

Jeffreys, H., 1939, Theory of probability, Clarendon Press, Oxford. Reprinted in 1961 by Oxford University Press.

Kleinert, H., 1989, Gauge fields in condensed matter, Vol. II, Stresses and Defects, World Scientific Pub.

Kurosh, A., 1955, The theory of groups, 2nd edn., translated from the Russian by K. A. Hirsch, two volumes, Chelsea Publishing Co.

Lastman, G.J., and Sinha, N.K., 1991, Infinite series for logarithm of matrix, applied to identification of linear continuous-time multivariable systems from discrete-time models, Electron. Lett., 27(16), pp. 1468–1470.

Leibniz, Gottfried Wilhelm, 1684 (differential calculus), 1686 (integral calculus), *in:* Acta eruditorum, Leipzig.

Lévy-Leblond, J.M., 1984, Quantique, rudiments, InterEditions, Paris.

Ludwik, P., 1909, Elemente der Technologischen Mechanik, Verlag von J. Springer, Berlin.

Malvern, L.E., 1969, Introduction to the mechanics of a continuous medium, Prentice-Hall.

Marsden J.E., and Hughes, T.J.R., 1983, Mathematical foundations of elasticity, Dover.

Means, W.D., 1976, Stress and strain, Springer-Verlag.

Mehrabadi, M.M., and S.C. Cowin, 1990, Eigentensors of linear anisotropic elastic materials, Q. J. Mech. Appl. Math., 43, pp. 15–41.

Mehta, M.L., 1967, Random matrices and the statistical theory of energy levels, Academic Press.

Minkowski, H., 1908, Die Grundgleichungen für die elektromagnetischen Vorgänge in bewegten Körper, Nachr. Ges. Wiss. Göttingen, pp. 53-111.

Moakher, M., 2005, A differential geometric approach to the geometric mean of symmetric positive-definite matrices, SIAM J. Matrix Analysis and Applications, 26 (3), pp. 735–747.

Moler, C., and Van Loan, C., 1978, Nineteen dubious ways to compute the exponential of a matrix, SIAM Rev., Vol. 20, No. 4, pp. 801–836.

Morse, P.M., and Feshbach, H., 1953, Methods of theoretical physics, McGraw-Hill.

Murnaghan, F.D., 1941, The compressibility of solids under extreme pressures, Kármán Anniv. Vol., pp. 121–136.

Nadai, A., 1937, Plastic behavior of metals in the strain-hardening range, Part I, J. Appl. Phys., Vol. 8, pp. 205–213.

Neutsch, W., 1996, Coordinates, de Gruiter.

Newcomb, S., 1881, Note on the frequency of the use of digits in natural numbers, Amer. J. Math., 4, pp. 39–40.

Newton, Sir Isaac, 1670, Methodus fluxionum et serierum infinitarum, English translation 1736.

Nowacki, W., 1986, Theory of asymmetric elasticity, Pergamon Press.

Ogden, R.W., 1984, Non-linear elastic deformations, Dover.

Oprea, J., 1997, Differential geometry and its applications, Prentice Hall.

Pflugfelder, H.O., 1990, Quasigroups and loops, introduction, Heldermann Verlag.

Pitzer, K.S. and Brewer, L., 1961, Thermodynamics, McGraw-Hill.

Poirier J.P., 1985, Creep of crystals High temperature deformation processes in metals, ceramics and minerals. Cambridge University Press.

Powell, R.W., Ho, C.Y., and Liley, P.E., 1982, Thermal conductivity of certain metals, *in:* Handbook of Chemistry and Physics, editors R.C. Weast and M.J. Astle, CRC Press.

Richter, H., 1948, Bemerkung zum Moufangschen Verzerrungsdeviator, Z. amgew. Math. Mech., 28, pp. 126–127

Richter, H., 1949, Verzerrungstensor, Verzerrungsdeviator und Spannungstensor bei endlichen Formänderungen, Z. angew. Math. Mech., 29, pp. 65–75.

Rinehart, R.F., 1955, The equivalence of definitions of a matric function, Amer. Math. Monthly, 62, pp. 395–414.

Rodrigues, O., 1840, Des lois géométriques qui régissent les déplacements d'un système solide dans l'espace, et de la variation des coordonnées provenant de ses déplacements considérés indépendamment des causes qui peuvent les produire., J. de Mathématiques Pures et Appliquées, 5, pp. 380–440.

Rougée, P., 1997, Mécanique des grandes transformations, Springer.

Schwartz, L., 1975, Les tenseurs, Hermann, Paris.

Sedov, L., 1973, Mechanics of continuous media, Nauka, Moscow. French translation: Mécanique des milieux continus, Mir, Moscou, 1975.

Segal, G., 1995, Lie groups, *in:* Lectures on Lie Groups and Lie Algebras, by R. Carter, G. Segal and I. Macdonald, Cambridge University Press.

Soize, C., 2001, Maximum entropy approach for modeling random uncertainties in transient elastodynamics, J. Acoustic. Soc. Am., 109 (5), pp. 1979–1996.

Sokolnikoff, I.S., 1951, Tensor analysis - theory and applications, John Wiley & Sons.

Srinivasa Rao, K.N., 1988, The rotation and Lorentz groups and their representations for physicists, John Wiley & Sons.

Sylvester, J.J., 1883, On the equation to the secular inequalities in the planetary theory, Phil. Mag., (5) 16, pp. 267–269.

Taylor, S.J., 1966, Introduction to measure and integration, Cambridge University Press.

Taylor, A.E., and Lay, D.C., 1980, Introduction to functional analysis, Wiley.

Terras, A., 1985, Harmonic analysis on symmetric spaces and applications, Vol. I, Springer-Verlag.

Terras, A., 1988, Harmonic analysis on symmetric spaces and applications, Vol. II, Springer-Verlag.

Thomson, W.T., 1950, Transmission of elastic waves through a stratified solid, J. Appl. Phys., 21, pp. 89–93.

Truesdell C., and Toupin, R., 1960, The classical field theories, *in:* Encyclopedia of physics, edited by S. Flügge, Vol. III/1, Principles of classical mechanics and field theory, Springer-Verlag, Berlin.

Ungar, A.A., 2001, Beyond the Einstein addition law and its gyroscopic Thomas precession, Kluwer Academic Publishers.

Varadarajan, V.S., 1984, Lie groups, Lie algebras, and their representations, Springer-Verlag.

Yeganeh-Haeri, A., Weidner, D.J., and Parise, J.B., 1992, Elasticity of α-cristobalite: a silicon dioxide with a negative Poisson's ratio, Science, 257, pp. 650–652.

Index